PLEASE STAMP DATE DUE, BOTH BELOW AND ON CARD

DATE DUE	DATE DUE	DATE DUE	DATE DUE
APR 0 9 2013			

GL-15

Raman Spectroscopy, Fullerenes and Nanotechnology

RSC Nanoscience & Nanotechnology

Series Editor:
Professor Paul O'Brien, *University of Manchester, UK*
Professor Sir Harry Kroto FRS, *University of Sussex, UK*
Professor Harold Craighead, *Cornell University, USA*

Titles in the Series:
1: Nanotubes and Nanowires
2: Fullerenes: Principles and Applications
3: Nanocharacterisation
4: Atom Resolved Surface Reactions: Nanocatalysis
5: Biomimetic Nanoceramics in Clinical Use: From Materials to Applications
6: Nanofluidics: Nanoscience and Nanotechnology
7: Bionanodesign: Following Nature's Touch
8: Nano-Society: Pushing the Boundaries of Technology
9: Polymer-based Nanostructures: Medical Applications
10: Molecular Interactions in Nanometer Layers, Pores and Particles: New Findings at the Yoctovolume Level
11: Nanocasting: A Versatile Strategy for Creating Nanostructured Porous Materials
12: Titanate and Titania Nanotubes: Synthesis, Properties and Applications
13: Raman Spectroscopy, Fullerenes and Nanotechnology

How to obtain future titles on publication:
A standing order plan is available for this series. A standing order will bring delivery of each new volume immediately on publication.

For further information please contact:
Book Sales Department, Royal Society of Chemistry,
Thomas Graham House, Science Park, Milton Road, Cambridge,
CB4 0WF, UK
Telephone: +44 (0)1223 420066, Fax: +44 (0)1223 420247, Email: books@rsc.org
Visit our website at http://www.rsc.org/Shop/Books/

Raman Spectroscopy, Fullerenes and Nanotechnology

Maher S. Amer
Russ Engineering Center, Wright State University, Dayton, OH 45435, USA

RSCPublishing

RSC Nanoscience & Nanotechnology No. 13

ISBN: 978-1-84755-240-2
ISSN: 1757-7136

A catalogue record for this book is available from the British Library

© Maher S. Amer 2010

All rights reserved

Apart from fair dealing for the purposes of research for non-commercial purposes or for private study, criticism or review, as permitted under the Copyright, Designs and Patents Act 1988 and the Copyright and Related Rights Regulations 2003, this publication may not be reproduced, stored or transmitted, in any form or by any means, without the prior permission in writing of The Royal Society of Chemistry or the copyright owner, or in the case of reproduction in accordance with the terms of licences issued by the Copyright Licensing Agency in the UK, or in accordance with the terms of the licences issued by the appropriate Reproduction Rights Organization outside the UK. Enquiries concerning reproduction outside the terms stated here should be sent to The Royal Society of Chemistry at the address printed on this page.

The RSC is not responsible for individual opinions expressed in this work.

Published by The Royal Society of Chemistry,
Thomas Graham House, Science Park, Milton Road,
Cambridge CB4 0WF, UK

Registered Charity Number 207890

For further information see our web site at www.rsc.org

Dedication

To my teachers, students, friends, and family.

Preface

Two fundamental discoveries have recently started a new era of scientific research: the discovery of fullerenes and the development of single-molecule imaging capabilities. The discovery of fullerenes with their unique properties, highly versatile nature, and many potential applications in materials science, chemistry, physics, opto-electronics, biology, and medicine has launched a new branch of interdisciplinary research known as "nanotechnology." This technology revolutionized the multibillion-dollar field of opto-electronics and is a key to wireless communications, remote sensing, and medical diagnostics, and still has a lot to offer. The development of single-molecule imaging and investigating capabilities provided the means for studying the reactions of complex material systems, and biological molecules in natural systems.

The real importance of these discoveries is that they, synergized together, put forward the platform for what can be called "the next industrial revolution" in human history: "nanotechnology." Just as the quantum mechanics work of the 1930s led to the electronic material revolution in the 1980s, and as the fundamental work in molecular biology in the 1950s gave rise to the current biotechnology, it is believed that the emerging work in nanotechnology has the potential to fundamentally change the way people live within the next two decades. The ability to manipulate matter on the atomic level and to manufacture devices from the molecular level up will definitely have major implications. Among the advances and benefits foreseen for nanotechnology implementation are inexpensive energy generation, highly efficient manufacturing, environmentally benign materials, universal clean water supplies, atomically engineered crops resulting in greater agricultural productivity, radically improved medicines, unprecedented medical treatments and organ replacement, greater information storage and communication capacities, and increased human performance through convergent technologies. This means that nanotechnology is expected to revolutionize manufacturing and energy production, in addition to healthcare, communications, utilities, and definitely defense. Hence, nanotechnology will transform labor and the workplace, medical system, transportation, and power infrastructures. In short, nanotechnology will transform life, as we currently perceive it.

Here, we discuss nanotechnology and its attributes based on the observation that it represents a domain in which conventional materials perform in an unconventional way. We will try to explain the phenomenon – the nanophenomenon – based on our current state of knowledge and will try to predict its potentials and challenges.

Nanotechnology is based on certain building blocks that include fullerenes (in spheroidal, cylindrical, and sheet forms), nano-crystals, and nanowires, and on characterization and imaging techniques capable of interrogating such building blocks. This book focuses on fullerenes (as a major family of building blocks), Raman spectroscopy (as a powerful investigative spectroscopic technique), and how both contribute to the advancement of our state of knowledge as far as nanotechnology is concerned. The book consists of four chapters. In Chapter 1, we introduce and discuss nanotechnology. Based on the fact that social studies indicate that the majority of the public is not aware of the nature of nanotechnology, we felt obligated to start the book with a "layperson" level of introduction to nanotechnology. We basically show that, as far as nature is concerned, nanotechnology is over 3 billion years old, and as far as humankind is concerned nanotechnology was practiced several thousand years ago. On a more scientific level of discussion, we explore nanotechnology and define nanodomain as the domain in which a system becomes a thermodynamically small system that can no longer be treated or described by classical thermodynamic equations of state originally observed and developed for bulk or large systems. We show that once the size of a system is on the order of certain length-scales it becomes thermodynamically inhomogeneous and its thermodynamic potentials and functions become indefinable. In Chapter 1, we discuss such length-scales and show that materials behavior at such length-scales is unconventional and represents what can be termed as the "nano-behavior." We explain in this chapter how nano-behavior is related to the system's size as compared to thermodynamic inhomogeneity and not just to the system's physical dimensions. We conclude Chapter 1 with a discussion of interesting nanophenomena recently observed in optical, electric, thermal, and mechanical performance of nano-materials. We emphasize that, unlike conventional bulk systems, nanosystems are very sensitive to perturbation effects. We provide evidence and show examples to the fact that significant changes in the behavior of a nanosystem can be observed as the result of minute perturbation fields affecting the system.

Chapter 2 is dedicated to Raman spectroscopy as an investigative technique. In this chapter, we discuss all aspects of Raman scattering phenomenon: its theory, and instrumentation. We also discuss the concept of symmetry and its significance to the Raman scattering phenomenon. Most importantly, we show that Raman spectroscopy is an extremely powerful technique that can provide essential information not only regarding the structure and properties of the system under investigation but also regarding the behavior of a system in response to various perturbation effects.

Preface   ix

In Chapter 3, we discuss the carbon-based building blocks – the fullerenes. In this chapter we present the history of fullerenes; their original predictions, and their initial discovery. We classify fullerenes based on their dimensionality into three classes: zero-dimensional fullerenes, one-dimensional fullerenes, and two-dimensional fullerenes. For each class we discuss the structure as well as the production and purification methods. For zero-dimensional fullerenes, we discuss C_{60}, C_{70}, and larger or giant fullerenes. For one-dimensional fullerenes, we cover single-walled, double-walled, and multi-walled carbon nanotubes. For two-dimensional fullerenes, we discuss "graphene" in single- and multi-sheet forms.

Chapter 4 is devoted to the properties of all types of fullerenes discussed in Chapter 3. We start with Raman scattering in fullerenes and what information it provides regarding their structure, performance, and response to perturbation effects. We consider nanophenomenon related to solvent effects on fullerenes, and, more interestingly, we discuss fullerene effects on solvents. The structuring induced into liquid solvents due to fullerene interaction and its effect on the solvent properties is covered. We also discuss fullerene under pressure effects, their Raman response and mechanical properties. We conclude Chapter 4 and the book by an overview and concluding remarks regarding the current state of knowledge, the future potentials, and challenges that must be faced.

Maher S. Amer
Dayton, Ohio

Contents

Chapter 1 Nanotechnology, the Technology of Small Thermodynamic Systems   1

     1.1 Introduction   1
     1.2 Origins of Nanotechnology   1
     1.3 What Nanotechnology Is   4
         1.3.1 What Can Nanotechnology Do For Us?   5
         1.3.2 Where did the Name "Nano" Came From?   5
         1.3.3 Does Every Nanosystem Have To Be So Small?   6
         1.3.4 How and Why do the Properties of Matter Change by Entering the Nano-domain?   7
         1.3.5 Has Nanotechnology Been Used Before?   7
         1.3.6 Why did it Take us so Long to Realize the Importance of Nanotechnology?   9
     1.4 Back to the Science   11
     1.5 Large Systems and Small Systems Limits   11
     1.6 Scales of Inhomogeneity   14
         1.6.1 Thermal Gravitational Scale   14
         1.6.2 Capillary Length   14
         1.6.3 Tolman Length   15
         1.6.4 Line Tension (τ) and the (τ/σ) Ratio   16
         1.6.5 Correlation Length (ξ)   17
     1.7 Thermodynamics of Small Systems   19
     1.8 Configurational Entropy of Small Systems   21
     1.9 Nanophenomena   26
         1.9.1 Optical Phenomena   26
         1.9.2 Electronic Phenomena   30

	1.9.3	Thermal Phenomena	35
	1.9.4	Mechanical Phenomena	37
References			38

Chapter 2 Raman Spectroscopy; the Diagnostic Tool — **43**

- 2.1 Introduction — 43
- 2.2 Raman Phenomenon — 44
- 2.3 General Theory of Raman Scattering — 44
- 2.4 Raman Selection Rules — 47
  - 2.4.1 Vibration Modes and the Polarizability Tensor — 47
- 2.5 Symmetry — 50
  - 2.5.1 Identity (E) — 51
  - 2.5.2 Center of Symmetry (i) — 51
  - 2.5.3 Rotation Axes (C_n) — 52
  - 2.5.4 Planes of Symmetry (σ) (Mirror Planes) — 53
  - 2.5.5 Rotation Reflection Axes (S_n) (Improper Rotation) — 53
  - 2.5.6 Symmetry Elements and Symmetry Operations — 55
- 2.6 Point Groups — 56
  - 2.6.1 Point Groups of Molecules — 56
  - 2.6.2 Point Groups of Crystals — 60
- 2.7 Space Groups — 62
  - 2.7.1 Screw Axis (n_p) — 63
  - 2.7.2 Glide Planes — 65
  - 2.7.3 Space Groups in One- and Two-dimensional Space — 66
- 2.8 Character Table — 69
  - 2.8.1 Symmetry Operations and Transformation of Directional Properties — 69
  - 2.8.2 Degenerate Symmetry Species (Degenerate Representations) — 73
  - 2.8.3 Symmetry Species in Linear Molecules — 74
  - 2.8.4 Classification of Normal Vibration by Symmetry — 74
  - 2.8.5 Raman Overtones and Combination Bands — 79
  - 2.8.6 Molecular and Lattice Raman Modes — 79
- 2.9 Raman from an Energy Transfer Viewpoint — 81
- 2.10 Boltzmann Distribution and its Correlation to Raman Lines — 83
- 2.11 Perturbation Effects on Raman Bands — 85
  - 2.11.1 Strain Effects — 85
  - 2.11.2 Heat Effects — 86
  - 2.11.3 Hydrostatic Pressure Effects — 88
  - 2.11.4 Structural Imperfections Effects — 90
  - 2.11.5 Chemical Potentials Effects — 92
- 2.12 Resonant Raman Effect — 95

Contents

2.13	Calculations of Raman Band Positions	95
2.14	Polarized Raman and Band Intensity	96
2.15	Dispersion Effect	99
2.16	Instrumentation	101
Recommended General Reading		106
References		106

Chapter 3 Fullerenes, the Building Blocks **109**

3.1	Overview	109
3.2	Introduction	109
3.3	Fullerenes, the Beginnings and Current State	110
3.4	Zero-dimensional Fullerenes: The Structure	117
	3.4.1 Structure of the [60] Fullerene Molecule	123
	3.4.2 Structure of the [70] Fullerene Molecule	126
3.5	Production Methods of Fullerenes	129
	3.5.1 Huffman–Krätschmer Method	129
	3.5.2 Benzene Combustion Method	131
	3.5.3 Condensation Method	132
3.6	Extraction Methods of Fullerenes	133
3.7	Purification Methods of Fullerene	137
3.8	Fullerene Onions	140
3.9	One-dimensional Fullerene: The Structure	143
	3.9.1 Single-walled Carbon Nanotubes (SWCNTs)	143
	3.9.2 Multi-walled Carbon Nanotubes (MWCNTs)	155
	3.9.3 Production of Carbon Nanotubes	158
3.10	Two-dimensional Fullerenes – Graphene	161
References		168

Chapter 4 The Nano-frontier; Properties, Achievements, and Challenges **182**

4.1	Introduction	182
4.2	Raman Scattering of Fullerenes	183
	4.2.1 Raman Scattering of C_{60} Molecules and Crystals	183
	4.2.2 Raman Scattering of C_{70}	189
	4.2.3 Raman Scattering of Single-walled Carbon Nanotubes	190
	4.2.4 Raman Scattering of Double- and Multi-walled Carbon Nanotubes	197
	4.2.5 Raman Scattering of Graphene	201
	4.2.6 Thermal Effects on Raman Scattering	208
4.3	Fullerene Solubility and Solvent Interactions	215
	4.3.1 Solvent Effects on Fullerenes	221
	4.3.2 Fullerene Effects on Solvents	225

	4.4 Fullerenes under Pressure	229
	4.5 Overview, Potentials, Challenges, and Concluding Remarks	236
	References	240
Appendix 1	**Character Tables for Various Point Groups**	**259**
Appendix 2	**General Formula for Calculating the Number of Normal Vibrations in Each Symmetry Species**	**267**
Appendix 3	**Polarizability Tensors for the 32 Point Groups including the Icosahedral Group**	**272**
Subject Index		**276**

CHAPTER 1
Nanotechnology, the Technology of Small Thermodynamic Systems

1.1 Introduction

This chapter introduces nanotechnology, emphasizing the fact that it is more related to the thermodynamic behaviour of small systems than to the physical dimensions of the system. The importance of entropic forces in such systems will be considered and examples of how such forces can alter the behaviour of materials systems and enable them to exhibit unusual chemical, physical, electrical, optical, and mechanical properties will be given. The building blocks of nanotechnology will be identified and discussed. Examples of biological and natural utilization of nanostructured system as well as recent engineering applications of such systems will be given.[1]

1.2 Origins of Nanotechnology

Almost 50 years ago, on December 29, 1959, Richard P. Feynman,[i] a great physicist and, later, a Nobel Laureate, gave a lecture at the annual meeting of the American Physical Society at California Institute of Technology, Pasadena, entitled "There's Plenty of Room at the Bottom, an Invitation to enter a new field of Physics." The lecture was published later[2] and was republished[3] again in 1992 as the topic it first introduced overwhelmingly caught the attention of many of the scientists, politicians, and the public across the globe.

---

[i] Richard P. Feynman (1918–1988) was born in New York City on the 11th May 1918. He studied at the Massachusetts Institute of Technology where he obtained his BSc in 1939 and at Princeton University where he obtained his PhD in 1942. He was Research Assistant at Princeton (1940–1941), Professor of Theoretical Physics at Cornell University (1945–1950), Visiting Professor and thereafter appointed Professor of Theoretical Physics at the California Institute of Technology (1950–1959). Feynman received the Nobel Prize in Physics in 1965.

The term *"nanotechnology"* was never used in Feynman's lecture; instead, Feynman spoke about *miniaturization*, emphasizing the important scientific and economic aspects of our ability to make things *small*. Small machines that are capable of making even smaller ones. In his own words "although it is a very wild idea, it would be interesting in surgery if you could swallow the surgeon". Not to be misunderstood, Feynman emphasized that such a vision necessitates an ability to manipulate materials systems on a *small* scale. Feynman would not have missed the obvious and logical fact that atoms and molecules behave differently when arranged in a *small* system compared to their behavior in *large* or *bulk* systems. In his famous lecture Feynman said:

"I can hardly doubt that when we have some control of the arrangement of things on a small scale we will get an enormously greater range of possible properties that substances can have."

The obvious reason for that was:

"...Atoms on a small scale behave like nothing on a large scale, for they satisfy the laws of quantum mechanics. So, as we go down and fiddle around with the atoms down there, we are working with different laws, and we can expect to do different things."

Hence, while Feynman did not explicitly speak about what is referred to nowadays as "nanotechnology," he pointed out a new and important domain of physics where matter is investigated on a *new* scale at which quantum effects are dominant.

The term nanotechnology was actually coined in 1974 by Norio Taniguchi (1912–1988), a professor at Tokyo Science University, Japan.[4] The term *nano* is Greek for "dwarf". Professor's Taniguchi's main interest was in high precision machining of hard and brittle materials. He pioneered the application of energy beam techniques, including electron beam, lasers, and ion beams, to ultra-precision processing of materials. In his famous paper entitled "On the Basic Concept of 'Nano-Technology'", Professor Taniguchi defined the field as:

"Nano-technology mainly consists of the processing of, separation, consolidation, and deformation of materials by one atom or by one molecule."

Professor Taniguchi was mainly using the term to describe possibilities in precision machining for the electronic industry to enable smaller and smaller devices down to the *nanometer* length scale. The prefix *nano* is known in the metric scale system to represent a billionth or 10^{-9} of a unit. In 1974, Professor Taniguchi was interested in precision machining down to the nanometer level, which requires an ability to manipulate materials on the atomic or molecular level.

In 1986, K. Eric Drexler reused and popularized the term "nanotechnology" in a much broader prespective describing a whole new manufacturing technology based on molecular machinery. The premise was that such molecular machinery does exist, by countless examples, in biological systems, and, hence, sophisticated, efficient, and optimized molecular machines can be produced. In a series of books,[5–8] Drexler described a number of possible molecular machinery suitable for a very wide range of applications. He also described "profiles of the possible" as well as "dangers and hopes" associated with the nanotechnology. Drexler was awarded a PhD in 1991. His work, indeed, triggered and inspired what is currently referred to as the nano-revolution. He adapted the viewpoint that although nanotechnology can be initially implemented by resembling biological systems, ultimately it could be based on pure mechanical engineering principles, rendering nanotechnology as a manufacturing technology based on the mechanical functionality of molecular size components. Such mechanical components, *i.e.*, gears, bearings, motors, and structural members, would enable programmable assembly with atomic precission.[9] Figure 1.1 shows the three scholars Richard Feynman, who first envisioned nanotechnology, Norio Taniguchi, who coined the term nanotechnology, and Eric Drexler, who popularized the term in a new perspective.

The pure mechanical viewpoint shaping Drexler's proposed vision led, however, to a long and heated debate between him and Richard Smalley. Richard Smalley – a professor of chemistry at Rice University who shared the 1996 Nobel Prize in Chemistry with Robert Curl, Jr. and Sir Harry Kroto for discovering the C_{60} molecule (fullerene [60]) – had very well founded reservations on applying pure mechanical engineering principles to nano-machinery, and on the premise of mechanical functionality of molecules.[10] The debate was indeed a significant controversy about nanotechnology's meaning and possibilities. Drexler, later, backed off of his position on the basis that his original ideas have been misunderstood.[11] Several analyses of the debate were published.[12,13] Unfortunately, the debate left a negative impression on public view of the technology and, to a large extent, deepened the wrong concept that nanotechnology is the technology by which to make tiny (bug-like) machines

Figure 1.1 The three scholars Richard Feynman, who first envisioned nanotechnology, Norio Taniguchi, who coined the term "nanotechnology", and Eric Drexler who popularized the term in a new perspective.

Figure 1.2 Science art illustrating the perception of nanobots. (© Coneyl Jay/Science Photo Library.)

capable of replicating themselves, working miracles, but that could run amok. Given the effective role media usually play on public viewpoint and understanding of science,[14] it was concluded recently that the media has contributed to bounding nanotechnology by representing the term as a technology that trades on ideas of wonder as well as risk.[15]

A good example that demonstrates the general misunderstanding of nanotechnology is the Winner illustration of the 2002 "Visions of Science Award" by Coneyl Jay (Figure 1.2). The illustration shows how the public, in general, imagined what nanotechnology is all about; a bug-like tiny machine injecting *stuff* in a red blood cell! Unfortunately, that was the general impression of nano-medicine.

Such a simplistic understanding of nanotechnology bred enormous public concerns and suspicion. Well founded and justified concerns regarding the impact of nanotechnology on scientific, economic, ethical, and societal aspects of humankind future were raised and are still being debated. We will discuss these important issues in later sections of this chapter. At this point, however, it is beyond doubt that what we decided to call a *dwarf (nano)* turned out to be a *giant*.

1.3 What Nanotechnology Is

Nanotechnology has been described as the next industrial revolution in human history. As with each of the previous industrial revolutions, it is expected to have a huge and long-term impact on all aspects of human life. In addition, and not surprisingly, the new technology has not been very well understood in some cases and has been misunderstood in many other cases. According to the USA National Science Foundation 2007 released statistics, most Americans (54%) have heard "nothing at all" about nanotechnology. In this section, we will address the nature of nanotechnology in very simple, even layperson, terms. As Albert Einstein pointed out, one can claim knowledge of a subject only when one is capable of explaining the subject to one's grandmother. Time has changed since the Einstein era and many of nowadays grandmothers have advanced degrees. While this makes it easier for new generations to claim knowledge, it might be the time to change the rule of knowledge claiming to state that one may claim knowledge of a subject only if one is capable of *correctly* explaining the subject to the public.

1.3.1 What Can Nanotechnology Do For Us?

Everything we deal with is either man-made (made by humans) or natural (made by nature). For example, a car is a man-made transportation means, while a horse is a natural one. Currently, cars are faster and much stronger than horses. However, cars are still not capable of sensing the danger down the road as horses do. Also, cars cannot take their passenger home while the passenger is asleep as horses do. In addition, horses are much safer to travel by since none of us has ever heard about an accident between two horses resulting in a rider's life loss! To this end, we can describe nanotechnology as a new level of knowledge that could enable us to bridge the gap between the capabilities of man-made and natural things. This would result in a new generation of regular size, and not tiny, cars capable of sensing the danger, driving home, and reducing or eliminating accidents due to operator errors. In addition, a more important and a major difference between man-made and natural things is their efficiency. Over billions of years, nature mastered the art of efficient design and operation. Humans, however, are still at the beginning of a learning curve in those regards. For example, the best gasoline car engine we currently make has an efficiency of 25–30%. Mechanical efficiency of athletes during running was measured to range between 47% and 62%.[16] Other species and natural processes can even reach higher efficiencies. This clearly demonstrates how crucial nanotechnology can be considering the energy crises our civilization is currently facing.

1.3.2 Where did the Name "Nano" Came From?

The word *nano* is Greek for *dwarf*. This word was actually used to indicate the length unit equal to one billionth of a meter (10^{-9} meter). To have a good idea of what this length actually is, let us consider a typical single human hair. This is about 50–100 micron, which is 50 to 100 millionth of a meter. Hence, a single human hair would be 50 to 100 thousand times larger than a nanometer. Atoms and molecules are typically measured by a unit called the ångström (Å), which is one tenth of a nanometer (nm), or one ten billionth of a meter. Fifty years ago, Feynman predicted, and more recently, many scientists observed that the behavior of material clusters on the 1–100 nm scale is essentially different from that of larger clusters that we currently use. It can be scientifically sound to say that on the nanometer level the laws of nature controlling materials behavior are different, hence new phenomena can be observed. This is where the name "nanotechnology" came from.

1.3.3 Does Every Nanosystem Have To Be So Small?

The answer to this question is absolutely not. In fact, most biological systems, including ourselves, are nanosystems. This is in the sense that these systems on a certain level operate according to nanophysics and nano-chemistry laws. In fact, humans in their daily life activities obey two different sets of laws; the traditional laws of physics that we already know and nanoscale laws that we are still exploring. For example, if one jumps up, one's body will follow the gravitational law that was first identified by Isaac Newton in the fifteenth century, and one will, according to this law, fall back down. However, as one breathes one's blood exchanges carbon dioxide for oxygen at the lungs and does the opposite at the cells according to a different set of laws that we refer to here as nano-laws. The blood component in charge of the exchange (known as hemoglobin) is much larger than a nanometer, and so are our body cells, and definitely we and all other breathing creatures are. A nanosystem, or more appropriately, a nanostructured system is a system that is made of components that operate according to nano-laws regardless of the system size. Good examples to illustrate this point are butterfly wings and opal stones (Figure 1.3). The beautiful colors of butterfly wings and opals are due to light reflection by nanostructures and are not due to pigmentation. Different colors are due to different nanostructures. The opal example tells us that nanostructures are not

Figure 1.3 Nanostructures responsible for the amazing colors in butterfly wings and opal.

limited to biological systems. In fact, over billions of years, nature has mastered nano-manufacturing techniques as the best and most efficient techniques to build sophisticated and efficient products.

1.3.4 How and Why do the Properties of Matter Change by Entering the Nano-domain?

A good example to illustrate how the properties of matter change as it enters the nano-domain is water. In its bulk form, water is a liquid that every one of us is familiar with. It is a colorless odorless liquid that is heavier than air. Hence, gravitational laws are in control of the system and it fills our lakes, seas, and oceans. As water evaporates, due to heat effects, in the form of single and tiny clusters of molecules, gravitational laws are no longer in charge. The bulk laws are actually overruled by the new *nano-domain* laws. Water in the form of tiny clusters becomes airborne. These tiny clusters of water can accumulate and form huge clouds containing enormous amounts of water but still can be transported by wind over very long distances. Controlled by weather conditions, the tiny clusters of water in the clouds can grow into bigger and bigger clusters until they depart the nano-domain and enter the bulk domain again in the form of water droplets. Once in the bulk domain, gravitational laws will take control again, and water droplets will fall as rain. To this end, it is very clear that nature has utilized nanotechnology, for billions of years, in transporting enormous amounts of water over great distances very efficiently. Interestingly, even with our current *advanced* technologies, as we like to call it, we are not capable of carrying out such a transportation operation as efficient, if at all. It might be wisely and timely to learn from nature.

1.3.5 Has Nanotechnology Been Used Before?

While nature has been utilizing nanotechnology in building biological systems, as we mentioned before, for almost 3.7 billion years, humans have also used nanotechnology before. The mysterious optical behavior of the famous Lycurgus Cup (AD 400, Figure 1.4) is a good example of nanotechnology effects on optical properties of matter. This Roman cup is made of ruby glass. When viewed in reflected light, for example in daylight, it appears green. However, when a light is shone into the cup and transmitted through the glass it appears red! Recently, this mysterious behavior was investigated and it was found that while the chemical composition of the glass of Lycurgus Cup is almost the same as that of modern glass, the fascinating optical behavior is totally due to gold nanoparticles within the cup glass.[17]

In addition, it was revealed recently[18–21] that the famous Damascus saber (14th–16th century era), with its traditional wavy patterns[ii] on the surface

---
[ii] It is believed that the naturally appearing wavy patterns on the surface of Damascus swords are the inspiring origin of the famous damask patterns used in fashion and decoration.

Figure 1.4 Lycurgus Cup (British Museum; AD fourth century). This Roman cup is made of ruby glass. When viewed in reflected light, for example in daylight, it appears green. However, when a light is shone into the cup and transmitted through the glass it appears red. (Reproduced with kind permission from Leonhardt, ref. 17. Copyright Nature Publishing Group, 2007.)

(Figure 1.5), owes its superior strength and performance to the presence of nanostructures in the form of tubes and wires within its alloy.

Nanotechnology, namely nano-medicine, was used even much earlier in human civilization, most probably unintentionally. Finely ground gold particles in the size range 10–500 nm can be suspended in water. Such suspensions were used for medical purposes in ancient Egypt over 5000 years ago. In Alexandria, Egyptian alchemists used fine gold particles to produce a colloidal elixir known as "liquid gold" that was intended to restore youth! It is an interesting coincidence to realize that the 2007 Medal of Science, the U.S.A.'s highest honor in the field, was awarded to the Egyptian-American chemist Professor Mostafa El-Sayed, of Georgia Institute of Technology, for his many outstanding contributions, among which using gold nanorods in cancer tumor treatment was the most recent.

More recently, in 1856, Michael Faraday (1791–1867) independently prepared colloidal gold, which he called "divided state of gold". Faraday's samples are still preserved in the Royal Institution. In addition, in 1890, the work of the German bacteriologist Robert Koch (1843–1910) showed that gold compounds inhibit the growth of bacteria. He was awarded the Nobel Prize for Medicine in

Nanotechnology, the Technology of Small Thermodynamic Systems 9

Figure 1.5 Typical wavy patterns on the surface of a Damascus saber. The structure was found to be due to carbon nanotubes in the saber alloy. (Reproduced with kind permission from Levin *et al.*, ref. 19. Copyright Wiley-VCH, 2005.)

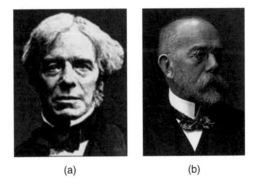

(a)   (b)

Figure 1.6 The two scholars who pioneered the preparation of gold nanoparticles and their medical applications: (a) Michael Faraday and (b) Robert Koch. (Photos courtesy of the Nobel Foundation.)

1905. Figure 1.6 shows these scholars who pioneered the preparation and medical applications of gold nanoparticles.

1.3.6 Why did it Take us so Long to Realize the Importance of Nanotechnology?

The development and progress of any technology is a matter of probability. Three main ingredients have to exist: a vision or a theory, characterization and diagnostic techniques, and a material system, or in other words, building blocks for experimental verification. In addition, economic, societal, and political environments have to be suitable to enable the synergy of the three main ingredients into one successful and sustainable endeavor. If we consider nanotechnology, Feynman's vision was born as a natural consequence of the successful theoretical physics developments in quantum mechanics during the 1930s. The vision was most probably delayed by a decade because of World War II. Feynman's vision, however, had to wait till the 1980s for the discovery of the scanning tunneling microscope in 1981 and the fullerene building blocks in 1985 for the triad to be complete. On one hand, scanning tunneling microscopy was the tool that allowed the required resolution and the manipulation capability of single atoms and molecules. On the other hand, fullerene discovery opened the door for the production and investigation of suitable material building blocks. Figure 1.7 shows the two scholars and Nobel Laureates Gerd Binnig and Heinrich Rohrer who invented the scanning tunneling microscope (STM), and Figure 1.8 shows the three scholars and Nobel

Figure 1.7 The two scholars and Nobel Laureates (a) Gerd Binnig and (b) Heinrich Rohrer who invented the scanning tunneling microscope (STM). (Photos courtesy of the Nobel Foundation.)

Figure 1.8 The three scholars and Nobel Laureates (a) Curl, (b) Smalley, and (c) Kroto who first discovered and isolated C_{60} fullerene molecules. (Photos courtesy of the Nobel Foundation.)

Laureates – Curl, Smalley, and Kroto – who first discovered fullerene molecules.

Before we proceed to our scientific discussions and equations, a crucial point has to be made very clear. While nanotechnology will definitely play a major role in the future of human civilization, no one should suffer the illusion that nanotechnology will provide solutions for all human-kind problems. It is unfortunate that some people in the scientific field, for different reasons, portrayed nanotechnology as capable of making every human on planet earth younger, more beautiful, and richer than what they currently are! In fact, as much as nanotechnology will provide solutions for our challenges, it will also impose new challenges on us. We will identify and discuss some of these challenges at the end of this book. Other challenges cannot be even predicted at the present.

1.4 Back to the Science

Nowadays, 50 years after Feynman's announcement of his vision, we know that it is not merely obeying the quantum mechanical laws that makes materials behave differently in small and large systems but also the system size dependence of the nature and relative importance of forces controlling the system. We now know that as the system size becomes smaller and smaller, gravitational forces, a major player in large systems, start to lose their control on the system and other forces, such as surface tension, van der Waals, and entropic or depletion forces, start to take control. The terms *small* and *large* or *nano* and *bulk* have been mentioned frequently so far in this chapter. It might be a good idea to start our discussion of nanotechnology by defining and differentiating between these important terms.

1.5 Large Systems and Small Systems Limits

It is well known that thermodynamics and statistical mechanics, our principal theoretical tools for understanding the physics and behavior of material systems, are mainly based on the assumption that the system under consideration is infinitely large and, therefore, the system is essentially uniform at equilibrium even if it has multiple phases. Uniformity in this context is the uniformity of thermodynamic functions in the system. Thermodynamic functions of common interest include pressure (p), temperature (T), chemical potential (μ), internal energy (U), and free energy (F).

If the uniformity, of thermodynamic functions, condition is not met then, regardless of the actual physical dimension of the system or its number of molecules, the system can no longer be considered a *large* system and has to be dealt with as a *small* system for which classical definitions of the aforementioned thermodynamic functions are no longer valid.

For small thermodynamic systems the *quasi-thermodynamic assumption*,[22] sometimes called the *point thermodynamic approximation*,[23] assumes that it is possible to define unique and useful thermodynamic functions for the system at a point (r). This led to the definition of local thermodynamic functions at any point (r) in the system for the three fields:[24,25] local pressure $p(r)$, local temperature $T(r)$, and local chemical potential $\mu(r)$, as well as for the three densities: local number density $\rho(r)$, local energy density $\phi(r)$, and local free energy density $\psi(r)$. Note that the density functions can be defined by describing, around point (r), a small volume (δV) that contains a number of molecules (δN), and has an internal energy (δU), and free energy (δF) as follows:

$$\rho(r) = \lim_{\delta V \to 0} \left(\frac{\delta N}{\delta V} \right) \tag{1.1}$$

$$\phi(r) = \lim_{\delta V \to 0} \left(\frac{\delta U}{\delta V} \right) \tag{1.2}$$

$$\psi(r) = \lim_{\delta V \to 0} \left(\frac{\delta F}{\delta V} \right) \tag{1.3}$$

In addition, the local temperature $T(r)$ can be defined by knowing the translation kinetic energy (δK) of molecules within the system as:

$$\frac{3k}{2} T(r) = \lim_{\delta V \to 0} \left(\frac{\langle \delta N \rangle}{\langle \delta V \rangle} \right) \tag{1.4}$$

It is also important to realize that the local chemical potential $\mu(r)$ is taken as the chemical potential (μ) of the homogeneous material with a density equal to the local density at point (r) and at the same temperature (T), i.e.:

$$\mu(r) = \mu[\rho(r), T(r)] \tag{1.5}$$

According to the potential distribution theorem by Widom,[26] the local chemical potential [Equation (1.5)] can be determined by an exact solution as follows:

$$\frac{\mu(r)}{kT} = \ln[\Lambda \rho(r)] - \ln \left\langle \exp \left[\frac{u(r)}{kT} \right] \right\rangle \tag{1.6}$$

where, Λ is the de Broglie wavelength. The average in Equation (1.6) can be determined by a particle insertion method that is very well explained by Widom.[26–28] While Equation (1.6) is considered to be exact, well-defined, and computable, recent studies raised questions regarding the ability to define local temperature on the nanoscale. With advances enabling temperature measurement on the nanoscale with what was referred to as a "nanothermometer",[29] theoretical studies addressed the fundamental question of how meaningfully temperature can be defined on the nanoscale. Special attention was given to the smallest system size for which local temperature can be said to exist and can be defined.[30] Fundamental issues regarding the definition of local temperature on such a small scale, and the agreement between theory and experiments, are still to be resolved.[31]

To this end, we know that a thermodynamically homogeneous system is large and its physics can be described by classical thermodynamics and statistical mechanics. Equations of state can be developed for such a system; hence, its physical behavior can be predicted. A thermodynamically inhomogeneous system, however, has to be treated as a small system and local thermodynamic functions have to be determined for such a system to enable an accurate description of its physics. This raises a question regarding how determinable local thermodynamic functions are and the scale limit, if any, at which such functions become accurately indeterminable.

Let us consider the case of the local energy density function $\phi(r)$ defined in Equation (1.2). We realized that the energy of a small sample of matter (δU) indeed depends on interactions between pairs of chemical species within the system volume (δV). The system energy also depends on interactions with groups of such species that are not within the system volume (δV). Therefore, there is no unique way of determining how much of the interaction energy with groups outside the system should be ascribed to (δV). Hence, $\phi(r)$ cannot be uniquely determined. Another good example to consider is the local number density $\rho(r)$ function defined in Equation (1.1). While determining the number of species δN within a volume element δV seems to be a straightforward issue, hence, $\rho(r)$ is usually considered a one-body function that is always well-defined, we realize that under certain circumstances, especially when δV is very small, thermal fluctuations effects would make $\rho(r)$ uniquely indeterminable.[32]

Returning to the $\phi(r)$ example just discussed, let us consider two cases to elucidate the effect of system size (δV) on its thermodynamic treatment. First, consider the simple case in which the strength of interactions within the system is on the same order as that with species outside the system. Here, the physical size (δV) of the system clearly plays a crucial role in determining the relative contribution of interactions with species outside the system volume to the system energy (δU). It could be straightforward to realize that as the system volume increases, the outside interactions relative contribution to the system energy is less relevant. Therefore, the effect of such interactions on the ability to accurately determine the system energy (δU) will diminish as the system physical size increases. This simple case highlights the point that even thermodynamics of small systems has its own limits of applicability that depend on the system physical dimensions and raises a serious persistent question regarding such limits.

Now, let us consider the more realistic case in which the outside interactions are substantially stronger than the within the system interactions. This could be due to an external field affecting the system of consideration or due to inhomogeneity in the system resulting from an interface or a membrane for example. This case, unfortunately, represents most of the nanostructured systems that we could encounter and, hence, is essential to the field of nanotechnology. In such a case, the physical dimensions of the system (δV) play a much less significant role in determining the relative contributions to the system's energy (δU). The system physical dimensions can be significantly large, and still its local thermodynamic functions cannot be uniquely determined;

hence, its physics cannot be accurately predicted. This, not so simple case, highlights the point that physical dimensions of a system can be irrelevant to its thermodynamic behavior, and again raises the same crucial persistent question regarding limits of applicability. It is important to note that here we used the $\phi(r)$ function as an example. The same argument applies to other local thermodynamic field and intensity functions mentioned above. To this end, our previous discussion imposes one important question regarding the extent to which the local thermodynamic functions in a small system in equilibrium can be uniquely defined. The answer to this question is crucial if the physics of inhomogeneous or small systems is to be accurately predicted.

1.6 Scales of Inhomogeneity

It has been shown that the capability of uniquely defining local thermodynamic functions for fields and densities in a thermodynamically small system depends on the scale of inhomogeneity in the system. Several characteristic scales have been identified and discussed in the literature.[33]

1.6.1 Thermal Gravitational Scale

Since all systems of interest on planet earth are under the effect of its gravitational field, it is important to realize the restriction such a gravitational field might impose on the use of classical thermodynamic. For a one-phase system, the characteristic length associated with the thermal effect of a gravitational field (l_g) is given by:

$$l_g = \frac{kT}{mg} \tag{1.7}$$

where m is the molecular mass and g is the gravitational field.

For nitrogen gas (N_2), the main constituent of the earth's atmosphere, the gravitational characteristic length is about 9000 m. This means that for earth atmosphere and within 9000 m above sea level, local thermodynamic potentials, such as pressure and chemical potentials, can be defined and will be functions only of local temperature and density:

$$p(r) = p[T(r), \rho(r)], \quad \text{and} \quad \mu(r) = \mu[T(r), \rho(r)] \tag{1.8}$$

1.6.2 Capillary Length

If we consider a two-phase system, a liquid and a gas, in equilibrium under a gravitational field, the liquid phase will be at the bottom and the vapor will be above it with an interface in between. Owing to thermal fluctuations, the interface will instantaneously depart from planarity, creating what is referred to as *capillary waves*. The characteristics of these capillary waves substantially depend on the system size and the strength of the gravitational field. The characteristic length that governs the propagation of the capillary waves is known as *the capillary length* (l_c) and is defined as:[25]

$$l_c^2 = \frac{2\sigma}{g(\rho_l - \rho_g)} \quad (1.9)$$

where ρ_l, and ρ_g are the (mass) density of the liquid and gas phases, respectively, and σ is the liquid surface tension. Capillary length is typically in the range of 10^{-3} m. For the water–air interface at 0 °C in earth gravitational field, it is 3.93 mm. Water surface waves with wavelength much larger than the capillary length (*e.g.*, sea waves) will be completely governed by gravity. As the wavelength, however, becomes comparable to the capillary length, the surface tension of the liquid starts to play more significant role.

Considering liquid–gas interfaces with a radius of curvature (R) (as in the case of a liquid droplet), if R is comparable with l_c, then both gravity and surface tension are important to consider when assessing the liquid droplet physical behavior. If, however, the liquid droplet is very small compared with l_c, then surface tension effects are dominant and gravity effects can be ignored in the thermodynamic treatment of the system.

Several investigations have addressed the uncertainty in defining local thermodynamic functions applicable for spherical liquid droplets with a radius of curvature (R) smaller than the capillary length.[32,34,35] The work, however, could be essentially reduced to the purely mechanical solution that was obtained by Laplace based on mechanical arguments, and does not directly enable the definition of local thermodynamic functions for such a system.

1.6.3 Tolman Length

Another interesting point to consider is the applicability of Laplace's equation in situations where the droplet or bubble radius is in the nanoscale. This is a point that should be of extreme interest in the field of nano-fluids. According to Laplace's equation, the excess pressure inside a liquid droplet or a bubble ($\Delta p = p^l - p^g$) can be expressed as:

$$\Delta p = p^l - p^g = \frac{2\sigma}{R} \quad (1.10)$$

Tolman,[36] however, showed that the surface tension (σ) decreases with decreasing the droplet radius over a wide range of circumstances and that it can be expressed as:

$$\sigma_R = \sigma_\infty \left(1 - \frac{2\delta}{R} + \cdots \right) \quad (1.11)$$

where, σ_R is the tension of a surface of a radius R, and σ_∞ is the tension of a planar surface. Here, δ is the *Tolman length*. Importantly, in Equation (1.11), the Tolman length is defined as a coefficient in an expansion in $1/R$ and, therefore, is not a function of R. In the literature,[37] other definitions of the Tolman length have been given in which it is defined as a function of R to account for deviations from the planar limit to all orders in $1/R$. Such definition was expressed as:

$$\delta = R_e - R \tag{1.12}$$

where R_e is the equimolar radius of the liquid droplet.[iii]

Defined as a coefficient in an expansion in $1/R_e$, the Tolman equation can yield an expression for the excess pressure inside a liquid droplet (Δp) as:

$$\Delta p = \frac{2\sigma}{R_e}\left(1 - \frac{2\delta}{R_e} + \cdots\right) \tag{1.13}$$

with the first term being the same as Laplace's equation.

The Tolman length has received considerable theoretical attention. However, some issues still remain completely unresolved.[38] While there is some agreement in the literature that it is not a sharp dependent of temperature, a huge controversy regarding its value exists. It has been shown more recently that the value of Tolman length sensitively depends on the interaction potentials in the liquid.[39] The discrepancy in its sign and its dependence on the interaction potential is still not understood. The most widely accepted value for the Tolman length ranges between 1 nm and a fraction of a nanometer. The important point to realize here, however, is that the surface tension, as a physical property of a liquid, cannot be uniquely defined as the droplet radius is on the order of Tolman length. This is a very clear example of the effect of entering the nanodomain on our understanding of materials behavior. A simple physical property, such as surface tension, first investigated by Leonardo da Vinci in the early fifteenth century and officially introduced by J. A. Segner in 1751, becomes completely unexplored and un-understood once the system size becomes on the order of Tolman length, *i.e.*, 1 nm or less.

1.6.4 Line Tension (τ) and the (τ/σ) Ratio

Another important area of nanotechnology covers self-assembly and formation of molecular monolayer films known as Langmuir–Blodgett films. The technique depends on the spreading of a substance, usually a polymer or nanoparticles in a solvent, on the surface of water to form a continuous ultra-thin film. Depending on the nature of spread solution, it may spread and form the desired film or may not spread and form a set of liquid lenses on the surface of the water. Thermodynamic investigation of the spreading (wetting) and non-spreading (non-wetting) regimes revealed another important scale of

---

[iii] For elaborated discussion, the reader is referred to ref. 37

inhomogeneity that is crucial for such a process. This length scale is the *line tension* (τ) to the surface tension (σ) ratio. Let us further clarify this point.

If three phases, α, β, and γ (in our case these three phases would be the water substrate, the spread solution, and air), meet at three surfaces, and if the three surfaces meet at a line, then the free energy (F) of such a system can be expressed as:[23]

$$F = V^\alpha \psi^\alpha + V^\beta \psi^\beta + V^\gamma \psi^\gamma + A^{\alpha\beta}\sigma^{\alpha\beta} + A^{\alpha\gamma}\sigma^{\alpha\gamma} + A^{\gamma\beta}\sigma^{\gamma\beta} + L^{\alpha\beta\gamma}\tau^{\alpha\beta\gamma} \quad (1.14)$$

where V denotes the phase volume measured to the equimolar dividing surface, A denotes the contact area between the phases, L is the length of the three phases contact line, and $\tau^{\alpha\beta\gamma}$ is the line tension. Notably, unlike the surface tension which is always positive, the line tension can be positive or negative. The ratio (τ/σ) is very important for the physics of small systems. For soap solutions, for example, this ratio is on the order of 20 nm. Hence, clearly, for droplets or bubbles with dimensions between 1 mm and 1 μm the physics of the system is mainly determined by surface tension and can be handled by ignoring both gravity and line tension. However, if the system dimensions are on the order of the (τ/σ) ration (20 nm or less) the effect of line tension cannot be disregarded and has to be accounted for in any attempt to understand the behavior of the system.

Another important length scale to consider, especially while dealing with nanoparticles, is the surface tension (σ, in $J\,m^{-2}$) to the bulk energy density (ϕ, in $J\,m^{-3}$) ratio (σ/ϕ). Such a ratio has a length scale usually in the nanometer range, or less, depending on the nature of the material. For example, the ratio for argon at its boiling point is in the range of 0.5 nm. The ratio for water at 25 °C drops down to about 0.3 Å.

Talking about line contributions, we can still consider the case where multiple lines (usually three as in the triple point in foams) of phase contacts meet in a point. This can hardly be considered tension; however, it contributes to the system's free energy independent of the system size.[23]

1.6.5 Correlation Length (ξ)

In any system, regardless of its physical size, fluctuations in one region of the system can influence other regions in the system. In such a situation, the two regions of the system are said to be *correlated*. The *correlation length* (ξ) is the distance or the range over which different regions, or points, in the system can be correlated. This means that if two points in the system are separated by a distance equal to or smaller that the system's correlation length then these two points will influence each other and any fluctuations or changes in one point will be felt by the other. However, if the two points are separated by a distance that is larger than the system's correlation distance, then the two points will not *feel* each other and each of them will behave as if the other does not exist. As we said before, point correlations in systems do exist regardless of the size of system. For example, correlations among the planets in our own and other solar systems do affect the planets trajectories, with a gravitational correlation length on the order of a few million miles. Tidal phenomenon on our planet is a direct result of the correlation between Earth and its moon. Correlations in the distribution of galaxies were also found to provide some important clues regarding the structure and evolution of the Universe on scales larger than individual galaxies.[40]

In a material system, however, correlation length, or molecular interaction range, is typically considered to be on the order of 1 nm. Recent studies, however, showed that correlation length in certain systems can extend in the range 7–10 nm.[41,42] The correlation length (ξ) in materials was also found to depend on the temperature (T) and to diverge at the system's critical point (T_c). Mathematically, it can be expressed as:

$$\xi \sim |t|^{-v} \text{ and } t = \frac{T - T_c}{T} \tag{1.15}$$

where v is known as a *critical exponent*.

Divergence of the correlation length at the critical point means that the correlation length becomes very large. Therefore, molecules very far apart become correlated, and the system's physical behavior, regardless of the system size, becomes completely different from what we are accustomed to. A few centuries age, and still in some cultures, this natural phenomenon would have been interpreted as *a miracle*, at best, or *the act of evil spirits*, at worst! Nowadays, we should realize that this is nanophysics in action.

Away from the critical point, where the correlation length extends to reach the physical dimension of a system, if we reduce the physical dimensions of a system to match its correlation length, then the same phenomenon can be observed. For a system in such a state, thermodynamic local functions cannot be uniquely defined. Hence, the system behavior cannot be mathematically expressed based on statistical mechanics principals or predicted.

Before we move onto the next point, a crucial issue needs to be emphasized. Let us consider a well-understood phenomenon such as capillarity, which was first investigated by Leonardo Da Vinci in the early fifteenth century and also was the subject of Einstein's first real paper in 1901.[43] The height (h) of a liquid column inside a tube of radius (r) is well known to depend on the liquid surface tension (γ), density (ρ), and its contact angle with the tube walls (θ). It can be mathematically expressed as:

$$h \approx \frac{2\gamma \cos \theta}{\rho g r} \tag{1.16}$$

For water in a glass tube at ambient conditions, Equation (1.16) can be reduced to:

$$h \approx \frac{14 \times 10^{-6}}{r} \text{ m} \tag{1.17}$$

where r is in meters. Hence, if the tube internal diameter is in the meter range, h will be in the micrometer range, and the phenomenon is hardly observed. Once the tube internal diameter is reduced into the millimeter range, h will be in the centimeter range, and the phenomenon becomes very observable. In this case, the system size enables surface tension effects to override gravitational effects and to dominate the physical behavior of the system. The question may arise as what if the tube internal diameter becomes in the nanometer range? Equation (1.17) would predict that h in this case should be about 14 km! This, in fact, is not true since once the system is confined to a size on the order of its correlation length (ξ), the correlation between water molecules and the silica molecules in the tube walls will take over and the a new set of equations, that still has not been developed, is needed to predict the system's physical behavior. Recent experimental investigations revealed unfamiliar physical phenomena and violation of established laws by water confined to nanometer-size pores.[44,45]

To this end, we have seen that our current understanding and ability to predict the physics of materials systems is based upon our ability to formulate thermodynamic equations of state for such systems. We have also discussed the fact that once thermodynamic inhomogeneity exists in a system at certain length scales, the system has to be treated as a thermodynamic small system that requires the definition of local thermodynamic functions. Most importantly, we have discussed different scales of inhomogeneity at which local thermodynamic functions cannot be uniquely defined. We have also shown that the physics of a system depends on where the system size lies in respect to these characteristic length scales. We find that:

$$\xi \sim \sigma/\phi \sim \delta \sim \tau/\sigma \sim 1 \text{ nm} \tag{1.18}$$

Once the system physical size is on the order of its correlation length, the system is thermodynamically small, or in other words, the system is nanostructured.

Interestingly, the correlation length in the Universe is on the order of millions of light years, which makes the Universe a nanostructured system! A nanosystem is not necessarily small after all.

1.7 Thermodynamics of Small Systems

As discussed above, our only theoretical tools to understand the physics of matter, *i.e.*, thermodynamics and statistical mechanics, are derived for large, or macroscopic, systems, and are very limited once the system size approaches one of the aforementioned scales of inhomogeneity. This point is not new and was very well recognized and tackled by Terrell L. Hill, in 1962. Hill is considered the father of the thermodynamics of small systems. He was the first to point out the difference in thermodynamic treatment between large systems, where the number of molecules in the system tend to infinity ($N \to \infty$), and small systems, where the number of molecules is limited.[46] The material system that caught his attention in the early 1960s was colloidal particles and latter in the late 1960s he applied his treatment to biological systems.[47] Hill's aim of the proposed thermodynamic treatment was to extend the range of validity of classical thermodynamic definitions and interrelations into "nonmacroscopic systems".[46,48] Notably, the "nonmacroscopic systems" term used in the early 1960s is what we refer to as "nanosystems" nowadays. Another interesting point to realize is that Feynman's invitation to explore small systems announced in his famous lecture in December 1959 was already recognized, most probably independently, by Hill who submitted his first thermodynamic treatment of small systems in January 1962.

In his treatment of thermodynamic small system, Hill proposed to utilize "correction terms" to account for interactions that cannot be ignored as the system size hits the "*small*" boundaries. By this, Hill showed that thermodynamic equations can be generalized so that they will be valid for small systems.

To explain, let us consider a large one-component system. The Gibbs free energy (G) of such a system can be expressed as:

$$G = Nf(p, T) \tag{1.19}$$

where N is the number of particles (molecules) in the system, and f is a function of pressure (p) and temperature (T) only. The chemical potential (μ) of the molecules in such a system can be expressed as:

$$\mu = \left(\frac{\partial G}{\partial N}\right)_{p,T} \tag{1.20}$$

In the large thermodynamic system limit where $N \to \infty$, the system free energy $G \to Nf$, and hence the chemical potential $\mu \to f(p,T)$, or, in other words, becomes a function of pressure and temperature only. Now, if the system is small enough, Hill proposed the addition of correction terms to the original thermodynamic expression to ensure its validity. For example, the expression for Gibbs free energy should include such correction terms as:

$$G = Nf(p, T) + a(p, T)N^{\frac{2}{3}} + b(T) \ln N + c(p, T) \tag{1.21}$$

where a, b, and c are functions of pressure and temperature.

Here, Hill added the correction terms $N^{2/3}$ to account for surface effects that are known to contribute to the free energy of the system on that order, $\ln N$ to account for phonon confinement effects, and a size independent term (c) to account for point contributions as mentioned in Section 1.6.3. He also pointed out that if line (or second-order surface) contributions to the free energy of the system were to be accounted for, a new term of order $N^{1/3}$ should also be included. In addition, if surface contributions to the free energy of a two-dimensional system (a single graphene sheet is a system of current interest, for example) were to be considered, a correction term of the order $N^{1/2}$ should also be included. Hence, Equation (1.21) can be generally written with as many correction terms as needed to fully account for all possible contributions. For example, an equation such as:

$$G = Nf(p,T) + a(p,T)N^{\frac{2}{3}} + b(T)N^{\frac{2}{3}} + c(T)\ln N + d(p,T) \qquad (1.22)$$

is meant to account for volume, surface, line, phonon confinement, and point contribution to the system.

Assuming that the f, a, b, c, and d functions are completely defined, and that they are strictly functions of pressure and temperature only, Equation (1.22) can be used to describe thermodynamic functions in physically small systems as defined by a limited number of molecules (N).

Another major point of Hill's treatment of thermodynamics of small systems is noting that while the effect of the system's environment can be safely neglected in thermodynamic treatment of a large system the same cannot be done for a small system. The environment of a thermodynamic small system plays a major role in its physical behavior and must be accounted for in any thermodynamic treatment of the system.

Fluctuations are also a crucial point that need to be considered in thermodynamic treatment of small systems. In a system containing N particles, fluctuations are usually on the order of $N^{-1/2}$. For a large system where $N \to \infty$, thermal fluctuations are negligible. However, for a small system of N in the range of 100, e.g., a hundred-atom cluster, fluctuations are on the order of 10%, which cannot be neglected. As Hill pointed out, it is important to note that the physics of small systems can be dominated by their fluctuations.

More recently, Hill commented on the relevance of his approach of thermodynamics treatment of small systems to nanosystems using the term *nanothermodynamics*,[49] and also derived general equations for energy fluctuations in small systems.[50]

1.8 Configurational Entropy of Small Systems

For large systems that would mix ideally at a certain temperature and pressure, the driving force for mixing is the configurational entropy. This is a well-known natural phenomenon that takes place in solids (copper and gold), liquids (water and methanol), and gas (oxygen and nitrogen) bulk systems. In such cases, the two components in the system, as we all know and expect, would mix to maximize the system's configuration entropy, hence reaching the state of equilibrium as we observe it. The interesting question would be "would these components mix in a small system configuration as they do in a large system configuration?" It is widely known that, based on statistical mechanics concepts, the state of equilibrium for a system is the most probable state of the system. Configurational states of a system are just the different configurations the system can assume. In other words, these are the different ways the system

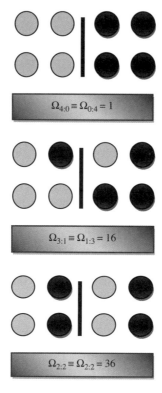

Figure 1.9 Representation of two small systems in contact.

can arrange its components while maintaining equilibrium (as defined by the minimum free energy state) under the effect of constant, temperature, pressure, and other thermodynamic fields. To address our question, let us consider the simplest case possible for mixing as controlled by configurational entropy. A system of four atoms of a component (a) and four atoms of a component (b) with an interface as shown in Figure 1.9. The number of ways (Ω) such eight atoms can be distributed among the available eight spatial positions can be calculated according to:

$$\Omega = \frac{N!}{N_a! N_b!} \quad (1.23)$$

where N_a, and N_b are the number of a and b species, respectively, and $N = N_a + N_b$, is the total number of atoms or molecules in the system, which is eight in our case.

The first configuration of the system to consider is the configuration where the atoms are not mixing and each species is isolated on one side of the interface. In this case the four a atoms are on one side and the four b atoms are on the other side of the interface. The number of ways the system can assume this configuration is clearly 1. We will express this as $\Omega_{4:0} = 1$. If one of the a species crosses the interface – note that this will require that one of the b species also crosses the interface in the opposite direction – it can be shown that this partial mixing situation configuration of the system can be assumed by 16 different ways, and hence, $\Omega_{3:1} = 16$. Now let us consider the case when two of the a species cross the interface to generate the total mixing situation where the composition will be homogeneous across the system. In this case, the number of spatial distribution of the atoms among available spatial configurations is 36, so $\Omega_{2:2} = 36$. Similar calculations can be reached for the remaining two configurations $\Omega_{1:3}$ and $\Omega_{0:4}$.

Now let us address the equilibrium state issue, which of the aforementioned five configurations would be the most probable configuration that the system would assume? Well, the total number of ways the system can distribute its eight constituents among the available eight spatial positions turned out to be $1 + 16 + 36 + 16 + 1 = 70$ [see also Equation (1.23)]. This means that the probability of the five different configurations we discussed should be 1/70, 16/70, 36/70, 16/70, and 1/70, respectively. This indicates that the two unmixed configurations would have a probability of 1.4% each, the two partially mixed configurations would have a probability of 22.8% each, and the totally mixed configuration would have a probability of 51%. The totally mixed situation in this case is the *most probable state* of the system. However, it is not the *only possible state* of the system. Clearly, from our example, a system made of only eight atoms would spend only 51% of its time in the totally mixed state and the other 49% of its time will be in the partially or unmixed states.

If the system is large in size ($N \to \infty$) the *most probable state* becomes *the only possible state*. Hence, the system would spend 100% of its time in such a state, which we refer to as the state of equilibrium. This is why if we lay a typical object (N is on the order of 10^{23}) on the table, the object stays there as long as no other external field causes it to move. It remains in its only possible state. For a small system, however, with a limited number of atoms, equilibrium is truly the most probable but not the only possible state. Other non-equilibrium states are possible too and, hence, it is possible, according to statistical mechanics principles, that a theoretical molecular assembly made of limited numbers of molecules, such as the one shown in Figure 1.10, would suddenly reverse its rotation direction or separate and assume a different configuration.

This by no means should mislead the reader into believing that molecular assemblies are not thermodynamically possible or stable. Nature has been designing and utilizing very successful and efficient *molecular machinery* for millennia. Bacteria and green plants run the *nano-machinery* that makes life possible using the very same clean, sustainable, and readily available source of power that we are still trying to harness – sunlight. Other forms of natural molecular machines utilize other forms of energy to function. Humans are a good example for fuel diversity. Several studies and reviews have described the concept and first principles of designing molecular machines or nano-machines similar to those perfected by nature.[51-61] Our previous discussion was meant to indicate that designing and utilizing nanostructured systems referred to as

24   Chapter 1

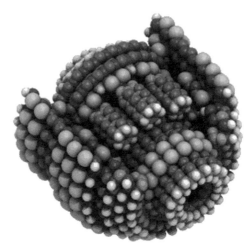

Figure 1.10   The MarkIII(k), a nanoscale planetary gear designed by K. Eric Drexler. A planetary gear couples an input shaft via a sun gear to an output shaft through a set of planet gears (attached to the output shaft by a planet carrier). The planet gears roll between the sun gear and a ring gear on the inner surface of a casing. This and other molecular assemblies are available on the World Wide Web at http://www.nanoengineer-1.com/content/index.php?option = com_content&task = view&id = 40&Itemid = 50.

molecular machinery necessitates a deep consideration, appreciation, and understanding of the thermodynamic of such small systems. The subject of nanosystems is too sophisticated to be considered from a single discipline viewpoint. The new frontier of nanotechnology is interdisciplinary in nature and any advances in such a field require collaboration and deep appreciation of physics, chemistry, materials science, biology, in addition to several different disciplines of engineering. It is crucial to realize that nature in its 3.5 billion year quest to design efficient molecular machinery has designed molecular machines capable of harnessing Gibbs free energy changes and transforming such changes into any of the other forms of energy, including mechanical work. Different nano-machines designed by nature can harness changes in the Gibbs free energy of the system resulting from changes in any of the thermodynamic potentials affecting the system. These include changes in mechanical force or pressures, changes in length or volume, changes in temperature, changes in chemical potential or numbers of a chemical species, changes in the oxidative state or numbers of electrons of chemical moieties, changes due to the absorption of electromagnetic radiation, or even changes in the extent of spatial order (the entropy) of the system. Figure 1.11 demonstrates all possible (observed and supposed) energy conversions that living organisms (molecular machines based upon protein molecules) can perform. Bold arrows indicate the energy conversions that have been observed.[61]

It is interesting to realize that natural polymer molecules (in the form of proteins) are not the only moieties capable of assembling molecular machines.

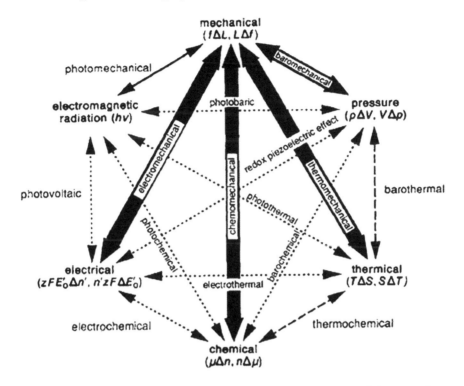

Figure 1.11 The (observed and possible) energy conversions that living organisms (molecular machines based upon protein molecules) can perform. Bold arrows indicate the energy conversions that have been observed. (Reproduced with kind permission from D. Urry, ref. 61. Copyright VCH, 1993.)

Certain manmade polymers[55] and other smaller molecules[62] can be utilized as well. With so many energy conversion possibilities, tremendous number of building blocks, and a deep understanding of the physics of small systems, nanotechnology can definitely benefit humankind if wisely utilized.

To summarize our discussions in previous sections, it is clear that thermodynamic functions and equations of state as based upon fundamentals of statistical mechanics are our most powerful and only tool to understand and be able to predict the physics of matter. Such tools, however, have limits of applicability once certain length scales of inhomogeneity are approached. As discussed earlier, such length scales can be approached regardless of the physical size of the system. Once such boundaries of inhomogeneity length scales are approached, the system becomes a nanostructured system. In nanostructured systems, forces that usually have negligible effects on the physics of the system become dominant and, basically, control the system's behavior, leading to the observation of new, and unusual, physical phenomena. In addition, we briefly discussed the principles and basic concepts of nano-machines. The important message to remember here is that nano-machines, as a product of nanotechnology, are not about making miniature versions of our machines based on the same concepts of design we currently utilize. In fact, nano-machines should be based upon concepts utilized by nature, the same thermodynamic concepts determining the behavior of small systems. The following section gives several examples to demonstrate different physical phenomena observed for typical material systems once they approach the nano-domain.

1.9 Nanophenomena

Several phenomena have been recently observed that can be attributed to the behavior of matter in the nanodomain. In this section we mention some of such very interesting phenomena and their immediate impact on the development of our technology.

1.9.1 Optical Phenomena

We mentioned before how nature utilized nanostructure in butterflies wings and opal stones to develop amazing colors. In addition, gold nanoparticle solutions have been shown to yield different colors depending upon the gold particle size and shape.[63–65] Figure 1.12 shows solutions of nano-gold particles of different sizes in water. Interestingly, for gold particles of 2 nm size (far left) the particle size is too small to affect the light interaction with the solution, resulting in a solution with optical properties similar to those of water. As the gold particles increase in size, their interaction with the light becomes significant, causing the observed different colors of the solution. It is widely

Figure 1.12 Monodispersed gold nanoparticles of different sizes (2–200 nm) in water. Note the different colors due to different light interactions, which are dominated by particle size. (Courtesy of Ted Pella, Inc., available on the World Wide Web at http://www.tedpella.com/gold_html/goldsols.htm.)

Nanotechnology, the Technology of Small Thermodynamic Systems 27

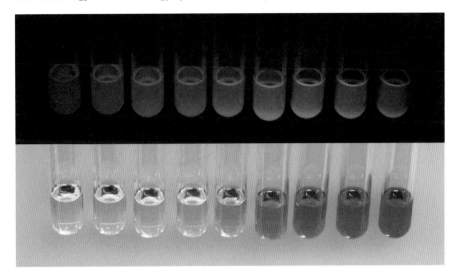

Figure 1.13 Cadmium selenide (CdSe) quantum dots of different sizes observed in visible light (bottom) as well as dark light (ultraviolet radiation) (top). (Courtesy of the University of Wisconsin-Madison, Materials Research Science and Engineering Center. Available on the World Wide Web at: http://www.mrsec.wisc.edu/Edetc/background/quantum_dots/index.html.)

known that light interaction with the solutions in this case is strongly affected by the electronic structure of the solution. Figure 1.12 is a good example of particle size effect on electronic structure as revealed by its interaction with light.

Such an optical nanophenomenon is not limited to metallic particles. Semiconducting nanoparticles (sometimes referred to as quantum dots) also show very interesting dependence of their electronic structure on the particle size, as reflected by their interaction with visible light and electromagnetic radiation in general.[66–71] Figure 1.13 shows nanoparticles of cadmium selenide (CdSe) of different sizes in solution observed under radiation with visible light (bottom) as well as dark light (ultraviolet radiation) (top). Clearly, from the photo, the size of the quantum dot affects its electronic structure as reflected by the different solution colors. More interestingly, as shown in the top photo, irradiation with ultraviolet radiation (dark light) causes the solution to irradiate different colors in the visible range. This is another interesting example of an optical nanophenomenon with potential applications in solar energy harvesting.[72–76]

It was also interesting to realize that not only particle size but also particle shape affects its optical properties and controls the system absorption behavior. It was found that gold nanorods interact differently with light based upon the rod length to diameter (aspect) ratio. Depending upon their aspect ratio, gold nanorods absorb light with a maximum located at different wavelengths in the visible and the near-infrared spectra range. Figure 1.14 shows the maximum

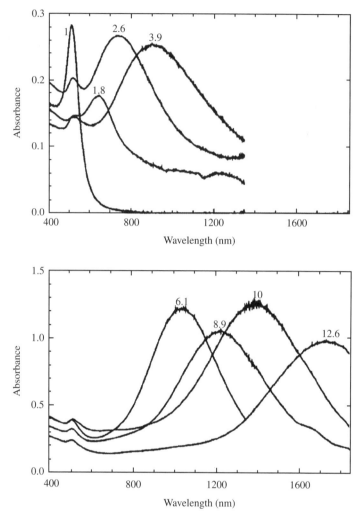

Figure 1.14 Absorption spectra for gold nanorods of different aspect ratios. Note the shift in maximum absorption wavelength at the different aspect ratios denoted at the top of each curve. Aspect ratio 1 denotes gold nanospheres. (Reproduced with kind permission from van der Zande et al., ref. 77. Copyright American Chemical Society, 1999.)

absorption wavelength for gold nanorods of different aspect ratios.[77] Notably, gold nanospheres (denoted by an aspect ratio 1 in Figure 1.14) absorb light at a different wavelength as well.

In addition, the maximum absorption wavelength was also shown to depend on the orientation of the gold nanorods in respect to the polarization direction of the light. The maximum wavelength of light absorption for light polarized parallel to the nanorods axis ($\lambda_{max}^{\parallel}$) differs from that for a light polarized in a direction normal ($\lambda_{max}^{\perp}$) to the nanorods axis direction.[78] The phenomenon was related to a resonance effect in the particle surface plasmons activity, The surface plasmon resonance (SPR) effect that, in turn, depends on the geometry of the nanorods was found to control the position of a maximum in the optical adsorption spectrum of such nanoparticles. Figure 1.15 shows the dependence of the ($\lambda_{max}^{\parallel}$) on the aspect ratio of the gold nano-rods. It has been shown that ($\lambda_{max}^{\parallel}$) strongly shifts linearly to longer wavelengths as the rod aspect ratio increases while ($\lambda_{max}^{\perp}$) shifts to lower wavelengths as the rod aspect ratio increases, reaching a plateau for rods with an aspect ratio higher than five.[78]

Optical properties of nanospheres with core–shell structures have also been investigated.[65,79–85] Various types of nanoparticles with a core–shell structure could be prepared using different methods. Figure 1.16 shows the different types of nanoparticles with shell–core structures that have been produced and investigated. They range from short molecules tethered to surface-modified core particles (Figure 1.16a) to multi-shell (onion-like) structures (Figure 1.16e).

Different types of materials (*i.e.*, metallic, dielectric, *etc.*) were also used to make such nanoparticles. Interestingly, the optical absorption of these nanoparticles showed dependence on the relative dimensions of the core and shell as well as the exact structure of the particle. It was reported that the observed optical resonance and maximum absorption wavelength can be varied over hundreds of nanometers in wavelength across the visible and into the infrared region of the spectrum. The work of Halas *et al.*[65,79–84] in this area pointed out the crucial possibility of engineering such structures on the nanoscale to produce a new class of matter with well-designed optical properties for several applications. Figure 1.17(a) shows the theoretical calculations for the resonance optical absorption spectra of a 60 nm silica core with gold shells of different thicknesses. Figure 1.17(b) shows the calculated optical resonance wavelength for the same nanoparticles.[79] Intriguingly, the resonance wavelength increases (shifts from the visible into the infrared spectral region) as the shell thickness is reduced from 20 to 5 nm. Figure 1.18 shows the different colors obtained from gold nanoshell particles with different shell–core relative dimensions dispersed in water, in support of the theoretical calculation results shown in Figure 1.17.

In addition, self-assembled nanoparticle monolayers exhibit optical absorption spectra that not only depend on the properties of the individual particles but also on their mutual interactions, as well as on their interaction with their environment.[86,87] With their optical properties depending on the shape, size, composition, and their exact structure in addition to their interaction with each other and their environment, nanoparticles represent a very fertile ground and a virtually unexplored frontier for future investigations. Figure 1.19(a) shows a schematic drawing of a multilayer film assembled using a layer-by-layer technique of silica-coated gold nanoparticles embedded in cationic polyelectrolyte on a glass substrate (1 = glass substrate, 2 = cationic polyelectrolyte, 3 = nanoparticles), and (Figure 1.19b) photographs of transmitted (top) and reflected (bottom) colors from the multilayer thin films with varying silica shell thickness.

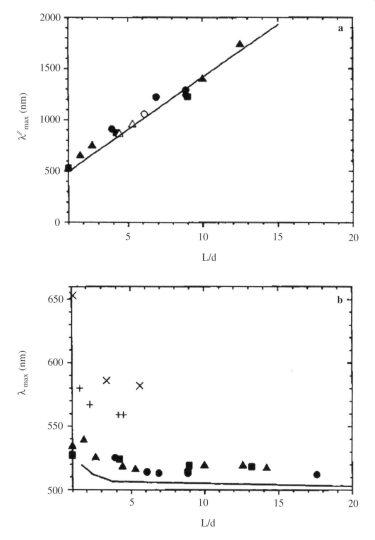

Figure 1.15 Dependence of the plasmon absorption maxima for (a) longitudinal ($\lambda_{max}^{\parallel}$) and (b) transverse ($\lambda_{max}^{\perp}$) polarizations conditions on the gold nanorod aspect ratio for gold nanorods dispersed in water. Filled squares, circles, and triangles represent experimental data obtained from rods with diameters of 12, 16, and 20 nm, respectively. Open symbols indicate samples with a narrow size distribution. The + and × symbols represent rods with diameters of 86 and 120 nm, respectively. (Reproduced with kind permission from van der Zande, ref. 78. Copyright American Chemical Society, 2000.)

1.9.2 Electronic Phenomena

It is very well known that electronic and optical properties of matter are very well correlated. The previously mentioned unusual and versatile optical

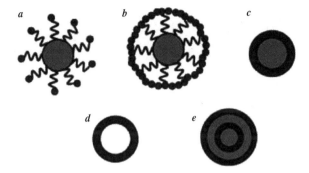

Figure 1.16 Variety of core–shell particles produced and investigated: (a) surface decorated particles, (b) heavier surface decoration, forming a shell around the core, (c) solid shell around a core, (d) quantum bubbles, and (e) multilayered particles.

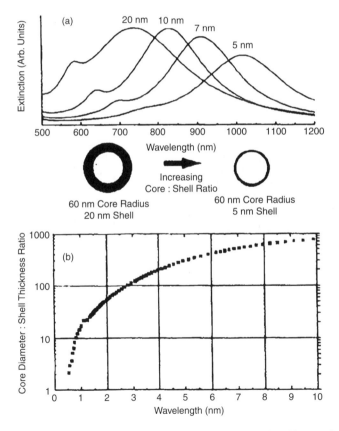

Figure 1.17 (a) Theoretically calculated optical resonance of gold nanoshells of different thicknesses on a 60 nm silica core. (b) Calculated optical resonance wavelength for the same samples in (a). (Reproduced with kind permission from Oldenburg *et al.*, ref. 79. Copyright Elsevier 1998.)

Figure 1.18 Photograph showing colors of gold nanoshell particles with different shell thicknesses dispersed in water. Core:shell ratio decreases from left to right. (Photo by C. Rodloff. Courtesy of Professor Halas's group of Rice University, USA.)

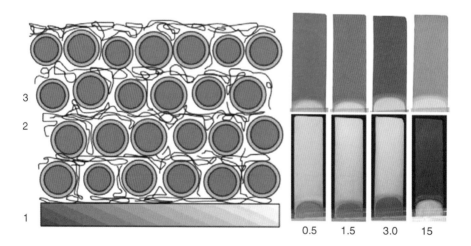

Figure 1.19 (Left) Schematic drawing of a multilayer film assembled using the layer-by-layer technique of silica-coated gold nanoparticles embedded in cationic polyelectrolyte on a glass substrate (1 = glass substrate, 2 = cationic polyelectrolyte, 3 = nanoparticles). (Right) Photographs of transmitted (top) and reflected (bottom) colors from the multilayer thin films with varying silica shell thickness. (Reproduced with kind permission from L. Liz-Marzán, ref. 86. Copyright Elsevier 2004.)

phenomena observed for nanoparticles are simply the result of an unusual and versatile electronic structure of such a new class of matter, or of such old matter observed in a new domain. The electronic structure of small metal clusters has been an active and fertile field of investigation for over two decades now.[88–91]

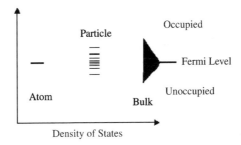

Figure 1.20 Electronic structure of a metal as an atom, a particle, and in the bulk state.

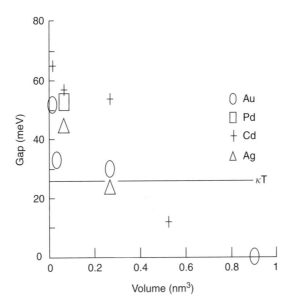

Figure 1.21 Conduction gap observed in small clusters of gold (Au), palladium (Pd), cadmium (Cd), and silver (Ag) as a function of cluster volume. (Adapted with kind permission from Vinod et al., ref. 93. Copyright Elsevier 1998.)

There has been good progress in understanding the nature of small metal clusters and a realization that they cannot be simply treated as minute elements of a block of a metal. This is based on research findings showing that the conduction band present in a bulk metal will be absent and instead there would be discrete states at the band edge.[92] Figure 1.20 illustrates the electronic band structure of metals in the bulk form and in the atomic form, passing through the nanodomain.

The very well known conductive nature of bulk metals is no longer a given fact as metals assume smaller and smaller clusters. Energy gaps in the electronic structure of conductor clusters with a value dependent of the cluster size becomes observable.[93] Figure 1.21 shows the conduction gap observed in small clusters of gold (Au), palladium (Pd), cadmium (Cd) and silver (Ag) as a function of cluster volume. The energy bandgap for all tested metals increased with the reduction of metal cluster volume.

The dependence of electronic structure, as reflected in the size of the energy bandgap, on matter cluster size was not limited to metal clusters. In fact a similar phenomenon was observed for semiconductor clusters as well.[94–101] The traditional band energy gap in semiconductors was also found to depend on the cluster size. Figure 1.22 shows the energy bandgap for different semiconductors as measured by the wavelength of the absorption threshold. It can be inferred from the experimental results shown in Figure 1.22 that the energy bandgap remains size independent till the cluster becomes smaller than a critical size, below which the energy bandgap increases sharply. Such a critical size value apparently depends on the energy gap in the bulk state of the semiconductor. The smaller the bulk-state energy bandgap of the semiconductor the larger the critical size of the cluster. It is also clear from the results that the smaller the bulk state energy bandgap of the semiconductor is the more sensitive the size dependence becomes as the semiconductor cluster becomes smaller than its critical size.

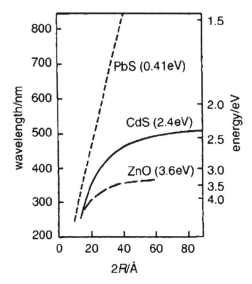

Figure 1.22 Size-dependence of the wavelength of the absorption threshold in semiconductors. Values in parentheses represent the energy bandgap of the semiconductor in the bulk state. (Reproduced with kind permission from A. Henglein, ref. 99. Copyright American Chemical Society 1989).

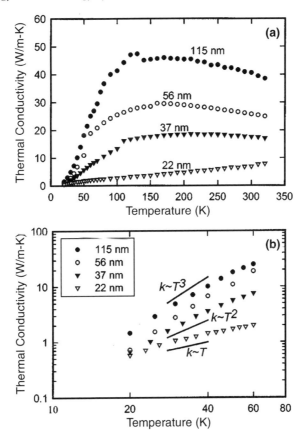

Figure 1.23 (a) Measured thermal conductivity of different diameter Si nanowires. The number beside each curve denotes the corresponding wire diameter. (b) Low temperature experimental data on a logarithmic scale. Also shown are T^3, T^2, and T^1 curves for comparison. (Reproduced with kind permission from Li et al., ref. 107. Copyright American Institute of Physics 2003.)

1.9.3 Thermal Phenomena

Given the very well known fact that thermal properties of matter are direct functions of its thermodynamic behavior, it is not surprising that once matter is in the nanodomain its thermal properties would become unusual. For example, heat capacity as the first derivative of the thermodynamic function (enthalpy) is expected to have a system size dependence based upon our previous discussions of the thermodynamic behavior of small systems and how it becomes sensitive to system size once that latter entered the nanodomain. In addition, a phenomenon such as thermal conductivity that mainly depends on electronic and phononic behavior of the system should be expected to exhibit anomalous behavior in systems approaching or in the nanodomain. In fact, recent investigations have confirmed such expectations.[31,102–110]

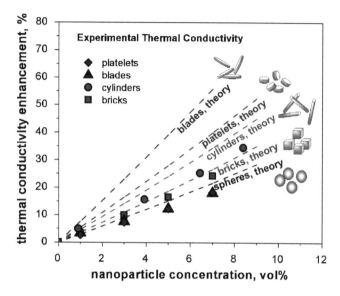

Figure 1.24 Experimentally measured thermal conductivity of alumina nanofluids in EG/H$_2$O compared to predictions of theoretical models. (Reproduced with kind permission from Timofeeva et al., ref 102. Copyright American Institute of Physics 2009.)

Recent theoretical studies have shown that as a solid enters the nanodomain its thermodynamic constants, such as Debye temperature, and specific heat capacity, are no longer constant but become size dependent. These quantities were also shown to change with the system temperature and the nature of the chemical bonds involved. While surface effects and quantum confinement explanations are most appealing, the exact mechanisms of such intriguing phenomenon and the correlation among these quantities, however, are still to be clarified. The fact that matter on the nanometer scale can exist in one-dimensional forms such as nanowires and nanotubes adds an appealing dimension to thermal transport in this class of matter. For example, the thermal conductivity of individual single crystalline intrinsic silicon nanowires with diameters of 22, 37, 56, and 115 nm was recently measured over a temperature range of 20–320 °K and was reported to be more than two orders of magnitude lower than the bulk value.[107] Figure 1.23(a) shows the measured thermal conductivity for silicon nanowires of different diameters, ranging between 22 and 115 nm. The dependence of the thermal conductivity of the nanowires on the wire diameter is very clear. Figure 1.23(b) shows the logarithm of thermal conductivity of the same nanowires in the low temperature range plotted against the logarithm of temperature. Interestingly, the data show that the behavior of the large diameter nanowires fits the well-known Debye T^3 law quite well in this temperature range. However, for the smaller diameter wires, 37 and 22 nm, the power exponent decreases as the diameter decreases and deviation from the Debye T^3 law is clear.

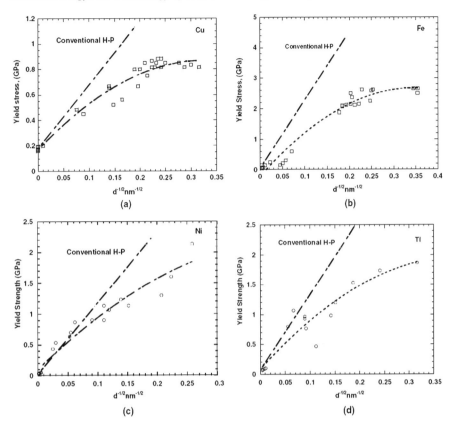

Figure 1.25 Plots showing the trend of yield stress with grain size for different metals as compared to the conventional Hall–Petch (H-P) response: (a) copper, (b) iron, (c) nickel, and (d) titanium. (Reproduced with kind permission from Meyers et al., ref. 116. Copyright Elsevier 2006.)

Another interesting nanophenomenon is the observed effect of alumina nanoparticles on the thermal conductivity of fluids once suspended in them. Experimental data obtained for alumina nanoparticles of different shapes in a fluid consisting of equal volumes of ethylene glycol (EG) and water showed an enhanced thermal conductivity.[102] Figure 1.24 shows the measured enhancement in fluid thermal conductivity as a function of nanoparticle concentration for different nanoparticle shapes of alumina. Clearly, the experimental results agree well with the predictions of developed theoretical models.[111]

1.9.4 Mechanical Phenomena

The mechanical behavior of small systems is another anomalous phenomenon that has attracted the attention of many researchers and is still under investigation.[112–118] For example, nanostructured metals showed an interesting deviation from the historically accepted Hall–Petch relationship expressed as:

$$\sigma_y = \sigma_o + kd^{-\frac{1}{2}} \qquad (1.24)$$

where σ_y is the yield strength, σ_o is the friction stress, k is a constant, and d is the grain size. The deviation from such a relationship was first reported by Chokshi et al.[119] in 1989 for nanostructured copper. Such a phenomenon has since been referred to as the inverse Hall–Petch effect and has been reported for many other nanostructured metals with grain size ranging between 1000 and 100 nm.[116] Figure 1.25 shows experimental plots depicting the trend of yield stress with grain size for different metals as compared to the conventional Hall–Petch response: (a) copper, (b) iron, (c) nickel, and (d) titanium.

In addition, while a reduction in grain size leads to an increase in ductility in conventional metals, ductility was found to be small for systems with a grain size less than 25 nm.[120] Other anomalous behavior has been reported for nanostructured materials regarding their creep, fatigue, as well as their strain rate sensitivity.[118]

References

1. V. Borisenko and S. Ossicini, *What is What in the Nanoworld: A Handbook on Nanoscience and Nanotechnology*, VCH Verlagsgesellschaft Mbh, 2008.
2. R. Feynman, *Eng. Sci.*, 1960, 23(No. 5, February), pp. 22–26.
3. R. Feynman, *J. Microelectromech. Systems.*, 1992, **1**, 60.
4. N. Taniguchi, *Proc. Intl. Conf. Prod. Eng. Part II* Society of Precision Engineering, Tokyo, Japan, 1974, **vol. II**.
5. K. Drexler and R. Whitaker, *Engines of Creation: [The Coming Era of Nanotechnology]*, Anchor, 1986.
6. K. Drexler, C. Peterson and G. Pergamit, *Unbounding the Future*, Morrow New York, 1991.
7. K. Drexler, *Nanosystems Molecular Machinery, Manufacturing, and Computation*, John Wiley and Sons Inc, New York, 1992.
8. J. Hall and K. Drexler, *Nanofuture: What's Next for Nanotechnology*, Prometheus Books, 2005.
9. K. E. Drexler, *Proc. Natl. Acad. Sci. U.S.A.*, 1981, **78**, 5275.
10. R. M. Baum, "Nanotechnology, Drexler and Smalley make the case for and against "molecular assemblers", *Chem. Eng. News*, 2003, **81**, 37–42. Available at http://pubs.acs.org/cen/coverstory/8148/8148counterpoint.html (accessed 30 June 2009).
11. K. Drexler and R. Whitaker, *Engines of Creation:[The Coming Era of Nanotechnology]*, Anchor, 2006.
12. O. Bueno, *HYLE–Int. J. Philos. Chem.*, 2004, **10**, 83.

13. D. Berube and J. D. Shipman, *Technol. Society Mag., IEEE*, 2004, **23**, 22.
14. M. Bucchi, *Science and the Media: Alternative Routes in Scientific Communication*, Routledge, 1998.
15. S. Kaplan and J. Radin, working paper, University of Pennsylvania, May 2009.
16. H. Kyröläinen, P. V. Komi and A. Belli, *Scand. J. Med. Sci. Sports.*, 1995, **5**, 200.
17. U. Leonhardt, *Nat. Photon.*, 2007, **1**, 207.
18. W. Kochmann, M. Reibold, R. Goldberg, W. Hauffe, A. Levin, D. Meyer, T. Stephan, H. Müller, A. Belger and P. Paufler, *J. Alloys Compds.*, 2004, **372**, 15.
19. A. A. Levin, D. C. Meyer, M. Reibold, W. Kochmann, N. Pätzke and P. Paufler, *Crystal Res. Technol.*, 2005, **40**, 905.
20. C. Srinivasan, *Curr. Sci.*, 2007, **92**, 279.
21. M. Reibold, P. Paufler, A. A. Levin, W. Kochmann, N. Pätzke and D. C. Meyer, in *Physics and Engineering of New Materials*, ed. A. Pucci and K. Wandelt, Springer Verlag, 2009, **vol. 127**, p. 305.
22. S. Ono and S. Kondo, in *Encyclopedia of Physics*, ed. S. Fluegge, Springer Verlag, Berlin, 1960, **vol. 10**, p. 134.
23. J. Rowlinson, *Chem. Soc. Rev.*, 1983, **12**, 251.
24. R. B. Griffiths and J. C. Wheeler, *Phys. Rev. A*, 1970, **2**, 1047.
25. J. Rowlinson and F. Swinton, *Liquids and Liquid Mixtures*, Butterworths, London, 1969.
26. B. Widom, *J. Phys. Chem.*, 1982, **86**, 869.
27. B. Widom, *J. Chem. Phys.*, 1963, **39**, 2808.
28. B. Widom, *J. Chem. Phys.*, 1965, **43**, 3892.
29. Y. Gao and Y. Bando, *Nature*, 2002, **415**, 599.
30. M. Hartmann, G. Mahler and O. Hess, *Phys. Rev. Lett.*, 2004, **93**, 080402.
31. D. G. Cahill, W. K. Ford, K. E. Goodson, G. D. Mahan, A. Majumdar, H. J. Maris, R. Merlin and S. R. Phillpot, *J. Appl. Phys.*, 2003, **93**, 793.
32. R. Evans, *Mol. Phys.*, 1981, **42**, 1169.
33. J. Rowlinson, *J. Chem. Thermodynam.*, 1993, **25**, 449.
34. P. Tarazona, U. Marconi and R. Evans, *Mol. Phys.*, 1987, **60**, 573.
35. R. Evans, D. C. Hoyle and A. O. Parry, *Phys. Rev. A*, 1992, **45**, 3823.
36. R. C. Tolman, *J. Chem. Phys.*, 1949, **17**, 333.
37. J. Rowlinson and B. Widom, *Molecular Theory of Capillarity*, Dover Publications, 2003.
38. E. Blokhuis and J. Kuipers, *J. Chem. Phys.*, 2006, **124**, 074701.
39. Y. A. Lei, T. Bykov, S. Yoo and X. C. Zeng, *J. Am. Chem. Soc.*, 2005, **127**, 15346.
40. S. M. Fall, *Rev. Mod. Phys.*, 1979, **51**, 21.
41. M. S. Amer and M. T. Abdu, *Philos. Mag. Lett.*, 2009, **89**, 615.
42. B. M. Ginzburg and S. Tuichiev, *Russ. J. Appl. Chem.*, 2008, **81**, 618.
43. A. Einstein, *Annal. Phys.*, 1901, **4**, 513.

44. S. Chen, F. Mallamace, C. Mou, M. Broccio, C. Corsaro, A. Faraone and L. Liu, *Proc. Natl. Acad. Sci. U.S.A.*, 2006, **103**, 12974.
45. F. Mallamace, M. Broccio, C. Corsaro, A. Faraone, U. Wanderlingh, L. Liu, C. Mou and S. Chen, *J. Chem. Phys.*, 2006, **124**, 161102.
46. T. L. Hill, *J. Chem. Phys.*, 1962, **36**, 3182.
47. T. Hill, *Thermodynamics for Chemists and Biologists*, Addison-Wesley Educational Publishers Inc, 1968.
48. T. Hill, *Thermodynamics of Small Systems*, Dover Publications, 2002.
49. T. L. Hill, *Nano Lett.*, 2001, **1**, 111.
50. T. L. Hill and R. V. Chamberlin, *Nano Lett.*, 2002, **2**, 609.
51. J. D. Badjic, V. Balzani, A. Credi, S. Silvi and J. F. Stoddart, *Science*, 2004, **303**, 1845.
52. R. Ballardini, V. Balzani, A. Credi, M. T. Gandolfi and M. Venturi, *Acc. Chem. Res.*, 2001, **34**, 445.
53. V. Balzani, A. Credi and M. Venturi, *Nano Today*, 2007, **2**, 18.
54. V. Balzani, A. Credi and M. Venturi, *Molecular Devices and Machines: Concepts and Perspectives for the Nanoworld*, Wiley-VCH Verlag, Weinheim, 2008.
55. J. R. Banavar, M. Cieplak, T. X. Hoang and A. Maritan, *Proc. Natl. Acad. Sci. U.S.A.*, 2009, **106**, 6900.
56. D. Chowdhury, *Resonance*, 2007, **12**, 39.
57. G. Comtet, G. Dujardin, A. J. Mayne and D. Riedel, *J. Phys.: Condens. Matter*, 2006, **18**, S1927.
58. E. R. Kay and D. A. Leigh, *Nature*, 2006, **440**, 286.
59. A. B. Kolomeisky and M. E. Fisher, *Annu. Rev. Phys. Chem.*, 2007, **58**, 675.
60. R. Lipowsky, Y. Chai, S. Klumpp, S. Liepelt and M. Müller, *Physica A (Amsterdam)*, 2006, **372**, 34.
61. D. Urry, *Angew. Chem. Int. Ed. Engl.*, 1993, **32**, 819.
62. C. Manzano, W. H. Soe, H. S. Wong, F. Ample, A. Gourdon, N. Chandrasekhar and C. Joachim, *Nat. Mater.*, 2009, **8**, 576.
63. M. M. Alvarez, J. T. Khoury, T. G. Schaaff, M. N. Shafigullin, I. Vezmar and R. L. Whetten, *J. Phys. Chem. B*, 1997, **101**, 3706.
64. S. Link and M. A. El-Sayed, *J. Phys. Chem. B*, 1999, **103**, 4212.
65. E. Prodan, C. Radloff, N. J. Halas and P. Nordlander, *Science*, 2003, **302**, 419.
66. L. Xu, L. Wang, X. Huang, J. Zhu, H. Chen and K. Chen, *Physica E (Amsterdam)*, 2000, **8**, 129.
67. M. Chu, X. Shen and G. Liu, *Nanotechnology*, 2006, **17**, 444.
68. M. Hines and P. Guyot-Sionnest, *J. Phys. Chem.*, 1996, **100**, 468.
69. S. Il, I. Yun, I. Ki and J. In, *J. Korean Phys. Soc.*, 2008, **52**, 1891.
70. D. Talapin, J. Nelson, E. Shevchenko, S. Aloni, B. Sadtler and A. Alivisatos, *Nano Lett*, 2007, **7**, 2951.
71. S. Jung, H. Yeo, I. Yun, S. Cho, I. Han and J. Lee, *J. Nanosci. Nanotechnol.*, 2008, **8**, 4899.
72. I. Robel, V. Subramanian, M. Kuno and P. Kamat, *J. Am. Chem. Soc*, 2006, **128**, 2385.

73. S. Tomic, N. Harrison and T. Jones, *Proc. NSOD*, IEEE, Newark, DE, 2007, p. 81.
74. P. Yu, K. Zhu, A. Norman, S. Ferrere, A. Frank and A. Nozik, *J. Phys. Chem. B*, 2006, **110**, 25451.
75. A. Kongkanand, K. Tvrdy, K. Takechi, M. Kuno and P. Kamat, *J. Am. Chem. Soc*, 2008, **130**, 4007.
76. S. Tomi, N. Harrison and T. Jones, *Opt. Quantum Electronics*, 2008, **40**, 313.
77. B. van der Zande, L. Pages, R. Hikmet and A. van Blaaderen, *J. Phys. Chem. B*, 1999, **103**, 5761.
78. B. van der Zande, M. Bohmer, L. Fokkink and C. Schonenberger, *Langmuir*, 2000, **16**, 451.
79. S. Oldenburg, R. Averitt, S. Westcott and N. Halas, *Chem. Phys. Lett.*, 1998, **288**, 243.
80. E. Prodan, P. Nordlander and N. J. Halas, *Nano Lett.*, 2003, **3**, 1411.
81. C. M. Aguirre, T. R. Kaspar, C. Radloff and N. J. Halas, *Nano Lett.*, 2003, **3**, 1707.
82. C. Charnay, A. Lee, S.-Q. Man, C. E. Moran, C. Radloff, R. K. Bradley and N. J. Halas, *J. Phys. Chem. B*, 2003, **107**, 7327.
83. S. Lal, S. E. Clare and N. J. Halas, *Acc. Chem. Res.*, 2008, **41**, 1842.
84. N. J. Halas, *Proc. Natl. Acad. Sci. U.S.A.*, 2009, **106**, 3643.
85. S. Katele, S. Gosavi, J. Urban and S. Kulkarni, *Curr. Sci.*, 2006, **91**, 1038.
86. L. Liz-Marzán, *Mater. Today*, 2004, **7**, 26.
87. J. Schmitt, P. Machtle, D. Eck, H. Mohwald and C. A. Helm, *Langmuir*, 1999, **15**, 3256.
88. G. Schmid and L. Gade, *Clusters and Colloids From Theory to Applications*, VCH, Weinheim, 1994.
89. L. De Jongh, *Physics and Chemistry of Metal Cluster Compounds: Model Systems for Small Metal Particles*, Kluwer Academic Publishers, Dordrecht, Boston, 1994.
90. W. A. de Heer, *Rev. Mod. Phys.*, 1993, **65**, 611.
91. C. Rao and A. Cheetham, *J. Mater. Chem.*, 2001, **11**, 2887.
92. P. V. Kamat, *J. Phys. Chem. B*, 2002, **106**, 7729.
93. C. Vinod, G. Kulkarni and C. Rao, *Chem. Phys. Lett.*, 1998, **289**, 329.
94. A. Henglein, *Ber. Bunsen-Ges. Phys. Chem.*, 1995, **99**, 903.
95. A. Henglein and D. Meisel, *Langmuir*, 1998, **14**, 7392.
96. J. H. Hodak, A. Henglein and G. V. Hartland, *J. Chem. Phys.*, 1999, **111**, 8613.
97. A. Henglein, *J. Phys. Chem. B*, 2000, **104**, 2201.
98. M. Haase, H. Weller and A. Henglein, *J. Phys. Chem.*, 1988, **92**, 4706.
99. A. Henglein, *Chem. Rev.*, 1989, **89**, 1861.
100. A. Henglein, B. G. Ershov and M. Malow, *J. Phys. Chem.*, 1995, **99**, 14129.
101. T. Linnert, P. Mulvaney, A. Henglein and H. Weller, *J. Am. Chem. Soc.*, 1990, **112**, 4657.

102. E. V. Timofeeva, J. L. Routbort and D. Singh, *J. Appl. Phys.*, 2009, **106**, 014304.
103. Y. Li, G. Li, S. Wang, H. Gao and Z. Tan, *J. Therm. Anal. Calorim.*, 2009, **95**, 671.
104. G. Parthasarathy, *Mater. Lett.*, 2007, **61**, 3208.
105. J. S. Kurtz, R. R. Johnson, M. Tian, N. Kumar, Z. Ma, S. Xu and M. H. W. Chan, *Phys. Rev. Lett.*, 2007, **98**, 247001.
106. M. X. Gu, C. Q. Sun, Z. Chen, T. C. A. Yeung, S. Li, C. M. Tan and V. Nosik, *Phys. Rev. B*, 2007, **75**, 125–403.
107. D. Li, Y. Wu, P. Kim, L. Shi, P. Yang and A. Majumdar, *Appl. Phys. Lett.*, 2003, **83**, 2934.
108. D. Li, Y. Wu, R. Fan, P. Yang and A. Majumdar, *Appl. Phys. Lett.*, 2003, **83**, 3186.
109. J. Rupp and R. Birringer, *Phys. Rev. B*, 1987, **36**, 7888.
110. L. Shi, D. Li, C. Yu, W. Jang, D. Kim, Z. Yao, P. Kim and A. Majumdar, *J. Heat Transfer*, 2003, **125**, 881.
111. R. L. Hamilton and O. K. Crosser, *Ind. Eng. Chem. Fundam.*, 2002, **1**, 187.
112. H. Gleiter, *Prog. Mater Sci.*, 1989, **33**, 223.
113. Y. Zhang, N. Tao and K. Lu, *Acta Mater.*, 2008.
114. S. Habelitz, S. J. Marshall, G. W. Marshall and M. Balooch, *Arch. Oral Biol.*, 2001, **46**, 173.
115. A. Desai and M. Haque, *Sens. Actuators, A*, 2007, **134**, 169.
116. M. Meyers, A. Mishra and D. Benson, *Prog. Mater. Sci.*, 2006, **51**, 427.
117. K. Kumar, H. Van Swygenhoven and S. Suresh, *Acta Mater.*, 2003, **51**, 5743.
118. C. Suryanarayana, *Int. Mater. Rev.*, 1995, **40**, 41.
119. A. H. Chokshi, A. Rosen, J. Karch and H. Gleiter, *Scripta Mater.*, 1989, **23**, 1679.
120. C. C. Koch, *Nanocryst. Mater.*, 2003, **18**, 9.

CHAPTER 2
Raman Spectroscopy; the Diagnostic Tool

2.1 Introduction

This chapter is devoted to Raman spectroscopy as a powerful investigating tool. Raman theory, selection rules, and its dependence on species symmetry will be covered. The ability of Raman spectroscopy to interrogate materials systems on their molecular level and to measure crucial thermodynamic functions (such as temperature and pressure) as well as to determine their structures will be demonstrated.

In 1922, Chandrasekhara Venkata Raman (Figure 2.1), an Indian professor of physics, at the University of Calcutta, working on light interaction with liquids published the first of a series of papers with his collaborator K. S. Krishnan. The first paper entitled "Molecular Diffraction of Light" ultimately led to his famous discovery on 28 February 1928. A week earlier, two Russian professors – G. S. Landsberg and L. I. Mandelstam – working at Moscow State University on light interaction with crystals since 1926 reported, independently, the same phenomenon. The observed phenomenon indicated that when light is scattered by matter a small percentage of the scattered light will have a frequency that is different from the frequency of the incident light. The observed shift in the scattered frequency is the result of a combination between the frequency of the incident light and the frequency of molecular motion of the interacting material.

In Russian literature, the phenomenon is known as *combinatorial scattering of light*. In the rest of the world, the phenomenon is known as the *Raman effect*. C. V. Raman received the Nobel Prize in Physics in 1930. Seventy years after its discovery, in 1998, the Raman effect was designated by the American Chemical Society as a National Historical Landmark in recognition of its significance as a tool for analyzing materials systems.

(a) (b)

Figure 2.1 (a) C. V. Raman and (b) L. I. Mandelstam, the two scientists who independently discovered the Raman effect.

2.2 Raman Phenomenon

The Raman effect, or phenomenon, can be defined as an inelastic scattering of light by matter. When a monochromatic light is scattered by matter, two types of interaction take place and result into two distinctive types of scattered light. One type of interaction does not involve energy transfer or exchange between the incident light photon and the molecules, or atoms, of matter. Hence, the scattered photon will have the same energy, or frequency, as the incident light. This type of scattering is elastic in nature and referred to as Rayleigh scattering. The second type of interaction involves energy exchange between the incident photon and the material's molecules. Hence, the scattered photon will have a new frequency, or energy, which is simply equal to the sum or the difference between the frequencies of the incident photon and the natural frequency of the thermally excited and kinetically active species in the material. This type of scattering is inelastic in nature and is referred to as Raman scattering.

2.3 General Theory of Raman Scattering

Despite the fact that Raman scattering is a quantum phenomenon and its detailed description requires quantum theory, the existence of the effect and a good qualitative description of the phenomenon can be predicted and described based upon classical electromagnetic theory. In this book, we will basically depend on the classical description of the Raman effect. We will, however, refer to quantum effects as needed to better explain specific points.

Using the classical theory of scattering, the Raman effect can be explained as follows; let us consider a monochromatic light (an electromagnetic) wave propagating in the z-direction (Figure 2.2) with an oscillating electric field

Raman Spectroscopy; the Diagnostic Tool

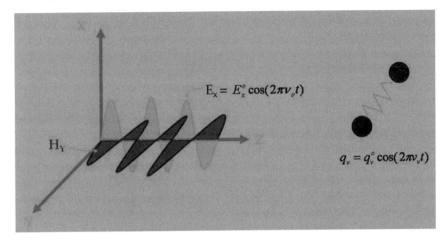

Figure 2.2 Illustration of an electromagnetic wave (monochromatic light) interacting with a vibrating molecule.

(in the x-direction). The amplitude of the electric field (E_x) at any time (t) can be expressed as:

$$E_x = E_x^o \cos(2\pi v_o t) \qquad (2.1)$$

where E_x^o is the maximum amplitude of the electric field, and v_o is the frequency of the electromagnetic wave (this is the incident light).

Let us also consider a diatomic molecule (to represent a material) with a natural vibration frequency (v_v). The normal vibration vector (q_v) can be expressed as a function of time as:

$$q_v = q_v^o \cos(2\pi v_v t) \qquad (2.2)$$

When the electromagnetic wave (the monochromatic light) interacts with the molecule, the light will polarize the electrons of the molecule inducing a dipole moment (μ):

$$\mu = \alpha E \qquad (2.3)$$

where α is known as the polarizability tensor, a second rank tensor that we will discuss later.

The induced dipole will oscillate and, hence, emit light (the scattered light) at three different frequencies. These three different frequencies can be determined by expanding the polarizability tensor (α) as a Taylor's series in (q_v) as:

$$\alpha = \alpha^o + \left(\frac{d\alpha}{dq_v}\right) q_v + \cdots \qquad (2.4)$$

Then, substituting from Equations (2.1) and (2.3), the dipole moment can be expressed as:

$$\mu = E_x^o \alpha^o \cos(2\pi v_o t) + E_x^o \left(\frac{d\alpha}{dq_v}\right)_o q_v^o \{\cos[2\pi(v_o - v_v)t] + \cos[2\pi(v_o + v_v)t]\} \tag{2.5}$$

Clearly, from Equation (2.5), the three frequencies the scattered light will have are v_o, (v_o-v_v), and (v_o+v_v). In other words, the scattered light will have three components at three different frequencies – one that has the same frequency as the incident light (v_o) and two components that have a frequency shifted from that of the incident light frequency by an amount equal to the natural frequency of the molecular vibration $(v_o \pm v_v)$. In the case of scattered light at the same frequency there is, clearly, no energy loss or gain due to interaction with the material. This type of scattering is known as *elastic scattering* or *Rayleigh scattering*. The other two components would have experienced a gain or loss of energy equal to the energy of the molecular vibration; this is referred to as *inelastic scattering* or *Raman scattering*. To further distinguish between the two types within the Raman scattering, (*i.e.*, the component that gained energy and the one that lost energy), they are given two different names: the light that is scattered with lower energy (v_o-v_v), is known as Stokes lines, while the light that is scattered with higher energy (v_o+v_v) is known as anti-Stokes lines. Stokes and anti-Stokes lines are fundamentally important for measuring local temperature and temperature distribution within a materials system as we will discuss later in this chapter.

Figure 2.3 depicts a generic spectrum showing both Rayleigh and Raman scattering lines.

Notably, the intensity of Rayleigh scattering is typically six orders of magnitude higher than that of Raman scattering. Hence, experimental setups need to be employed to filter the Rayleigh intensity to make the Raman lines experimentally observable. In a typical Raman spectrum, the scattered light

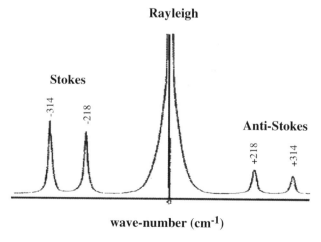

Figure 2.3 Illustration of a generic spectrum showing Rayleigh and Raman bands.

Raman Spectroscopy; the Diagnostic Tool

intensity is recorded on the ordinate while the frequency of the scattered light is recorded on the abscissa. The intensity is usually given in arbitrary units while the frequency is expressed in wavenumbers ($\tilde{v}$). Wavenumbers are reported in units of (cm^{-1}), and are calculated according to Equation (2.6):

$$\tilde{v} = \frac{1}{\lambda_o} - \frac{1}{\lambda_v} \qquad (2.6)$$

where the subscripts o and v are used to denote the incident light and the vibration mode, respectively; λ is the wavelength in centimeters and, as is well known, is related to the frequency (v) through the speed of light ($c = 3 \times 10^{10}$ cm s^{-1}) according to:

$$\lambda = \frac{c}{v} \qquad (2.7)$$

Once abscissa is reported in wavenumbers, $\tilde{v}$ for the Rayleigh line will always have a zero value. The Raman lines will be shifted by the wavenumber value of the mode they represent.

2.4 Raman Selection Rules

Since we have already mentioned that each Raman line[i] represents a mode, it is time to discuss vibration modes in a material system and which of such modes will be Raman active. This is usually referred to as the selection rules.

According to Equation (2.5), the Raman scattering will occur only under the condition that:

$$\left(\frac{d\alpha}{dq_v}\right) \neq 0 \qquad (2.8)$$

As we mentioned before, α is the polarizability tensor for any vibration mode. If the mode satisfies the aforementioned condition [Equation (2.8)] the mode will be Raman active. This means that the mode will produce a Raman peak in the spectrum of the material. In other words, it can be said that a vibration mode will be Raman active if the rate of change of polarizability with vibration (evaluated at the equilibrium position) is not zero.

2.4.1 Vibration Modes and the Polarizability Tensor

While the motion of atoms in matter (or actually the motion of nuclei as the atoms' centers of mass) seems to be random, analysis by classical mechanics showed that such seemingly random and complex motion is indeed compounded of a definite number of so-called *normal modes*. In each of such normal modes,

---

[i] The terms Raman line, Raman band, and Raman peak are usually used interchangeably in the field.

all the atoms (or nuclei) move with the same characteristic *normal frequency*. The number of such normal modes in any particular case can be easily deduced. Since every atom has three degrees of freedom (the number of coordinates required to fully specify the position of the atom), a molecule, with N atoms will have $3N$ degrees of freedom. However, six of these degrees of freedom represent translation and rotation motions around the three main Cartesian axes. This leaves $3N - 6$ as the number of possible *vibrational* degrees of freedom, known as the *normal vibration modes* or *phonons*. If the molecule is linear, the number of normal vibration modes becomes $3N - 5$ since rotation around the molecular axis is not accounted for.

If we consider a simple linear molecule such as carbon dioxide (CO_2), which is chemically represented as O=C=O, the molecule has four normal vibration modes as shown in Figure 2.4.

These four vibration modes have three frequencies that are designated v_1, v_2, and v_3 (Figure 2.4); v_1 is the frequency of the mode representing symmetric stretch of the oxygen atoms around the carbon atom; v_3 is the frequency of the mode representing the asymmetric stretch of the aforementioned species, v_2 is the frequency of two modes representing the in-plane, and out-of-plane bending. We will discuss later in this chapter the phenomenon of different modes having the same frequency. The question we would like to raise now is "which of the four normal vibration modes of CO_2 is Raman active?" For simplicity, consider the two axial stretch modes (v_1 and v_3). If we graphically represent the axial component of the polarizability (α_{ii} assuming that the molecular axis is in the i direction) as a function of the vibration vector ($\mathbf{q_v}$), it should resemble the curves shown in Figure 2.5. Clearly, from Figure 2.5, Equation (2.8) is satisfied for the symmetric stretch mode (v_1), and is not satisfied for the asymmetric stretch mode (v_3). This means that the symmetric stretch mode in CO_2 is Raman active while the asymmetric stretch mode is not.

It is more difficult to answer our question for other vibration modes using the condition in Equation (2.8). This is a consequence of the simplicity of the

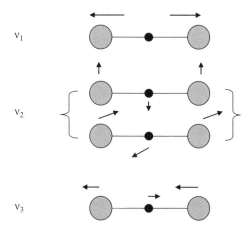

Figure 2.4 Illustration of vibration modes in a carbon dioxide molecule.

Raman Spectroscopy; the Diagnostic Tool

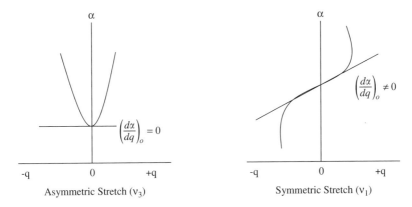

Figure 2.5 Illustration of axial polarizability of CO_2 molecules as a function of vibration vector of symmetric and asymmetric stretch modes.

general theory resulting in Equation (2.5). Another consequence of the simplicity of Equation (2.5) is that it predicts both Rayleigh and Raman scattered radiation to have the same polarization direction (relative to a fixed set of coordinate axes) as that of the incident light, which is not correct. In fact, since polarizability (α) is a tensor, the angular dependence and polarization of both Rayleigh and Raman scattering will depend on the tensorial properties of α. Equation (2.3) should actually be rewritten as:

$$\begin{vmatrix} \mu_x \\ \mu_y \\ \mu_z \end{vmatrix} = \begin{vmatrix} \alpha_{xx} & \alpha_{xy} & \alpha_{xz} \\ \alpha_{yx} & \alpha_{yy} & \alpha_{yz} \\ \alpha_{zx} & \alpha_{zy} & \alpha_{zz} \end{vmatrix} \begin{vmatrix} E_x \\ E_y \\ E_z \end{vmatrix} \quad (2.9)$$

The polarizability tensor is usually symmetric (i.e., $\alpha_{ij} = \alpha_{ji}$), and, according to the quantum theory of Raman spectroscopy, if one of the polarizability tensor components changes during the vibration, the mode is Raman active. If we plot $1/\alpha_i$ (where i represents the direction) we can construct a three-dimensional body known as *the polarizability ellipsoid*. According to the quantum theory of Raman scattering, if the shape, size, or orientation of the polarizability ellipsoid changed during vibration, the mode is Raman active. Figure 2.6 shows the polarizability ellipsoid for the different vibration modes of CO_2. Clearly, the symmetric stretch causes the polarizability ellipsoid to change size, hence it is Raman active. While the other two modes seem to change the ellipsoid shape or size, the fact that the ellipsoid is identical at the two extremes of the vibration mode indicates that the modes are not Raman active.

Plotting the polarizability ellipsoid for different vibration modes of molecules would rapidly become a trivial method to determine Raman activity. Consequently, more efficient analysis methods have been developed utilizing symmetry operations and group theory. Such methods enable better prediction of the Raman activity of different materials. To discuss such methods, we need to start with the fundamental symmetry operations and how they apply to materials systems.

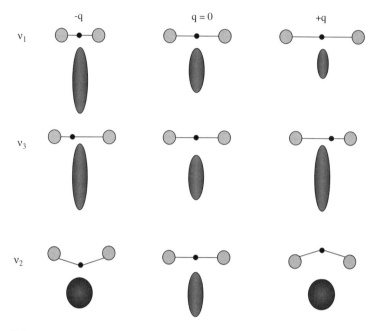

Figure 2.6 Polarizability ellipsoid for different normal vibration modes of CO_2.

2.5 Symmetry

Symmetry is an amazing natural concept. It can be observed in almost everything around us including, not surprisingly, ourselves. Everyone knows that our right half is indistinguishable from our left half if reflected in a mirror. Nature used symmetry billions of years before man and must have taught mankind how to understand symmetry and use it. The German mathematician Hermann Weyl (1885–1955) said that through symmetry man always tried to perceive and create order, beauty, and perfection. The correlation between symmetry and beauty would become clear if one was ever moved by the symmetry, or beauty, of a snowflake, a crystal, or a flower. Ancient Greeks used the term symmetry to describe "proportionality" or similarity in arrangements of parts. They applied such understanding to create the most beautiful sculptures for humanity. Symmetry and beauty can also be heard in music and poetry. In the broadest sense, symmetry can be defined as the opposite of chaos and randomness. This definition of symmetry draws attention to the deep relationship between symmetry and entropy – a subject that is beyond the scope of this book.[ii]

The mathematical formulation of symmetry was rigorously developed in the nineteenth century. According to Hermann Weyl, an object is called

---

[ii] Several inspiring articles and books have been published recently on the relationship between symmetry and entropy. Interested readers are referred to the Recommended General Reading section at the end of this chapter.

symmetrical if it can be changed "*somehow*" to obtain the same object. In this section we will explain what is meant by "*somehow*", and discuss the mathematical description of symmetry as related to molecules and crystals, the building blocks of materials. The mathematical description of symmetry is concerned with the correspondence of positions on opposite sides of a point, a line, or a plane. Mathematicians realized that at most, five different elements of symmetry are needed to fully describe the correspondence of two point positions. In other words, one would need at most five different elements of symmetry that once operated separately on a point the point can be moved to a new indistinguishable position. To correlate this concept more closely with molecules, one may note that every molecule would possess one or more symmetry elements that once operated the molecule will assume a new configuration indistinguishable from the original one. Hence, the "*somehow*" turns out to be simply the operation of any of the following five symmetry elements.

2.5.1 Identity (E)

The identity symmetry element exists in everything in the universe. It is usually given the symbol (E) for the German word *Einheit*[iii] meaning unity. Loosely, the word can be translated as "the same" or "identical".

2.5.2 Center of Symmetry (i)

A center of symmetry is a point in space that occupies a midpoint on a line connecting two indistinguishable positions. The center of symmetry is also known as *inversion center*, hence, the designation "i". If one connects a line from an atom in a molecule, or, generally speaking, a site or position in space, through a center of symmetry, extending the line for the same distance should lead to an equivalent indistinguishable atom, or position. For example, the carbon atom in a CO_2 molecule (Figure 2.4) occupies a center of symmetry. If we consider any of the oxygen atoms, connecting a line from that oxygen to the carbon and extending the line an equidistance, will lead us to the second oxygen atom in the molecule that is indistinguishable from the one we started at. Figure 2.7 illustrates the concept of a center of symmetry. Note that a center of symmetry of a molecule may, or may not, be occupied by an atom. For example, both CO_2 and C_2H_2 molecules possess a center of symmetry. While that center of symmetry is occupied by a carbon atom in the case of CO_2, it is unoccupied in the case of ethyne, and actually lies on the midpoint between the two carbon atoms. As mentioned above, if we operate the center of symmetry operation on one of the oxygen atoms in CO_2 we will move that atom to the second oxygen atom position. Now what if we operate the center of symmetry on the same oxygen atom twice? This will bring the oxygen atom back to its original (*i.e.*, identical) position. Clearly, then, operating a center of symmetry

---

[iii] In crystallography and spectroscopy the reader will come across much German nomenclature due to the ground-breaking work done by German scientists in these fields.

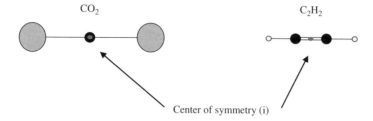

Figure 2.7 Illustration of the center of symmetry in two linear molecules, CO_2 and C_2H_2.

element twice (this is mathematically expressed as i^2) is equivalent to the identity element of symmetry (E). Mathematically this can be expressed as $i^2 \equiv E$.

2.5.3 Rotation Axes (C_n)

If one could rotate a molecule (clockwise or anticlockwise) about an axis into a new configuration[iv] that is indistinguishable from the original configuration, the molecule is said to possess a rotation axis of symmetry (C_n).[v] This symmetry element is also referred to as proper rotation axis. The subscript "n" is the rotation order describing the angle of rotation required to reach an indistinguishable configuration of the molecule. The rotation angle can be $2\pi/n$ ($360°/n$). For crystals, due to space filling requirements, n can only assume the values 1, 2, 3, 4, and 6. For molecules, however, n may take any integer value (1, 2, 3...∞). For example, a CO_2 molecule possesses two two-fold rotation axes (C_2) normal to the molecular axis as shown in Figure 2.7. As shown in the figure, if the molecular axis is in the x-direction, rotation of 180° ($2\pi/2$) about either the y-axis or the z-axis will bring the molecule into an indistinguishable configuration. It is also important to note that for such linear molecules, the molecular axis represents a (C_∞) rotation axis. It is also simple to observe that a benzene molecule possesses a six-fold rotation axis (C_6) normal to its plane. If one rotates a molecule that possesses a two-fold rotation axis 180° about that axis, this operation can be expresses as C_2^1. Similarly, rotation of 60° about a six-fold rotation axis is expressed as C_6^1. Also, a rotation of 120° about a six-fold axis is expressed as C_6^2. Both operations will bring the molecule into an indistinguishable configuration. A rotation of 360° about any rotation axis can be expressed as C_n^n. Such symmetry operation will bring the molecule into the original (identical) configuration. Hence, it is equivalent to the identity symmetry element (E). Mathematically this can be expressed as $C_n^n \equiv E$. Figure 2.8 illustrates the rotation axes symmetry elements possessed by a AB_4 type planar molecule.

[iv] Here, we use configuration to describe the equilibrium configuration of the molecule meaning that at vibration vector $q=0$.
[v] C is for cyclic.

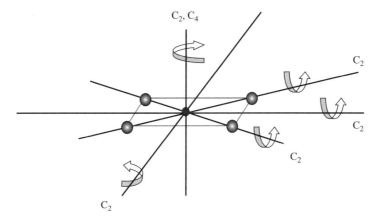

Figure 2.8 Illustration of the rotation axes symmetry elements possessed by an AB_4 planar molecule.

2.5.4 Planes of Symmetry (σ) (Mirror Planes)

If a molecular configuration can be divided by a plane into two parts that are mirror images of each other, then the molecule possesses a symmetry plane (σ). This symmetry element is also known as *mirror plane*. If the molecule possesses two symmetry planes intersecting in a line, the intersection line will be a rotation axis. Three different types of symmetry planes have been distinguished: vertical (σ_v), horizontal (σ_h), and diagonal (σ_d); σ_v is used to denote symmetry planes intersecting in a rotation axis. If the symmetry planes are bisecting the angle between two successive two-fold axes, the planes are denoted as diagonal planes (σ_d). Horizontal symmetry planes (σ_h) are usually those in the plane of planar molecules. Some texts use primes to distinguish different types of symmetry planes. Usually, symmetry planes in a plane of planar molecules are designated as σ. Symmetry planes out of the molecular plane are designated as σ'. Figure 2.9 illustrates the three different types of symmetry planes possessed by a planar molecule of the type AB_4. In addition, reflection in a symmetry plane twice (σ^2) results in the original configuration, hence $\sigma^2 = E$.

2.5.5 Rotation Reflection Axes (S_n) (Improper Rotation)

For some molecules, an indistinguishable configuration can be reproduced by rotation to a certain degree ($360°/n$) about an axis and then reflection through a reflection plane that is perpendicular to the rotation axis. Such symmetry element is denoted as a rotation reflection axis, or improper rotation axis (S_n). The symbol S is, again from German, from the word *Spiegel*, meaning a mirror; n is the order of the axis. If we consider an ethane molecule

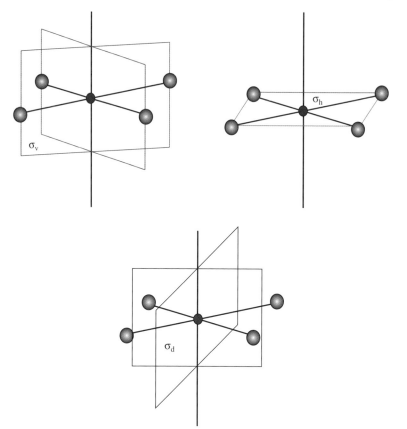

Figure 2.9 Different types of symmetry planes possessed by an AB$_4$ planar molecule.

(H$_3$–C–C–H$_3$), the positions of any two of the indistinguishable hydrogen atoms can be exchanged by rotating 60° about a vertical axis, then reflecting in a horizontal plane as shown in Figure 2.10. Such a symmetry element is denoted as (S$_6$). Note that neither a C$_6$ nor a σ_h are present on their own. An improper rotation axis can also be operated in stages exactly like rotation axes. The operation described in Figure 2.10(a) is expressed as S$_6^1$. Repeating the operation again results in S$_6^2$. Figure 2.10(b) shows that an S$_6^2$ is indeed equivalent to a C$_3$ rotation. One can clearly understand this by realizing that in a S$_6^2$ we rotate a total of 120° (equivalent to a C$_3$) and reflect in the same horizontal plane twice ($\sigma^2 \equiv E$). Improper rotation axes also have the characteristic that S$_n^n \equiv E$, if n is even. Another unique characteristic of improper rotation axes is that S$_n^{\frac{n}{2}} = i$, if n is even and $n/2$ is odd (S$_6^2$ for example). Understanding such characteristics of symmetry elements enables easier determination of all symmetry elements of a molecule. For example, if a molecule possesses a S$_6$ symmetry element, it must have an inversion center (i). The opposite is not necessarily true.

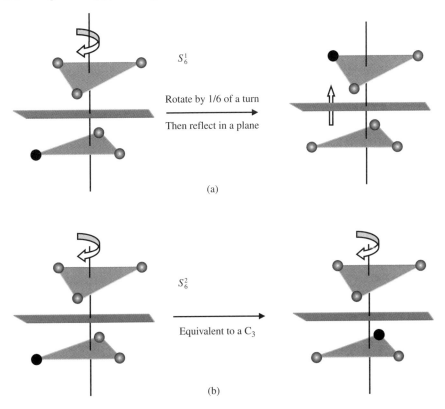

Figure 2.10 (a) Improper reflection (S_6^1) in an ethane molecule. Carbon atoms are not shown for clarity. One of the hydrogen atoms was made distinguishable for illustration purpose. (b) Illustration of the equivalency of improper rotation S_6^2 to a C_3 symmetry element.

2.5.6 Symmetry Elements and Symmetry Operations

It is time to distinguish between two important concepts: symmetry elements and symmetry operations. A symmetry element is a geometrical property that could be possessed by a molecule based on the exact shape of the molecule. A symmetry operation is the operation of applying an action (based on a symmetry element) that results in an indistinguishable configuration of the molecule. There are only five different symmetry elements but more than five symmetry operations. For example, a six-fold improper rotation axis (S_6) is a symmetry element that is possessed by benzene. Such symmetry element generates different symmetry operations. In the first instant, one would say that the generated symmetry operations are six in the form of S_6^j where, j can take any value between 1 and 6. This answer is not very accurate since, as we discussed before, the operation S_6^2 is equivalent to C_3, the operation S_6^3 is equivalent to i, the operation S_6^4 is equivalent to S_2^2, and S_6^6 is equivalent to E. This leaves the symmetry element S_6 capable of generating only two unique symmetry operations, namely, S_6^1, and S_6^5. This difference between point symmetry elements and operations should be clearly in mind when we try to classify molecules into point groups according to their symmetry operations.

2.6 Point Groups

Mathematically, a set of operations (A, B, C, *etc.*) forms a group if the following four rules are obeyed:

1. The product of any two members of the group and the square of any member are also members of the group.
2. There must be an identity element in the group.
3. Combination of members must be associative, *i.e.*, (AB)C should equal A(BC).
4. Every member must have an inverse that is also a member of the group, *i.e.*, if A is a member, then A^{-1} must also be a member, realizing that $AA^{-1} = E$ (the identity operation).

Hence, sets of point symmetry operations can be grouped into groups according to their satisfactory compliance with the four rules listed above. Such groups of symmetry operations are known as "point groups". It was found that all possible point symmetry operations can be grouped into 32 point groups.

2.6.1 Point Groups of Molecules

Point groups are given symbols made of a capital letter and, usually, two subscripts: a number and a lowercase letter, C_{2v} for example. The number indicates the order of the principal axis of the molecule. The principal axis is taken as the highest order axis, and usually defines the vertical direction. The capital letter is "D" (for dihedral) if an *n*-fold principal axis is accompanied by *n* two-fold axes at a right angle to it; otherwise, the letter will be a "C" (for cyclic). The lower case subscript letter is "h" if a horizontal plane is present. If *n* vertical planes are present, the letter is v for a C group, and d for a D group. It is important to note that h takes precedence over v or d. If no vertical or horizontal planes are present, the subscript letter is omitted. In addition, special point group symbols are reserved for molecules with certain shapes. If the molecule has a tetrahedral shape, the group symbol is "T_d". The group symbol is "O_h" for octahedral molecules and "I_h" for molecules with dodecahedral or icosahedral shapes. Such notation is known as *Shönflies notation* rule. Table 2.1 lists all possible 32 point groups for molecules.

Assigning a molecule to a certain point group describing its symmetry is the first and most important step in understanding and predicting the molecule's spectroscopic response. In 1956, Zeldin proposed a systematic method to assign molecules to their point groups.[1] Zeldin's method does not require the listing of all symmetry operations of the molecule. However, it focuses on major symmetry

Table 2.1 The 32 point groups for molecules.

Point group	Symmetry operations	Simple description	Example
C_1	E	No symmetry	Lysergic acid
C_s	E σ_h	Planar, no other symmetry	Hypochlorous acid
C_i	E, i	Inversion center	1,2-Dichloroethane
C_2	E C_2		Hydrogen peroxide
C_{2h}	E C_2 i σ_h	Planar with inversion center	*trans*-1,2-Dichloroethylene
C_{2v}	E, C_2, σ_h, σ_v	Angular or see-saw	Water
C_3	E, C_3	Propeller	Triphenylphosphine
C_{3h}	E, C_3, C_3^2, σ_h, S_3, S_3^5		Boric acid
C_{3v}	E $2C_3$ $3\sigma_v$	Trigonal pyramidal	Ammonia
C_4	E, C_4		None
C_{4h}	E, σ, C_4, i		$C_4H_4Cl_4$?

Table 2.1 Continued.

Point group	Symmetry operations	Simple description	Example
C_{4v}	E, $2C_4$, C_2, $2\sigma_v$, $2\sigma_d$	Square pyramidal	Xenon oxytetrafluoride
C_6	E, C_6		None
C_{6h}	E, σ, C_6, i		None
C_{6v}	E, 6σ, C_6		None
$C_{\infty v}$	E, C_∞ σ_v	Linear	Hydrogen chloride
D_2	E, $3C_2$		None
D_{2h}	E, 3σ, $3C_2$, i	Planar with inversion center	Ethylene
D_{2d}	E, $2S_4$, $3C_2$, $2\sigma_d$	90° Twist	Allene
D_3	E, C_3, $3C_2$	Triple helix	Tris(ethylenediamine) cobalt(III) cation
D_{3h}	E, $2C_3$, $3C_2$, σ_h $2S_3$ $3\sigma_v$	Trigonal planar or trigonal bipyramidal	Boron trifluoride
D_{3d}	E $2C_3$ $3C_2$ i $2S_6$ $3\sigma_d$	60° Twist	Cyclohexane
D_4	E, C_4, $4C_2$		Cyclobutane

Table 2.1 Continued.

Point group	Symmetry operations	Simple description	Example
D_{4h}	E $2C_4$, C_2 $2C_2'$ $2C_2''$ i $2S_4$ σ_h $2\sigma_v$ $2\sigma_d$	Square planar	Xenon tetrafluoride
D_{4d}	E $2S_8$ $2C_4$ $2S_8^3$ C_2 $4C_2'$ $4\sigma_d$	45° Twist	Dimanganese decacarbonyl
D_{5h}	E $2C_5$ $2C_5^2$ $5C_2$ σ_h $2S_5$ $2S_5^3$ $5\sigma_v$	Pentagonal	C_{70} fullerene
D_{5d}	E $2C_5$ $2C_5^2$ $5C_2$ i $3S_{10}^3$ $2S_{10}$ $5\sigma_d$	36° Twist	Ferrocene (staggered rotamer)
D_{6h}	E $2C_6$ $2C_3$ C_2 $3C_2'$ $3C_2''$ i $3S_3$ $2S_6^3$ σ_h $3\sigma_d$ $3\sigma_v$	Hexagonal	Benzene
$D_{\infty h}$	E, C_∞ $\infty\sigma$ ∞C_2 i S_∞	Linear with inversion center	Carbon dioxide (O=C=O, 116.3 pm)
T_d	E $8C_3$ $3C_2$ $6S_4$ $6\sigma_d$	Tetrahedral	Methane (108.70 pm)
O_h	E $8C_3$ $9C_2$ $6C_4$ i $6S_4$ $8S_6$ $3\sigma_h$ $6\sigma_d$	Octahedral or cubic	Cubane

Table 2.1 Continued.

Point group	Symmetry operations	Simple description	Example
I_h	E $12C_5$ $12C_5^2$ $20C_3$ $15C_2$ i $12S_{10}$ $12S_{10}^3$ $20S_6$ 15σ	Icosahedral	C_{60}

elements possessed by the molecule. Figure 2.11 shows a flowchart scheme usually used to assign molecules to point groups. Importantly, for some of the possible molecular point groups, no known molecules are listed as example. This should not be a source of confusion regarding the validity of these point groups. For example, while it had been generally believed that no molecules assume the icosahedral (I_h) symmetry,[2] three molecules have been known to defy such belief: the borohydride anion[3] $(B_{12}H_{12})^{2-}$, dodecahedrane[4] $(C_{20}H_{20})$, and the molecule that is the subject of this book, [60] fullerene[5] (C_{60}).

2.6.2 Point Groups of Crystals

To consider spectroscopic investigation of solid state materials, it is important to shed some light on the distinction between point groups as applied to molecules and point groups as applied to crystals. The difference basically comes from the crystallographic restriction theorem. The theorem is based on the observation that due to space filling restriction in solid crystals the order of rotation axes (*n*) should be restricted to only four values: 2, 3, 4, and 6. With such restriction in mind, applying the five point symmetry elements on the seven possible crystal lattices (triclinic, monoclinic, orthorhombic, tetragonal, rhombohedral, hexagonal, and cubic) results, again, in 32 point groups that are referred to as crystallographic point groups or crystal classes. Table 2.2 lists the 32 crystallographic point groups using Shönflies notation. Another notation that is also used to denote point groups is known as *Hermann–Mauguin notation*. It is named after the German crystallographer Carl Hermann and the French mineralogist Charles Mauguin. This notation is sometimes called *international notation* and is used mostly in crystallography. Table 2.2 also lists the corresponding Hermann–Mauguin notation for the 32 crystallographic point groups. A set of graphical symbols was also developed for symmetry elements. While not widely used in spectroscopy, the graphical symbol system is widely used in crystallography for stereographic representation of the 32 point groups. Table 2.3 shows the graphical symbols for the different symmetry elements and their corresponding *international* notation. For completeness, Table 2.4 shows the stereographic presentation of the 32 crystallographic point

Raman Spectroscopy; the Diagnostic Tool

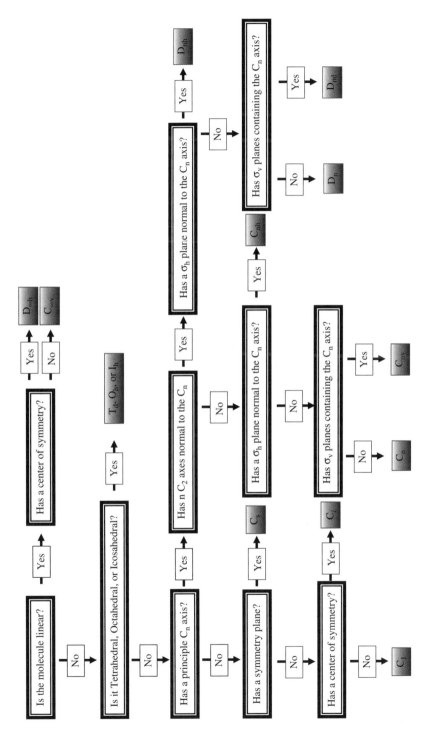

Figure 2.11 Flow chart for assigning molecules to point groups according to symmetry elements.

Table 2.2 The 32 crystallographic point groups, corresponding international notations, and related crystal systems.

Point group (Shönflies notation)	Symmetry operations	Hermann–Mauguin notation	Related crystal system
C_1	E	1	Triclinic
C_s	E, σ_h	m	Monoclinic
C_i	E, i	$\bar{1}$	Triclinic
C_2	E, C_2	2	Monoclinic
C_{2h}	E, C_2, i, σ_h	2/m	Monoclinic
C_{2v}	E, C_2, σ_h, σ_v	mm2	Orthorhombic
C_3	E, C_3	3	Rhombohedral
C_{3h}	E, C_3, C_3^2, σ_h, S_3, S_3^5	$\bar{6}$	Hexagonal
C_{3v}	E, $2C_3$, $3\sigma_v$	3m	Rhombohedral
C_4	E, C_4	4	Tetragonal
C_{4h}	E, σ, C_4, i	4/m	Tetragonal
C_{4v}	E, $2C_4$, C_2, $2\sigma_v$, $2\sigma_d$	4mm	Tetragonal
C_6	E, C_6	6	Hexagonal
C_{6h}	E, σ, C_6, i	6/m	Hexagonal
C_{6v}	E, 6σ, C_6	6mm	Hexagonal
D_2	E, $3C_2$	222	Orthorhombic
D_{2h}	E, 3σ, $3C_2$, i	mmm	Orthorhombic
D_{2v}	E, $2S_4$, $3C_2$, $2\sigma_d$	$\bar{4}$2m	Tetragonal
D_3	E, C_3, $3C_2$	32	Rhombohedral
D_{3h}	E, $2C_3$, $3C_2$, σ_h, $2S_3$, $3\sigma_v$	$\bar{6}$m2	Hexagonal
D_{3v}	E, $2C_3$, $3C_2$, i, $2S_6$, $3\sigma_d$	$\bar{3}$m	Rhombohedral
D_4	E, C_4, $4C_2$	422	Tetragonal
D_{4h}	E, $2C_4$, $5C_2$, i, $2S_4$, σ_h, $2\sigma_v$, $2\sigma_d$	4/mmm	Tetragonal
D_6	E, C_6, $6C_2$, $6\sigma_d$	622	Hexagonal
D_{6h}	E, $2C_6$, $2C_3$, $6C_2$, i, $3S_3$, $2S_6^3$, σ_h, $6\sigma_d$	6/mmm	Hexagonal
S_4	E, S_4, C_2	$\bar{4}$	Tetragonal
S_6	E, S_3, i	$\bar{3}$	Rhombohedral
T	E, $4C_3$, $3C_2$,	23	Cubic
T_h	E, 3σ, $4C_3$, $3C_2$, i	m$\bar{3}$	Cubic
T_d	E, 6σ, $4C_3$, $3C_2$	$\bar{4}$3m	Cubic
O	E, $3C_4$, $4C_3$, $6C_2$	432	Cubic
O_h	E, 9σ, $3C_4$, $4C_3$, $3C_2$, i	m$\bar{3}$m	Cubic

groups and their corresponding international notation as correlated to the seven crystal systems.

2.7 Space Groups

If one raises the question "why are the previous 32 groups called *point* groups?" the answer will be: because they are based on five symmetry elements, each of which requires a fixed *point* during its operation. Operations about a proper or improper rotation axis require that the axis is at a fixed point, reflection through a plane requires that the plane position is fixed at a point, a center of symmetry, itself, is a fixed point. This answer leads, logically, to two more questions: Are there symmetry elements that do not require a fixed point for

Table 2.3 Symbols for the symmetry elements used in stereographic representation of the 32 point groups.

Symmetry element	Symbol in stereogram	International symbol
$E(\equiv C_1)$	None	1
$i\ (\equiv S_1)$	None	$\bar{1}$
σ	Solid line or bold circle	m
C_2	⬤	2
C_3	▲	3
C_4	◆	4
C_6	⬢	6
S_2	As for mirror plane	$\bar{2}\ (\equiv m)$
S_3	△	$\bar{3}$
S_4	◇	$\bar{4}$
S_6	⬡	$\bar{6}\ (\equiv 3/m)$

operation? And would operation of such symmetry elements results in other symmetry groups? The answer for both questions is yes. This leads us to discuss two new space symmetry elements and how they produce 230 space groups once the three-dimensional solid state is considered.

Before discussing the two additional symmetry elements related to space symmetry, we should emphasize the translation operation. The translation operation along any of the unit cell principal axes is an operation to build three-dimensional crystals by repeating their building blocks (the unit cell) along the three principal axes of the crystal or space. From this viewpoint, translation operation is not exactly a symmetry operation. However, from a pure symmetry viewpoint, a translation operation can still be considered equivalent to the identity symmetry operation.

2.7.1 Screw Axis (n_p)

A screw axis (also known as a helical axis or twist axis) is a symmetry element that involves rotation about an axis followed by a translation along the same axis; n the rotation order, can be any allowed value according to the aforementioned crystallographic restriction theorem (*i.e.*, 2, 3, 4, and 6). Translation along the screw axis is measured as a fraction of the unit cell. Such fraction takes the value p/n, where p can assume any integer 1, 2, 3,...$(n-1)$. For example, 2_1 describes a screw axis symmetry operation in which an indistinguishable conformation (or an equivalent lattice site) is reached by rotating 180° about an axis then translating along that axis a distance that is half the unit cell. Similarly, a 4_3 describes a screw axis symmetry operation in which an indistinguishable

Table 2.4 The 32 crystallographic point groups in stereographic representation.

Crystal system	Point groups
Triclinic	1, $\bar{1}$
Monoclinic	2, m, $2/m$
Orthorhombic	222, $mm2$, mmm
Tetragonal	4, $\bar{4}$, $4/m$, 422, $4mm$, $\bar{4}2m$, $4/mmm$
Trigonal	3, $\bar{3}$, 32, $3m$, $\bar{3}m$
Hexagonal	6, $\bar{6}$, $6/m$, 622, $6mm$, $\bar{6}m2$, $6/mmm$

Table 2.4 Continued.

Crystal system	Point groups
Cubic	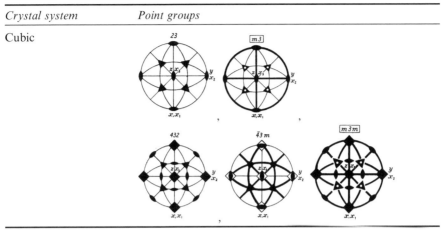

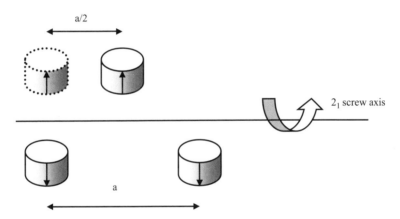

Figure 2.12 Operation of a two-fold screw axis 2_1.

conformation (or an equivalent lattice site) is reached by rotating 90° about an axis then translating along that axis a distance that is three fourth the unit cell in that direction. Figure 2.12 demonstrates the operation of a two-fold screw axis 2_1. Based upon the aforementioned restrictions on both n and p values, it is clear that only one two-fold screw axis is possible (2_1). Similarly, only two three-fold screw axes are possible (3_1, and 3_2). Interestingly, objects possessing a 3_1 screw axis and those possessing a 3_2 screw axis are mirror images (enantiomorphs) of each others. The same applies to 4_1 and 4_3, 6_1 and 6_5 as well as 6_2 and 6_4.

2.7.2 Glide Planes

A glide plane symmetry element involves reflection across a plane of symmetry followed by a translation parallel to that plane. Glide planes are usually

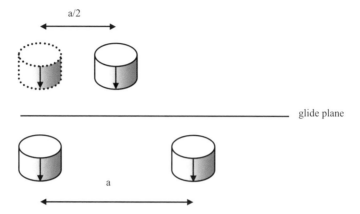

Figure 2.13 Operation of a glide plane symmetry element.

denoted as a, b, or c depending on which of the principal crystal axis the glide plane is along. If the glide is along a face diagonal it is denoted n, and if it is along a body diagonal it is denoted d. Figure 2.13 shows the operation of a glide plane symmetry element. Comparing the resulting configurations in Figures 2.12 and 2.13 should reveal the difference between a 2_1 screw axis and a glide plane. Notably, the two operations result in an object in the exact place but facing opposite directions.

As mathematical grouping of symmetry operations – resulting from the five point symmetry elements – resulted in 32 point groups, grouping of symmetry operations resulting from the seven symmetry elements – five point elements plus two space elements – results in 230 symmetry groups in a three-dimensional space. These groups are known as *space groups*. Space groups are most commonly denoted using the Hermann–Mauguin notations. Tables (similar to Table 2.2) exist to convert this notation into Shönflies notation. A Hermann–Mauguin notation for a space group consists of a set of four symbols. The first is a letter that describes the Bravais centering of the lattice (*i.e.*, primitive, body centered, face centered, *etc.*). Table 2.5 lists the symbols used to describe the different types of cells. The other three symbols describe the most prominent symmetry elements possessed by the crystal. The first of these symbols describes the symmetry of the principal axis of the cell. The next symbol describes the next major symmetry element, *etc.* For example, P2_1/m indicates a primitive cell with a 2_1 screw axis as the principal axis and a mirror plane perpendicular to the principal axis. P$mc2_1$ indicates a primitive cell with a mirror plane, a glide plane and a 2_1 screw axis.

2.7.3 Space Groups in One- and Two-dimensional Space

Realizing that space groups describe the symmetry of objects within space, and that they were developed based upon a number of symmetry operations available in that space, leads to the conclusion that the possible space groups

Raman Spectroscopy; the Diagnostic Tool

Table 2.5 Symbols for the six possible Bravais lattice centerings.[a]

Type	Symbol	Example
Primitive (lattice sites on corners only)	P	
Face centered (lattice sites on corners and all face centers)	F	
Side centered (lattice sites on corners and centers of two of the faces)	A, B, or C	C centering is shown
Body centered (lattice sites on corners and body center)	I	

[a]Primitive centering (P): lattice points on the cell corners only. Body centered (I): one additional lattice point at the center of the cell. Face centered (F): one additional lattice point at center of each of the faces of the cell. Centered on a single face (A, B, or C centering): one additional lattice point at the center of one of the cell faces.

depend on the dimensionality of the space in which they exist. Since, traditionally, materials scientists are inherently interested in three-dimensional space representing material systems, we have focused our discussion on three-dimensional space groups. It is important, however, to know that one- and two-dimensional space groups have also been investigated and are of great interest and applications in other fields such as theoretical physics, art, architecture, and the textile industry. In materials science, one- and two-dimensional liquid crystals as well as symmetry along a single polymer or a co-polymer molecule are subjects of increasing interest. In addition, as nanotechnology brings quasi one-dimensional (nanowires, nanotubes, one-dimensional crystals, *etc.*) and two-dimensional (graphene sheets, self-assembled layers, *etc.*) material systems to our attention, a brief discussion of space groups in these spaces is appropriate in this section.

2.7.3.1 *Space groups in One-dimensional Space (Linear Objects)*

Considering one-dimensional spaces, spatial restrictions make only certain symmetry elements and operations possible. In addition to translation along the one dimension of the space, rotation about an axis normal to the line is

Table 2.6 The seven possible one-dimensional space groups (Frieze groups).

Frieze group	Symmetry elements	Example
F1	Translation only	
F2	Translation, two-fold rotation	
F1m	Translation, transverse mirror	
F11m	Translation, longitudinal mirror	
F2mm	Translation, rotation, transverse and longitudinal mirrors	
F11g	Translation, glide	
F2mg	Translation, rotation, transverse mirror, glide	

possible as long as the rotation order is restricted to 180° (two-fold rotation axes only). We may also realize that inversion through a point on the line, reflection through a plane normal to the line, and glide through a plane passing through the line are also allowed. These allowed symmetry operations in one-dimensional space result in seven different one-dimensional space groups that are also known as *Frieze groups*. These groups can belong to only two types: one type with reflection and a second type with no reflection.

Symbols for one-dimensional space groups usually start with the letter "F" (for Frieze). In certain texts, the letter P (for primitive cells) is also used. The second symbol is a number, which is 2 if rotational symmetry does exist and is 1 if rotational symmetry does not exist. The third symbol is m if there is a transverse mirror, and it is 1 if no transverse mirror exists. The fourth symbol is m if a longitudinal mirror exists, and is a g if a longitudinal glide exists. The fourth symbol is omitted if no longitudinal symmetry exists. Table 2.6 lists the seven possible one-dimensional space groups (Frieze groups) and the symmetry elements possessed by each of them. Frieze groups are important in describing symmetry along polymer molecules.

2.7.3.2 *Space Groups in Two-dimensional Space (Wallpaper Groups)*

Considering a two-dimensional space restriction, only 17 space groups will be possible in such a case. Two-dimensional space groups are also referred to as *plane space groups* or *wallpaper* groups. The name wallpaper groups originates from the fact that these symmetry objects have been used for centuries as decorative art for textiles, architect, and – more recently – for wallpaper designs. The symbols for wallpaper groups start with a letter describing the cell centering (*i.e.*, P for primitive or C for centered) followed by an integer describing the highest order of rotation, followed by m, g, or 1 to reflect the existence of a mirror plane, a glide plane, or no symmetry element, respectively.

To conclude our discussions on space groups, it is important to realize that other types of space groups exist and have been investigated. Depending on the dimensionality of the space considered different space groups evolve and become relevant. For example, in physics and mathematics, when four-dimensional space is considered, in time evolving systems for example, 4895 space groups are possible. In addition, magnetic space groups can be constructed from ordinary three-dimensional space groups by considering magnetic spins associated with lattice points. Such magnetic space groups are also known as *colored space groups*, and 1651 of them are possible.

Now that we have discussed symmetry and symmetry groups in molecules and crystals in different spatial dimensionalities, it is important to revert to our main subject of spectroscopic properties of matter. According to Neumann's principle, the symmetry elements of any physical property (of a crystal) must include the symmetry element of the point group representing (that crystal). Realizing that Raman activity is part of the optical physical behavior of matter, let us discuss how the character of a point group governs its Raman activity.

2.8 Character Table

As we discussed earlier, a point group is, simply, a group of symmetry operations satisfying certain "membership" conditions. To gain full appreciation of Neumann's principle and be able to correlate the Raman activity of a species to the symmetry of its point group, mathematical representation of the symmetry operations is required. Hence a system was developed in which numbers were used to *represent* the effect of symmetry operations on directional properties. These numbers are called *representations*.

2.8.1 Symmetry Operations and Transformation of Directional Properties

Let us start with a simple example of how representations can be developed. If we consider a p_x electronic orbital, as shown in Figure 2.14, applying a C_2 symmetry operation on such orbital changes its orientation and, hence, switches the sign of the orbital. This can be mathematically expressed as:

$$C_2 p_x = -p_x \qquad (2.10)$$

Hence, the symmetry operation C_2 in this case can be represented by -1.

Let us apply another symmetry operation, σ_{yz}. As shown in Figure 2.15, the sign of the orbital is, again, switched, and it can be written that:

$$\sigma_{yz} p_x = -p_x \qquad (2.11)$$

Hence, the symmetry operation σ_{yz} in this case is represented by -1.

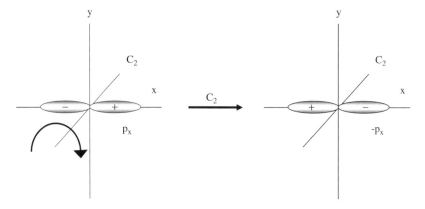

Figure 2.14 Transformation of p_x under a C_2 symmetry operation.

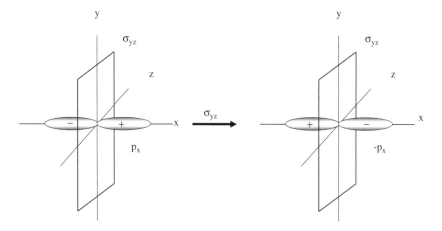

Figure 2.15 Transformation of p_x under a σ_{yz} symmetry operation.

If we apply the symmetry operation σ_{xz}, it is clear from Figure 2.16 that $\sigma_{yz}p_x = p_x$, and the representation for this operation is 1. Considering the identity operation (E), it will not be hard to realize that the representation for this operation is always 1.

What we considered in this example is a set of four symmetry operations: E, C_2, σ_{xz}, and σ_{yz}. These four symmetry operations, actually, form the point group (C_{2v}). Hence, we can mathematically represent the effect of the group operations on the electronic orbital p_x or, generally speaking, on anything with similar symmetry in the x-direction using a set of numbers as follows:

C_{2v}	E	C_2	σ_{xz}	σ_{yz}	
B_1	1	−1	1	−1	x

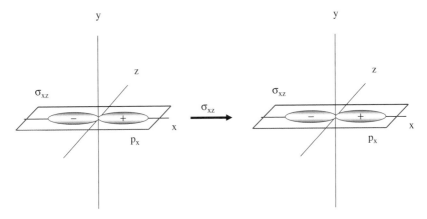

Figure 2.16 Transformation of p_x under a σ_{xz} symmetry operation.

In this case we say that the p_x is symmetric to E and σ_{xz} and it is anti-symmetric to C_2 and σ_{yz} in C_{2v} symmetry. The set of numbers is called a representation, and is usually given a letter name. This letter name describes the *symmetry species* of the representation as we will discuss later in this section. The letter name assigned for the p_x in the C_{2v} is B_1. We say that the x belongs to the B_1 representation of C_{2v}. We can also say that in a C_{2v} group, directional properties in the x-direction belong to the B_1 symmetry species. If we consider the effect of C_{2v} symmetry operations on the molecular orbital p_y, a different representation can be constructed, which is assigned the name B_2 as follows:

C_{2v}	E	C_2	σ_{xz}	σ_{yz}	
B_2	1	−1	−1	1	y

Similarly, considering the effect of the group symmetry operations on the molecular orbital p_z leads to the representation A_1 as follows:

C_{2v}	E	C_2	σ_{xz}	σ_{yz}	
A_1	1	1	1	1	z

Because all of the numbers in the A_1 presentation are positive numbers, it is a *totally symmetric* presentation.

To this end, we should emphasize the two conditions that make a set of numbers a representation. First, the numbers should represent the effect of the group symmetry operation on certain directional properties. Secondly, the numbers should multiply together in the same way as the group operations. Having said that, there is an additional set of numbers, called the A_2 representation, that satisfy the two aforementioned conditions. A_2 representation for the C_{2v} point group is symmetrical for rotation while anti-symmetrical for reflection is as follows:

C_{2v}	E	C_2	σ_{xz}	σ_{yz}
A_2	1	1	−1	−1

A table containing the full set of representations of a point group is known as the *character table* of the group.[vi] The character table provides essential information regarding the symmetry of point groups and directional properties of molecules belonging to such point groups. It also lists all the symmetry species available in a symmetry point group and shows the transformational properties of such symmetry species. For example, as we have just discussed, the A_2 symmetry species in C_{2v} is symmetrical for rotation and anti-symmetrical for reflection.

To this end, we know that molecules (and crystals) in their equilibrium conformations possess certain symmetry elements and operations that classify them into certain point groups. We also know that molecules translate, rotate, and vibrate in certain directions. During a vibration mode the locations of atoms change, and hence the symmetry of the molecule will be affected under certain symmetry operations. Hence, molecular motions and, specifically for our interest here, normal vibration modes can be assigned to certain symmetry species. The correlation between normal vibration modes and symmetry species emphasizes the most important role of the character table from a spectroscopic viewpoint. The character table provides information regarding the symmetry species (representations) that polarizability tensor components belong to. A vibration mode is Raman active if it belongs to the same symmetry species as one or more of the polarizability tensor components. Table 2.7 shows the character table of the C_{2v} point group. The table clearly shows that the diagonal components (α_{xx}, α_{yy}, and α_{zz}) of the polarizability tensor belong to the A_1 species while the off diagonal components, α_{xy}, α_{xz}, and α_{yz}, belong to the symmetry species A_2, B_1, and B_2, respectively. This means that vibration modes belonging to any symmetry species in a molecule with C_{2v} symmetry are Raman active. To complete our discussion of this point, we should mention that the total number of symmetry operations belonging to a group is denoted as the order of the group (g), and that the total number of symmetry species is denoted by (r). In our example (point group C_{2v}), $g = 4$ and $r = 4$.

Table 2.7 Character table of C_{2v} point group.

C_{2v}	E	$C_2(z)$	$\sigma_v(xz)$	$\sigma_v(yz)$	Raman activity
A_1	+1	+1	+1	+1	α_{xx}, α_{yy}, α_{zz}
A_2	+1	+1	−1	−1	α_{xy}
B_1	+1	−1	+1	−1	α_{xz}
B_2	+1	−1	−1	+1	α_{yz}

[vi] The name *character* was chosen because the numbers in the table are, actually, the character (sometimes called the trace, which is the sum of the matrix diagonal) of square matrixes known as irreducible representations of a symmetry operations.

Some general rules have also been developed for Raman activity as well. For example, if we go back to the CO_2 molecule (Figure 2.6), we realize that the stretch mode symmetric in respect to the molecule center of symmetry is Raman active while the stretch mode asymmetric in respect to the symmetry center is not. This is, in fact, a general rule that applies to all molecules with a center of symmetry.

A system has been developed to assign symbols to symmetry species (representations) that are indicative of their nature. A symmetry species that is totally symmetric for all symmetry operations is designated as A_1, a species that is symmetric for rotation but is anti-symmetric for reflection or rotation about a normal axis to the principal axis is designated as A_2. A species that is anti-symmetric for rotation about the principal axis is designated as B. If it is symmetric for rotation about a C_2 axis normal to the principal axis or for reflection it is designated as B_1, and if it is anti-symmetric in this regard it is designated as B_2. Also, in point groups possessing a σ_h, primes and double primes are usually used to designate symmetry with respect to the horizontal plane. A single prime (A′) indicates a symmetry while a double prime (A″) indicates anti-symmetry. In addition, in point groups possessing a center of symmetry (i), symmetry with respect to i is further designated by subscripts. A subscript g (from the German word "*gerade*," meaning even) is used to designate symmetrical transformation, while the subscript u (from the German word "*ungerade*," meaning uneven) is used to designate anti-symmetric transformation.

2.8.2 Degenerate Symmetry Species (Degenerate Representations)

Let us consider the effect of a C_4 symmetry operation on the directional properties of a molecular orbital p_y. As shown in Figure 2.17, a clockwise C_4^1

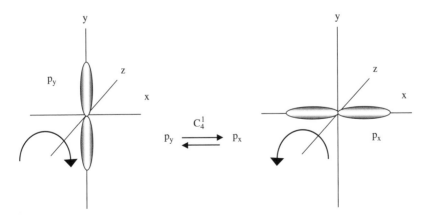

Figure 2.17 Doubly degenerate electronic orbital under C_4^1 symmetry operation.

operation will transform p_y into a p_x. Similarly, an anti-clockwise C_4^1 operation will change a p_x into a p_y orbital. In this case, the p_y and p_x under a C_4 symmetry operation are identical, and hence are called *degenerate*, meaning energetically equivalent. Therefore, in a point group possessing a C_4 operation (such as C_{4v}), both p_x and p_y orbitals belong to the same *doubly degenerate* symmetry species designated as E. It is important, however, to note that the p_x and p_y orbitals in a C_{2v} group are not energetically equivalent since they belong to different symmetry species, namely, B_1 and B_2, respectively.

As a general rule, it can be said that if a group possesses a rotation axis with 3 or higher order, some directional properties may convert into others as the result of a symmetry operations. If there is energy associated with the directional properties (such as orbital energies mentioned above, or energy of a vibration mode), these energies must be identical (degenerate). Hence, symmetry directly indicates that if a number of vibration modes are mixed by symmetry they must be degenerate, and should belong to the same degenerate symmetry species. A triple degeneracy is designated as F (or sometimes T), a quadruple degeneracy (four-fold) is designated as G, and a quintuple degeneracy (five-fold) is designated as H.

2.8.3 Symmetry Species in Linear Molecules

Linear molecules are a very important class of chemicals that includes diatomic gases and other molecules with great impact on our planet, such as carbon dioxide. Linear molecules belong to one of the two point groups, $C_{\infty v}$ or $D_{\infty h}$. A different set of symbols is used to designate symmetry species in linear molecules. Greek letters identical to the designation of electronic states in atomic bonding are used. The Greek letter σ (or capital Σ) is used to designate symmetry representations symmetrical with respect to the principal infinite axis (molecular axis). A superscript plus or minus signs (σ^+ or σ^-) are used to designate species that are symmetric or anti-symmetric with respect to a plane of symmetry through the molecular principal axis. The symbols π, Δ, and Π are used to designate degenerate species with the degree of degeneracy increasing in the same order. Subscripts g and u are also used to designate symmetry with respect to a center of symmetry as we discussed earlier. Table 2.8 shows the character table of the $D_{\infty h}$ point group to which the carbon dioxide molecule belongs.

2.8.4 Classification of Normal Vibration by Symmetry

For a molecule with N atoms, the $3N - 6(5)$ normal vibration modes (*phonons*) can be classified into symmetry species. A vibration mode would belong to a symmetry species if the mode's symmetry properties transform the same as those of the species under different symmetry operations. Different methods have been used to classify normal vibration modes into symmetry species. In this section we discuss what is considered the simplest method in which the classification process can be achieved by inspection for simple small molecules. The water molecule is a

Table 2.8 Character table of the $D_{\infty h}$ point group.

$D_{\infty h}$	E	$2C_\infty^\phi$	$2C_\infty^{2\phi}$	$2C_\infty^{3\phi}$	...	σ_h	∞C_2	$\infty \sigma_v$	$2S_\infty^{2\phi}$	$2S_\infty^{2\phi}$	...	$S_2 \equiv i$	Raman activity
Σ_g^+	+1	+1	+1	+1	...	+1	+1	+1	+1	+1	...	+1	$\alpha_{xx}+\alpha_{yy}, \alpha_{zz}$
Σ_u^+	+1	+1	+1	+1	...	−1	+1	−1	−1	−1	...	−1	
Σ_g^-	+1	+1	+1	+1	...	+1	−1	−1	+1	+1	...	+1	
Σ_u^-	+1	+1	+1	+1	...	−1	+1	−1	−1	−1	...	−1	
Π_g	+2	$2\cos\phi$	$2\cos 2\phi$	$2\cos 3\phi$	...	−2	0	0	$-2\cos\phi$	$-2\cos 2\phi$	...	+2	$(\alpha_{yz}, \alpha_{xz})$
Π_u	+2	$2\cos\phi$	$2\cos 2\phi$	$2\cos 3\phi$	...	+2	0	0	$+2\cos\phi$	$+2\cos 2\phi$	...	−2	
Δ_g	+2	$2\cos 2\phi$	$2\cos 4\phi$	$2\cos 6\phi$	...	+2	0	0	$-2\cos 2\phi$	$+2\cos 4\phi$	...	+2	$(\alpha_{xx}-\alpha_{yy}, \alpha_{xy})$
Δ_u	+2	$2\cos 2\phi$	$2\cos 4\phi$	$2\cos 6\phi$	...	−2	0	0	$-2\cos 2\phi$	$-2\cos 4\phi$	...	−2	
Φ_g	+2	$2\cos 3\phi$	$2\cos 6\phi$	$2\cos 9\phi$	...	−2	0	0	$-2\cos 3\phi$	$-2\cos 4\phi$	...	+2	
Φ_u	+2	$2\cos 3\phi$	$2\cos 6\phi$	$2\cos 9\phi$	...	+2	0	0	$+2\cos 3\phi$	$+2\cos 4\phi$	...	−2	
...													

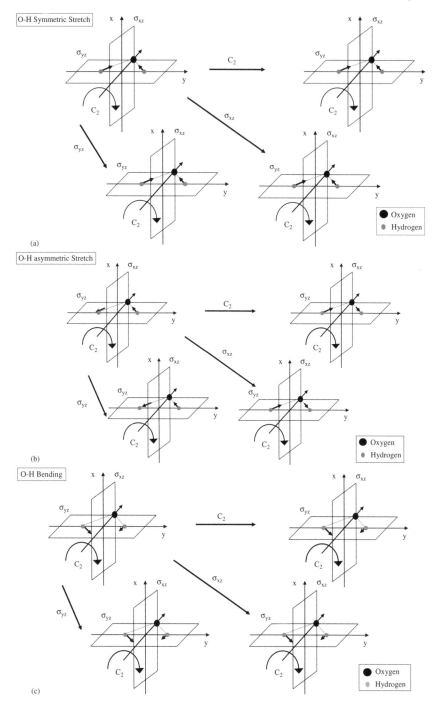

Figure 2.18 Normal vibration modes of water molecule and their transformation under symmetry operations of a C_{2v} point group: (a) O–H symmetric stretch, (b) O–H asymmetric stretch, and (c) O–H bending.

classical example for such a method. A water (H_2O) molecule is a tri-atomic, nonlinear molecule that belongs to the C_{2v} point group. Hence, in addition to the three translations (T_x, T_y, and T_z) along the three Cartesian axes, and the three rotations (R_x, R_y, and R_z) about them, the molecule has 3 [(3×3)–6] normal vibration modes. These three normal vibration modes are O–H symmetric stretch, O–H asymmetric stretch, and O–H in-plane bending (Figure 2.18a–c). By inspection, it can be shown that the O–H symmetric stretch mode is symmetric for the three symmetry operations of the group; hence it belongs to the totally symmetric species A_1. Inspection of the O–H in-plane bending mode also reveals that the mode is symmetric for all symmetry operations of the group, and also belongs to the A_1 species. Inspection of the O–H asymmetric stretch mode, however, reveals that the mode would transform symmetrically only for the in-plane mirror (σ_{yz}), but it would transform asymmetrically for the other symmetry operations and, hence, it belongs to the B_2 species. Therefore, we can say that, for a water molecule, two of the three normal vibration modes belong to the A_1 symmetry species, while the third belongs to the B_2 species. Consulting the character table for the C_{2v} point group (Table 2.7) reveals that different components of the polarizability tensor belong to the same symmetry species, hence, all the three normal vibration modes of the water molecule are Raman active.

The reader will realize that the three normal vibration modes of the water molecule belong to two of the four symmetry species available in the point group. A question might be raised regarding the other two symmetry species of the group, namely A_2 and B_1. Inspection of the translation motion along the three Cartesian axes reveals that T_z belongs to the totally symmetric species A_1, T_y belongs to B_2, while T_x belongs to B_1. Also, inspection of the three rotations about the three Cartesian axes reveals that R_z belongs to A_2, while R_y and R_x belong to B_1 and B_2, respectively. Realizing, as we mentioned before, that different polarizability tensor components belong to all four symmetry species in this case indicates that rotational modes in the water molecule are also Raman active. While rotational Raman spectroscopy is beyond the scope of this section, the reader is referred to several references where the subject has been discussed.[6,7] Table 2.9 lists the nine degrees of freedom (rotation, translation, and vibration modes) for a water molecule and the symmetry species they belong to.

While the inspection method for classifying vibration modes into symmetry species might be achievable in the case of a small simple molecule, the method becomes hard to apply as the molecular structure becomes more complicated. Thanks to the efforts of Herzberg,[8] the number of normal vibration modes in each symmetry species can be calculated using mathematical formulae (Herzberg's formulae) specific for each point group. For the C_{2v} point group, the Herzberg formulae are expressed as:

$$A_1 : \quad 3m + 2m_{xz} + 2m_{yz} + m_o - 1$$
$$A_2 : \quad 3m + m_{xz} + m_{yz} - 1$$
$$B_1 : \quad 3m + 2m_{xz} + m_{yz} + m_o - 2$$
$$B_2 : \quad 3m + m_{xz} + 2m_{yz} + m_o - 2,$$

Table 2.9 The nine degrees of freedom for a water molecule, and the symmetry species they belong to.

Degree of freedom	Symmetry species
x Translation	B_1
y Translation	B_2
z translation	A_1
x Rotation	B_2
y Rotation	B_1
z Rotation	A_2
O–H symmetric vibration	A_1
O–H Asymmetric vibration	B_2
O–H Bending vibration	A_1

and the total number of atoms (N) is given by:

$$N: \quad 4m + 2m_{xz} + 2m_{yz} + m_o$$

where, m is the number of sets of nuclei not on any symmetry elements, m_o is number of nuclei on all symmetry elements, m_{xy}, m_{yz}, and m_{xz} represent the number of sets of nuclei lying on the xy, yz, and xz planes, respectively, but not on any other rotation axis in these planes.

Considering our example for the water molecule, $m = 0$, $m_o = 1$ (the oxygen atom), $m_{xz} = 0$, $m_{yz} = 1$ (hydrogen atom), and the total number of atoms (N) equals 3. Hence, the number of normal vibration modes is 2, 0, 0, and 1 for the symmetry species A_1, A_2, B_1, and B_2, respectively, as was determined by inspection before. Character tables for the 32 molecular point groups as well as Herzberg formulae for them are listed in Appendices 1 and 2, respectively.

To this end, it should be clear that as a monochromatic light (a laser) interacts with a molecule each Raman active vibration mode will contribute a peak to the molecule Raman spectrum. Based upon the vibration frequency of the mode (v), and the frequency of the excitation laser (v_o), the Raman peak will appear at a certain wavenumber ($\tilde{v}$) as described in Equation (2.6). The frequency of a vibration mode (v) is actually scaled to the reduced mass of the atoms involved in the mode as well as the stiffness (spring constant) of the chemical bond between these atoms, and can be mathematically expressed as:

$$v \approx \sqrt{\frac{k}{m}} \quad (2.12)$$

where k is the bond spring constant and m is the reduced mass calculated as:

$$m = \frac{m_1 m_2}{(m_1 + m_2)} \quad (2.13)$$

where m_1 and m_2 are the atomic masses of the two chemically bonded atoms.

Raman Spectroscopy; the Diagnostic Tool

Based upon the symmetry of a chemical species, the number of its constituent atoms, their masses, as well as the nature and strength of chemical bonds connecting its atoms, a unique Raman spectrum identifying the species should be expected. Hence, Raman spectroscopy is a unique diagnostic technique capable of "finger printing" various chemical species. Figure 2.19 shows the Raman spectrum of liquid benzene and liquid toluene. Note the significant difference between the two spectra as the result of replacing one of the hydrogen atoms in benzene with a methyl (CH_3) group to create the toluene molecule.

2.8.5 Raman Overtones and Combination Bands

In a Raman spectrum of a certain molecule, one should expect to observe a Raman peak for each Raman active mode of the molecule. Additional modes, however, are commonly observed in a Raman spectrum. These modes are known as *combination* and *overtone* modes. A combination mode results when two Raman active modes combine together and contribute a new Raman peak into the spectrum. The position of the combination mode will be at (or very close to) the sum of the positions of the combined modes. An overtone mode will represent a higher order of a Raman active mode. A second-order mode will be at (or close to) twice the frequency of the original mode. A third-order will be at (or close to) three times the frequency, *etc*. For example,[9] Figure 2.20 shows the Raman spectrum of a graphitic (pyrocarbon) sample. The spectrum contains Raman bands at 1354, 1581, and 1620 cm^{-1} as the first-order Raman modes. These are known in the literature as the D, G, and D'-bands, respectively. The spectrum also exhibits Raman bands at 2708, 3185, and 3245 cm^{-1}, as the second-order Raman bands of the aforementioned first-orders, respectively. In addition,[vii] the spectrum exhibits a Raman peak around 2950 cm^{-1} that is a combination mode of the D-band and the G-band. Notably, in general,[viii] higher order Raman bands appear with much lower intensity. For this reason, second-order Raman peaks are commonly observed, while third-order Raman peaks are seldomly observed.

2.8.6 Molecular and Lattice Raman Modes

Now, we understand that based on the symmetry of a molecule or a crystal certain vibrational modes will be Raman active according to the selection rule discussed above. Importantly, in molecular crystals, where the crystal lattice sites are occupied by molecules, symmetry rules can be applied on two levels: the molecular level and the crystal level. Hence, in addition to Raman active

---

[vii] The origin of the 2450 cm Raman band in HOPG and carbon nanotubes was reported in 2005 as a second-order longitudinal optical mode resulting from a double resonance effect. Detailed discussion can be found in T. Shimada, T. Sugai, C. Fantini, M. Souza, L. G. Cançado, A. Jorio, M. A. Pimenta, R. Saito, A. Grüneis, G. Dresselhaus, M. S. Dresselhaus, Y. Ohno, T. Mizutani and H. Shinohara, *Carbon*, 2005, **43**, 1049.

[viii] The second order of the D-band (known as the G'-band or the 2D-band) is an exception for the general rule also due to its double resonance effect in graphite.

(a) **Benzene**

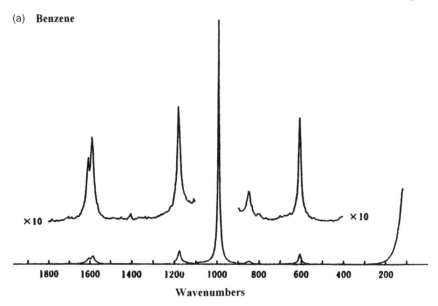

(b) **Toluene**

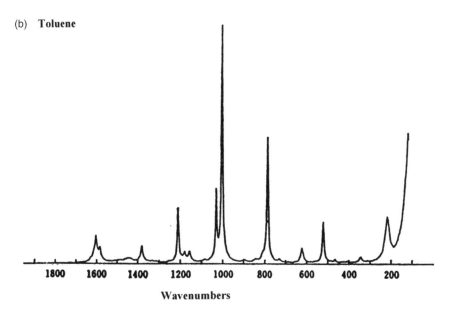

Figure 2.19 Raman spectra of (a) benzene and (b) toluene as examples of the fingerprinting capabilities of the Raman technique.

molecular modes, a new set of Raman active *crystal or lattice* modes can be observed. This is very common in polymer crystals as well as protein crystals. This point will be examined further as we discuss chemical and structural perturbation effects of Raman scattering in the next section.

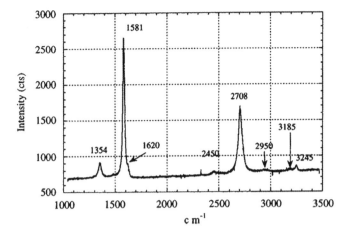

Figure 2.20 Raman spectrum of graphite showing Raman first order, combinations, and second order modes. (Reproduced with kind permission from Wilhelm et al., ref. 9. Copyright American Institute of Physics 1998.)

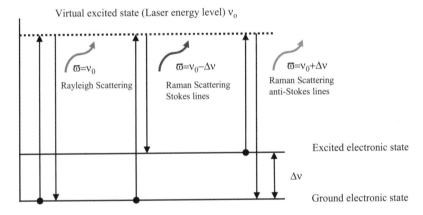

Figure 2.21 Ground and excited energy states describing Raman phenomenon from the energy transfer viewpoint.

2.9 Raman from an Energy Transfer Viewpoint

A different viewpoint from which to consider the light scattering phenomenon could be as follows: as a monochromatic light interacts with or excites matter, atoms or molecules in the matter could be considered as *virtually* excited from their equilibrium energy state into a higher excited energy state having the same energy as the excitation light ($E = h\nu$). The excited species would *instantaneously* descend into a lower energy state emitting light (the scattered light) having energy equal to the energy difference between two levels. Figure 2.21 shows a schematic of the process. As virtually excited species descend into a lower energy level, three possibilities should be considered. First, there is the possibility that a species descends back to its original ground energy state. In this case the emitted energy will be equal to the gained energy, and the scattered light will have the same frequency as the incident or exciting light. During this process no energy gain of loss would have taken place. The scattering process will be the elastic scattering known as Rayleigh scattering.

Secondly, we consider the possibility that a species excited from the equilibrium ground state descends into an excited state. In this case, the emitted energy will equal the gained energy minus the energy difference between the electronic ground and excited states. The scattered light will have a frequency equal to $(v_o - \Delta v)$ (Figure 2.21). In this case the scattered light (photon) has a lower energy than the incident photon. The scattering process is inelastic scattering, which we termed Raman scattering. Raman scattering in which the scattered photon has less energy than the incident photon is also known as Stokes-lines.

Thirdly, we should consider the possibility that a species excited from a high energy state descends into the equilibrium ground state. In this case, the emitted energy will equal the energy gained plus the energy difference between the electronic ground and excited states. The scattered light will have a frequency equal to $(v_o + \Delta v)$ (Figure 2.21). The scattered photon will have higher energy than the incident photon. The scattering process is, again, inelastic, or Raman scattering. This type of Raman scattering, where the scattered photon has more energy than the incident photon, is known as anti-Stokes lines.

While this explanation or description of the Raman phenomenon is very simple, and could be described as too simple, it serves an important purpose in explaining two crucial characteristics of the Raman phenomenon; first, it emphasizes that the Raman position depends on the relative position of the electronic energy states in the atom or molecule. Hence, any perturbation field – such as a strain field, a magnetic field, a thermal excitation, or a chemical potential resulting from a surrounding environment – that affects such an electronic structure should be expected to affect the Raman peak position. Secondly, the description makes it clear that Stokes and anti-Stokes lines are related to electronic state population, *i.e.*, distribution of atoms or molecules among available ground and excited electronic states in the system. Hence, the characteristics of Stokes and anti-Stokes lines should be correlated to the very well-known Boltzmann equilibrium distribution. We discuss these two important features in the following sections.

Before this discussion, however, it is important to point out an essential point regarding the nature of Raman scattering as just described. The name "Stokes" and "anti-Stokes" Raman lines came into use in memory of G. G. Stokes who pioneered work in fluorescence. This, as well as the simplified description of Raman scattering just discussed, might give the wrong impression that the phenomena of fluorescence and Raman scattering are cognate. In fluorescence, the incident photon of light is completely absorbed by the interacting molecule which then is actually raised into an excited electronic level. After a certain time spent in the excited level (lifetime), the molecule actually descends into a lower electronic state and, thereby, emits light with a frequency lower than that it originally absorbed. This mechanism is completely different from the Raman scattering mechanism. In Raman scattering, the incident light photon is never absorbed, but rather induces the molecule to undergo a vibrational or rotational transition. The molecule is never *actually excited* into a higher electronic state and this is why the term *virtually excited* was used. If the term *virtually excited* is still not clear, one may think of it as excited with a zero lifetime in the excited state! This also may help in differentiating Raman from fluorescence peaks. The event lifetime is usually reflected in the peak full width at half maxima (FWHM). Fluorescence peaks are always much broader than Raman peaks. In addition, an essential difference between the two phenomena becomes clear when considering the fact that fluorescence can be quenched by adding a species capable of absorbing the energy (by collision) from the molecule while it is in the excited state. Hence, the molecule will descend into the lower electronic state without emitting light. Raman scattering cannot be quenched since the molecule is never actually residing in the excited state.

2.10 Boltzmann Distribution and its Correlation to Raman Lines

In one of his great works, Boltzmann calculated the equilibrium (or most probable) distribution of a number of species among a number of energy levels such that the total energy of the system is constant. Following Boltzmann's notion, and using the electronic distribution among electronic states as an example, one may pose the question as: If a molecule has a ground electronic state (energy level) (ε_o) and a number of allowed excited (higher energy) states (ε_1, ε_2, ε_3, *etc.*), what would be the most probable (equilibrium) distribution of electrons among such electronic states at a constant total energy of the system (the molecule)? Boltzmann's answer for such an important question is known as Boltzmann's distribution which is given by the following equation:

$$\frac{N_i}{N} = \frac{e^{-\varepsilon_i/k_B T}}{Z(T)} \qquad (2.14)$$

The partition function $Z(T)$ is defined as:

$$Z(T) = \sum e^{-\varepsilon_i/k_B T} \qquad (2.15)$$

According to Boltzmann's distribution, at a certain temperature, most atoms will reside in the lowest possible energy state. Higher energy states will be populated with an exponentially decreasing number of electrons as the energy of the state increases. Bearing in mind the previously discussed origin of Stokes and anti-Stokes lines, their relative intensity should correlate to the local temperature of the molecule as follows:

$$\frac{I_S}{I_{AS}} = \frac{(v_o - v_v)^4}{(v_o + v_v)^4} \exp[{hv_v}/{k_B T}] \qquad (2.16)$$

where I_S is the Stokes line intensity, I_{AS} is the anti-Stokes line intensity, and all other symbols are the same as described before.

This adds another important utility to Raman scattering as a diagnostic tool: the ability to remotely measuring the local temperature of a system. It has to be added, however, that as evident from Equation (2.16), the I_S/I_{AS} intensity ratio for a vibrational mode increases with a power of four dependence on the frequency of the mode. Hence, from a practical viewpoint, as the wavenumber of a Raman band increases, its anti-Stokes line intensity decreases tremendously and becomes harder to quantify. This puts a practical limit on our ability to utilize many high wavenumber Raman lines for temperature measurements purposes. For good Raman scattering materials (such as diamond), the limit of this method, however, can still be useful for many applications and investigations. Figure 2.22 shows the temperature of a single-crystal diamond sample as obtained from the I_S/I_{AS} ratio [see Equation (2.16)] *versus* the reference temperature.[10a] From the figure, the temperature obtained from the non-contact Raman intensity method clearly agrees perfectly with the reference temperature up to around 750 K. Significant deviation (in the range of 75 K) can be observed at high temperatures. Such deviation (which can be denoted as error in the temperature measurements) is most plausibly due to electronic resonance and coupling effects at higher temperatures. This is the same obstacle faced by Einstein and Debye during their attempts to develop theoretical expressions describing the phonon contribution to heat capacity of crystals over a wide range of temperature. It is also important to note that the temperature beyond

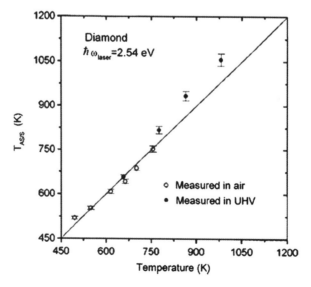

Figure 2.22 Single-crystal diamond sample temperature as obtained from the Stokes/anti-Stokes ratio versus the reference temperature. (Reproduced with kind permission from Cui *et al.*, ref. 10a. Copyright American Institute of Physics 1998.)

which deviation in the Raman measurements starts to occur does not only depend on the nature of the material but also on the system size (bulk or nano). As we will discuss in Chapter 4, Stokes/anti-Stokes temperature measurements for carbon nanotubes suffer from resonance effects even at room temperature.

2.11 Perturbation Effects on Raman Bands

Qualitatively speaking, any field that perturbs the electronic structure of a molecule or crystal will cause shifts in the Raman line position. Applying external strain, adding heat, interacting with a chemical environment, or even local perturbations resulting from structural imperfections such as vacancies, dislocations, substitution, and interstitial impurities are all examples of perturbation fields that have been shown to result in measurable Raman peak shifts. Such Raman peak shifts were investigated, and quantified. They contributed to the uniqueness of Raman spectroscopy as a diagnostic tool in several field.

2.11.1 Strain Effects

The effect of applied strain/stress[ix] on crystal symmetry and its phonon frequency are part of what is known as *morphic effects*. The term morphic effects was first coined by Hans Mueller of MIT in 1940 to denote symmetry changes in crystals induced by external generalized forces and leading to changes in the macroscopic properties of the system.[11]

Morphic effects and their impact on the Raman spectrum of the material have attracted the interest of many investigators. Expressions relating phonon frequencies to applied strain tensors as well as experimental results have been reported for various crystal classes. For a three-dimensional crystal, the components of a strain tensor were found to affect different vibrational modes differently based upon the phonon symmetry in respect to strain component direction. Some of the Raman active modes tend to shift to lower wavenumber (redshift), others tend to shift to higher wavenumbers (blue-shifts), while some modes did not show any shifts. Morphic effects in silicon single crystals were thoroughly investigated by Anastassakis *et al.*[12] in the 1970s and De Wolf[13] in the 1990s. Considering morphic effects on Raman spectra, it is important to realize that, owing to symmetry lowering under non-hydrostatic loading conditions (non-symmetrical strain tensors), degenerate Raman peaks most commonly split and start to appear as separate peaks in the spectrum. This should not be surprising because once the crystal symmetry was altered under loading effects vibration modes that used to have the same frequency and result in a degenerate Raman peak will have different frequencies and result in separate Raman peaks. Such ability of the Raman technique has been heavily utilized in

---

[ix] Since we mainly work within elasticity limits, stress and strain are linearly correlated. The reader should, however, note that it is the strain that affects the symmetry and, hence, it is the important quantity to focus on.

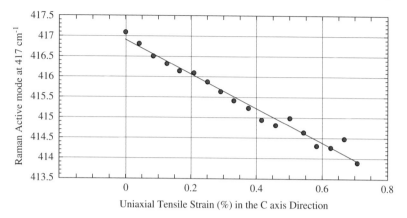

Figure 2.23 Effect of uniaxial tensile strain along the *c*-axis of a sapphire single crystal on Raman peak position.

investigating strain effects on single-crystal silicon wafers and profoundly benefited the electronic and MEMS (microelectromechanical systems) industry. It also enabled fundamental advancements in the field of laser machining of electronic materials. The work of Amer et al.[14] is a good example of the Raman technique capabilities in this field.

For a much simpler loading condition, the effect of applying simple uniaxial tensile strain along the *c*-axis of a sapphire (Al_2O_3) single crystal fiber on its Raman active mode at 417 cm^{-1} is demonstrated in Figure 2.23. The figure shows that this particular Raman mode shifts linearly to lower wavenumbers under the effect of uniaxial tensile strain.

In such a case, once caution has been taken to ensure that other non-zero component of the strain tensor will not dramatically affect the Raman peak position of the mode under consideration, the line shown in Figure 2.23 can be used as a calibration line for Raman mode position *versus* strain. Such calibration can be utilized to determine the axial strain level in strained fibers based on their measured Raman mode position. Figure 2.24 shows a strain profile measured along a 500-micron long graphite fiber embedded in epoxy. Such capability of the Raman technique paved the road for important investigations in solid mechanics and particularly in fibrous composites mechanics. The work of Young et al.,[15] Galiotis,[16] and Amer et al.[17,18] in the 1980s and 1990s are good examples of Raman applications in the field of composite mechanics.

2.11.2 Heat Effects

It was shown that heating or cooling the system would cause its phonons to soften or harden, respectively. This indicates that raising the system temperature would cause the system's normal vibration modes to shift to lower wavenumbers and *vice versa*. Plotting a Raman mode position for a

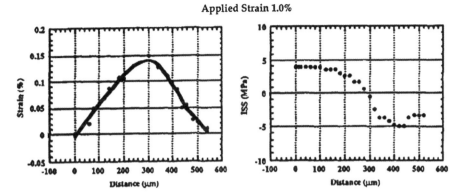

Figure 2.24 Strain profile measured along a 500-micron long graphite fiber embedded in epoxy (left), and calculated interfacial shear stress (ISS) profile (right). (Reproduced with kind permission from Amer et al., ref. 17. Copyright Elsevier 1996).

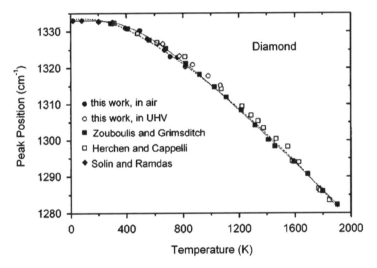

Figure 2.25 Diamond (1332 cm^{-1}) Raman peak position dependence on temperature as reported by several research groups (ref. 10b–d). (Reproduced with kind permission from Cui et al., ref. 10a. Copyright American Institute of Physics 1998.)

non-constrained crystal as a function of temperature would also yield a linear relation as shown in Figure 2.25 for single crystal diamond. Notably, the peak position–temperature relationship deviates from linearity at lower temperatures (around 800 K in this case). In addition, if the sample experiences a structural (phase) transition, such a linear relationship should be expected to change. Unpredicted electronic transitions (such as transition into a superconducting state) were also shown to cause anomalies and step-like changes in the Raman peak position–temperature relationship. Hence, it is crucial to note that such a linear relationship is applicable within well-defined temperature ranges that are different for different materials.

Occasionally, and incorrectly, the Raman peak position–temperature linear relationship has been generalized and used as a temperature calibration curve to estimate the system temperature. Most notably, this relationship between the Raman mode position and the system temperature cannot be used as a calibration curve for temperature determination, even in a calibrated range, without paying attention to the effects of constraints. While deviation from linearity, observed in Figure 2.25, was related to the relative importance of the effect of phonon–phonon and phonon–electron interactions on the Raman frequency, it is also important to note that physical constraints on the sample would also play a major role in the observed relationship between the Raman peak position and sample temperature. Hence, the effects of physical constraints, especially for thin films on solid substrates, should not be ignored in such investigations. The linear relationship shown in Figure 2.25, however, is not completely useless. If a Raman mode was calibrated against strain and temperature, within a practical range, a measure for the coefficient of thermal expansion of the material can, then, be deduced.

2.11.3 Hydrostatic Pressure Effects

Hydrostatic pressure effects on Raman spectra of materials systems both in liquid and solid states have been investigated.[19] For bulk systems,[x] Raman mode shifts under the effect of hydrostatic pressures have been used to investigate phase transitions in solids, coordination changes in molecular compounds, and to calculate compressibility of crystalline solids.

Regarding phase transitions, Figure 2.26 shows the dependence of solid H_2S Raman vibration modes on applied hydrostatic pressure in the range 0–20 GPa at room temperature (300 °K). H_2S has a structure similar to that of water. Hence it belongs to the C_{2v} point group, and has three molecular vibration modes: a symmetric stretch mode (v_1) around 2560 cm^{-1}, an asymmetric stretch mode (v_3), which is not observable at ambient pressures, and a bending mode (v_2) around 1160 cm^{-1}. As shown in Figure 2.26(a), the symmetric stretch mode (v_1) broadens and undergoes a redshift while the bending mode (v_2) is slightly affected by increasing the hydrostatic pressure. As the applied hydrostatic pressure reaches about 11 GPa, the asymmetric stretch mode (v_3) becomes very observable as a separate splitting peak that undergoes a redshift as the pressure is further increased. Simultaneously (at 11 GPa of applied pressure) another (v_2) mode appears around 1250 cm^{-1} which undergoes a blue-shift as the pressure is further increased. In addition to the aforementioned changes in the molecular Raman modes, five new lattice modes appear as the applied pressure exceeds 11 GPa, as shown in Figure 2.26(b), clearly indicating a phase

---

[x] Bulk systems will be used in this book to describe material systems that are not thermodynamically small and, hence, are not nanosystems.

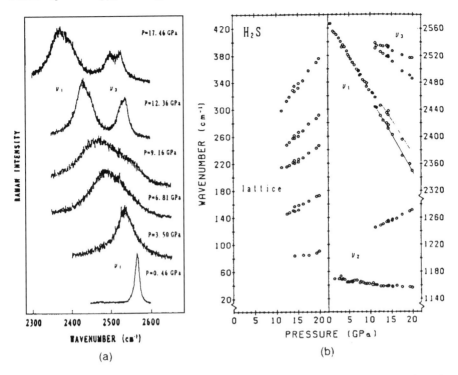

Figure 2.26 Effect of hydrostatic pressure on (a) Raman spectra and (b) peak position of H_2S. (Reproduced with kind permission from Pucci and Piccitto, ref. 19. Copyright Elsevier 1991).

transition. It is notable that all five lattice modes undergo a blue-shift with increasing pressure, a typical behavior of crystalline solids. More detailed Raman investigation were used to construct the complete pressure–temperature phase diagram of hydrogen sulfide, showing a liquid phase at room temperature and atmospheric pressure along with three different solid phases, as shown in Figure 2.27.[20,21]

Several empirical and semi-empirical correlations have been developed to correlate the symmetric stretch Raman mode frequency (v_1) to the corresponding chemical bond length. For example, Hardcastle and Wachs have developed an empirical correlation between the Raman symmetric stretching frequencies (v_1) of vanadium–oxygen bonds and their bond lengths, L, in vanadium oxide reference compounds[22] as follows:

$$v_1 = 21\,349\exp(-1.9176L) \tag{2.17}$$

where v_1 is in cm^{-1} and L is in Å.

By measuring the symmetric stretch Raman band positions as a function of hydrostatic pressure for vanadate-based compounds such as $TbVO_4$ and $DyVO_4$ and utilizing Equation (2.17), the V–O bond length could be obtained

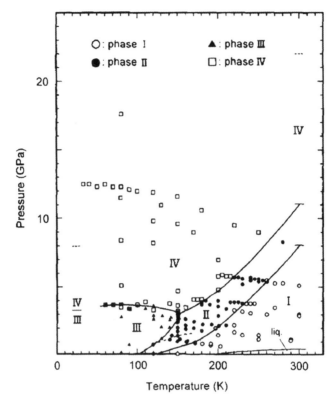

Figure 2.27 Pressure–temperature phase diagram of H_2S. Data points of each phase were determined by Raman measurements. (Reproduced with kind permission from Shimizu et al., ref. 20. Copyright American Chemical Society 1991.)

as a function of pressure. This allowed the calculation of crystal volume as a function of pressure, which is simply known as the $P-V$ equation of state (EOS). An EOS for a material system is very well known to be crucial for predicting the system's thermodynamic properties. Equations of state also provide important insights into the nature of molecular interactions within a system. Figure 2.28 shows the $P-V$ equation of state obtained from Raman spectroscopy for $TbVO_4$ and $DyVO_4$ compounds as reported by Chen et al.[23]

2.11.4 Structural Imperfections Effects

If we would like to predict the effect that structural imperfections might have on the Raman spectrum of a system, we will definitely need to consider two facts: first, that structural imperfections alter the system symmetry – if not on the local or molecular level, then definitely on the crystal or long range level – and, secondly, that there is always a strain field associated with such structural

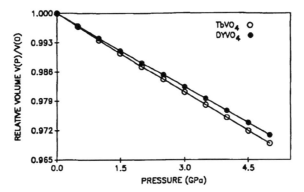

Figure 2.28 P–V equation of state obtained from Raman spectroscopy for TbVO$_4$ and DyVO$_4$ compounds. (Reproduced with kind permission from Chen et al., ref. 23. Copyright Applied Spectroscopy Society 1992.)

imperfections. Whether the imperfection is a point, linear, or planar imperfection, there will be always a symmetry alteration and a strain field associated with it, which will have a measureable effect on the Raman spectrum. For example, non-stoichiometry as point defects in nickel oxide (NiO) and barium titanate (BaTiO$_3$) are found to have a measureable effect on the characteristics of the Raman spectra of the compounds.[24,25] Non-stoichiometry in NiO$_{1+x}$ induces a new Raman active peak, the intensity of which depends linearly on the non-stoichiometry level as shown in Figure 2.29.

In addition, planar defects such as grain boundaries were found to induce a new Raman active peak around 1360 cm^{-1} in the spectrum of graphite. In their pioneering work, Tuinstra and Koenig,[26] back in the 1970s, compared the intensity ratio (R) of the defect induced mode around 1360 cm^{-1} (usually referred to as the D band) and the in-plane shear mode (E$_{2g}$) around 1575 cm^{-1} (usually referred to as the G band) ($R = I_{1360}/I_{1575}$) to grain size measurements (L_a) obtained from X-ray diffraction. They reported a linear correlation between the two quantities as shown in Figure 2.30. The work of Knight and White[27] on various carbon materials showed later that the linear dependence of R on L_a can be expressed as:

$$L_a (\text{Å}) = CR^{-1} \qquad (2.18)$$

over the extended range 25 < L_a < 3000 Å, and the constant C was reported to be 44 for excitation laser wavelength near 514.5 nm. Importantly, systematic investigation of the dependence of the constant C on laser excitation wavelength revealed that it does indeed depend on the excitation laser wavelength.[28] The dependence was found to be linear, and an empirical equation of the form:

$$C_{\lambda L} = -126 + 0.33\lambda_L \qquad (2.19)$$

was proposed for λ_L ranging between 400 and 700 nm.[29] The accuracy of the proposed relation was estimated by 10%. As we will discuss in Section 4.2

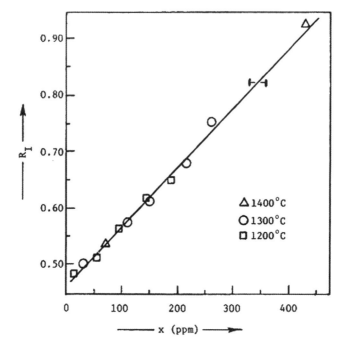

Figure 2.29 Effect of stoichiometry on the defect induced Raman band intensity (R_I) in NiO_{1+x}. Different symbols are for different samples quenched from different temperatures as shown in the figure. (Reproduced from ref. 25, courtesy of Professor B.C. Cornilsen, Michigan Technical University, USA.)

(Raman spectroscopy of fullerenes), one should be cautious in applying empirical relationships unless the experimental conditions are *exactly* the same as these used to establish the relationship.

2.11.5 Chemical Potentials Effects

A molecule placed in a fluid medium (gas or liquid) will have energy that is usually referred to as *cohesive* energy or *self*-energy (μ^i) associated with it. This energy is given by the sum of its interactions with all the surrounding molecules. The chemical potential of a molecule in a system at equilibrium (μ), which is the total free energy per molecule, is related to the molecule's *self*-energy (μ^i) according to the equation:

$$\mu = \mu^i + k_B T \ln X_i \tag{2.20}$$

where X_i is the mole fraction of the molecule species in the system. Notably, if a unary system is considered, $X_i = 1$, and the self-energy is equal to the chemical potential. In addition, since the chemical potential is a free energy,

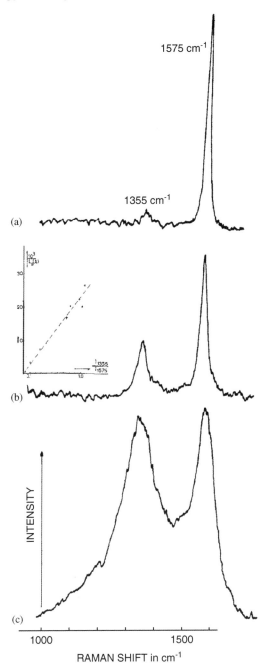

Figure 2.30 Raman spectra of HOPG (a), commercial graphite (b), and activated charcoal (c). Insert: intensity ratio of the D-band and the G-band correlated to the graphite crystal size (L_a) as measured using X-ray diffraction. (Reproduced with kind permission from Tuinstra and Koenig, ref. 26. Copyright, American Chemical Society 1970.)

Equation (2.20) indicates that (μ^i) is associated with the enthalpic contribution (energy interactions), and the term $k_B \ln X_i$ is associated with entropic contribution. This term results from the confinement of the molecule in the system and is referred to by various names, such as the ideal gas entropy, the configurational entropy, or the *confinement* entropy, the ideal solution entropy, the entropy of dilution, and the entropy of mixing.[30]

In general, molecules can *feel* the chemical potential (or the *presence*) of other chemical moieties near them within a certain range. This range is, in fact, the range of the chemical interaction that can develop between the molecule and neighboring chemical moiety. Such interactions usually have a range of one to a few atomic diameters, and hence are in the ångström to nanometer range. Longer range interactions also exist between macromolecules and surfaces. Their interaction range, however, rarely exceeds 100 nm. It is important to emphasize that in the previous statement we are not talking about strong chemical interaction resulting from primary chemical bonds, such as ionic, covalent, or metallic bonds, which would permanently change the nature of interacting moieties. Here, we are rather talking about *secondary* chemical interaction resulting from *weak* interactions between the two moieties on the level of hydrogen bonding and van der Waals and London interactions. These types of interactions affect the nature of the interacting moieties temporary, and their effect disappears once the interaction is no longer taking place. They have been known as *solvent* interactions. Solvent effects have been known to affect the Raman spectral characteristics of different systems such as proteins,[31] solid–liquid solutions,[32,33] polymers,[34] binary liquid solutions,[35] and aqueous foams.[36] Raman shifts in the range $5 \sim 20\, \text{cm}^{-1}$ due to solvent effects are typical. Figure 2.31 demonstrates the

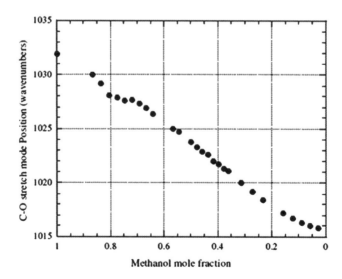

Figure 2.31 Changes in C–O stretch mode of methanol as a function of methanol mole fraction in a methanol–water mixture. (Reproduced with kind permission from Amer *et al.*, ref. 35. Copyright Elsevier 2006.)

solvent effect on Raman spectra. The figure shows the change in the C–O stretch mode position resulting from interaction of the methanol with water molecules in methanol–water binary mixtures with different compositions. Notably, a plateau can be observed in the Raman peak position trend, indicating a phase transition in the system.[35] Phase transitions in methanol–water binary systems have been observed through molecular simulation studies[37,38] and neutron diffraction spectroscopy.[39] The demonstrated ability of the Raman technique to probe structures on the nano and meso scales definitely adds to its power as a diagnostic technique excellent for nanotechnology investigations.

2.12 Resonant Raman Effect

In Section 2.9 we discussed the Raman phenomenon from an energy transfer viewpoint. One of the important aspects to be satisfied is that the energy level ($E = h\nu$) of the excitation laser used should be far from any excited energy level of the system being excited for Raman scattering. If the excitation laser used happened to have an energy that is the same or very close to the energy of an excited level in the material system, a resonance effect will take place. This phenomenon is known as the *resonance Raman* effect. Details of and selection rules for the resonance Raman effect are beyond the scope of this book. However, resonance Raman phenomenon can sometimes be used to observe and investigate very weak Raman bands due to the very strong intensity enhancement it produces.

2.13 Calculations of Raman Band Positions

Several quantum mechanics based methods have been developed to enable the calculation of the Raman spectrum of a chemical moiety. These methods are based upon empirical, semi-empirical, and density functional theory considerations. Several commercial and free codes have been written and are available on commercial and free download bases. While the theories and methods of molecular simulations used for such calculations are beyond the scope of this book, two points are important to emphasize and mention in this section because they can be correlated to the previous and the next sections of this chapter. First, it is important to know that calculated Raman peak positions are usually made for molecules in vacuum. Care must be taken to include the effect of the molecule environment (air is the most common to miss) if comparison with experimental results is intended. In other words, the chemical potential of molecules in the surrounding environment should not be ignored when comparison between calculated and measured Raman bands is essential. Secondly, it is also important to note that while accurate calculations of the Raman peak positions can be made, the intensity of the Raman band cannot be calculated. This is a consequence of the Raman activity conditions we discussed earlier in Sections 2.4 and 2.8. The Raman activity condition, in fact, classifies a normal vibration mode as Raman active with a probability = 1, or non-active with a probability = 0. The exact probability cannot be determined from the theory. Hence, the relative intensity cannot be calculated. Raman intensities, or to be exact, relative intensities will be discussed in the next section.

2.14 Polarized Raman and Band Intensity

The intensity of a Raman band is usually measured as the integration of the area under the Raman peak. This should be appropriately referred to as integrated intensity. This is different from the height of the Raman peak. Sometimes, however, the term intensity is used in the literature to refer to integrated intensity. As we discussed in the previous section, the intensity of a Raman band cannot be predicted based on theoretical aspects. Instead, experimental measurements are needed to determine such intensities. The intensity of a Raman active vibration mode, however, does depend on the symmetry species of the mode in context of the polarization direction of the exciting light relative to the molecule or crystal orientation. The relative Raman intensity of a mode (I) can be expressed as:

$$I \propto \left| e_i \mathcal{R} e_s \right|^2 \quad (2.21)$$

where e_i and e_s denote the electric vector of the incident and the scattered laser (polarization directions), respectively, while $\mathcal{R}$ represents the Raman scattering tensor of the particular Raman mode under consideration. The Raman tensor is a second rank tensor, which is the polarizability tensor we discussed earlier, and is specific for particular symmetry species in different point groups. Appendix 3 lists the Raman scattering tensors for different symmetry species in the 32 crystal point groups. Recent developments in single molecule spectroscopy and surface-enhanced Raman scattering enabled the experimental investigation of Raman scattering tensors for individual molecules[40] and biological assemblies.[41]

Note that each of the vectors and the tensor employed in Equation (2.21) has to be expressed in the same set of axes. Scattering tensors are customarily expressed in the crystal principal axes directions a, b, and c. Experimental measurements are more conveniently carried out in the space Cartesian axes system x, y, and z. Hence, it is necessary, in certain cases, to convert the crystal principal axes system into a space Cartesian system using the very well-known tensor rotation technique. Let us consider Equation (2.21) again and see how each of its right-hand side terms can be expressed in the space Cartesian axes system.

Let us consider a space Cartesian axes system as shown in Figure 2.32. For simplicity, we will consider an orthorhombic crystal oriented, as shown in the figure, with its c-axis along the z-axis, and a-axis having an angle θ with the x-axis. Using a polarizer, it is possible to control the polarization direction of incident light. If we send the incident laser along the positive direction of the z-axis (from up to down) polarized in the x direction, the electric field vector of the incident light (e_i) can be expressed as:

$$e_i = [1\ 0\ 0] \quad (2.22)$$

Raman Spectroscopy; the Diagnostic Tool

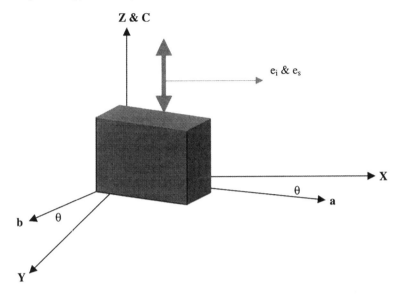

Figure 2.32 Schematic arrangement showing an orthorhombic crystal and laboratory axes considered in the Raman intensity analysis. (Reproduced with kind permission from Amer *et al.*, ref. 43. Copyright American Institute of Physics 2001.)

If we collect scattered light travelling along the negative direction of the z-axis, and only allow the x-polarized component of this scattered light to be recorded,[xi] the scattered light electric vector (e_s) can be expressed as:

$$e_s = \begin{bmatrix} 1 \\ 0 \\ 0 \end{bmatrix} \quad (2.23)$$

For the sake of demonstration, we consider an orthorhombic crystal with a D_{2h} point group symmetry and see how the intensity of an A_g Raman active mode would depend on the crystal orientation as the crystal is rotated about the c-axis as shown in Figure 2.32. For the A_g mode in a D_{2h} point group, the Raman tensor expressed in the crystal principal axes system (a, b, and c) takes the form:

$$\mathscr{R}_{Ag} = \begin{bmatrix} \alpha_a & 0 & 0 \\ 0 & \alpha_b & 0 \\ 0 & 0 & \alpha_c \end{bmatrix} \quad (2.24)$$

[xi] Such experimental setup in which the incident light travels along the positive z-axis, polarized in the x direction, while the collected scattered light travels along the negative z-axis polarized in the x direction is expressed as $zxx\bar{z}$ in what is referred to as Porto notations.

According to tensor rotation rules,[42] the Raman tensor expressed in the Cartesian axes system can be given as:

$$\mathscr{R}_{Ag}^{xyz} = \mathbf{\Phi}\mathscr{R}_{Ag}\bar{\mathbf{\Phi}} \tag{2.25}$$

where $\mathbf{\Phi}$ is the rotation matrix, and $\bar{\mathbf{\Phi}}$ is its transpose. For the geometry shown in Figure 2.32, the rotation matrix can be expressed as:

$$\mathbf{\Phi} = \begin{bmatrix} \cos\theta & \sin\theta & 0 \\ -\sin\theta & \cos\theta & 0 \\ 0 & 0 & 1 \end{bmatrix} \tag{2.26}$$

Substituting from Equations (2.22)–(2.26), the intensity Equation (2.21) can be rewritten as:

$$I \propto \left| \alpha_a \cos^2\theta + \alpha_b \sin^2\theta \right|^2 \tag{2.27}$$

Figure 2.33 shows experimental results[43] obtained from a single-crystal high-temperature superconductor (YBCO) as fitted to Equation (2.26). The dependence of Raman intensity on the experimental setup and the crystal or molecule orientation adds another utility to the technique as a diagnostic tool. Polarized Raman

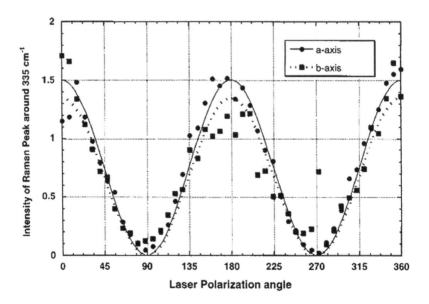

Figure 2.33 A_g Raman mode peak intensity for a single YBCO crystal as a function of crystal rotation angle (θ) about its c-axis. Solid line, θ measured from the a-axis direction, dashed line θ measured from the b-axis direction. Experimental data fitted to Equation (2.26). (Reproduced with kind permission from Amer et al., ref. 43. Copyright American Institute of Physics 2001.)

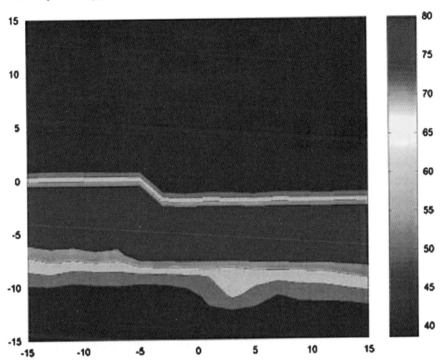

Figure 2.34 A 30×30 micron map of grain orientation in a 2.5-nm thick YBCO film as measured by polarized Raman spectroscopy. (Reproduced with kind permission from Amer *et al.*, ref. 43. Copyright American Institute of Physics 2001.)

has been utilized to investigate grain structure and to measure grain boundary mismatch angles in thin film coatings. Figure 2.34 shows a 30×30-micron map of the grain orientation in a 2.5 nm thick YBCO film as measured by polarized Raman spectroscopy. The map clearly shows a band with almost 40° misorientation resembling a twin boundary in the substrate that was reflected into the film structure. Such investigations have enabled important correlations between the film structure and its performance in high-temperature superconducting films.[44] Orientation measurement ability of polarized Raman spectroscopy has also contributed greatly in investigating and understanding liquid crystals.[45] We will discuss polarized Raman measurements of carbon nanotubes in Chapter 4, when Raman scattering in fullerenes is considered.

2.15 Dispersion Effect

Based on our discussion of the Raman phenomenon so far, one should expect that as the excitation laser wavelength changes, the Raman band wavenumber should remain the same. This is true and has been experimentally verified.

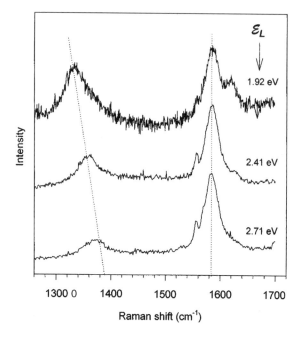

Figure 2.35 Raman spectra of poly(*para*-phenylene) (PPP), graphitized at 2400 °C, produced with three different excitation laser energies: 1.92, 2.41, and 2.71 eV. (Reproduced with kind permission from Mathews *et al.*, ref. 46. Copyright American Physical Society 1999.)

The position of certain Raman modes, however, shows dependence on the excitation laser wavelength. This phenomenon is known as the *dispersion* effect. The dispersion effect is clearly demonstrated in the disorder induced mode in graphite (the D band) around 1335 cm^{-1}. Figure 2.35 shows the Raman spectra of poly(*para*-phenylene) graphitized at 2400 °C. The three spectra were produced using laser excitation wavelengths of 647 (1.92 eV), 514.5 (2.41 eV), and 457.9 nm (2.71 eV).[46] Such dispersion behavior is a characteristic feature of the resonant behavior of the Raman D band in carbon materials.

Systematic investigation of the dispersion phenomenon in the D band of different carbon materials showed that the position of this disorder induced band increases (shows a blue-shift) linearly with increasing energy (E_L) (or reducing the wavelength) (λ_L) of the excitation laser.[47] The same behavior was observed for the second order (or the overtone) of the D band, usually referred to as the G′ band around 2700 cm^{-1}. The slope of the linear dependence of the D band position on the excitation laser energy was found constant regardless of the type of the carbon materials and almost equal to 50 cm^{-1} eV^{-1}. The slope for the overtone G′ band was found to be almost twice that for the D band at around a value of 100 cm^{-1} eV^{-1}. This should not be surprising since the G′ band is the second harmonic of the D band as we mentioned before. Figure 2.36 shows the dependence of the D band and its overtone G′ band (second-order

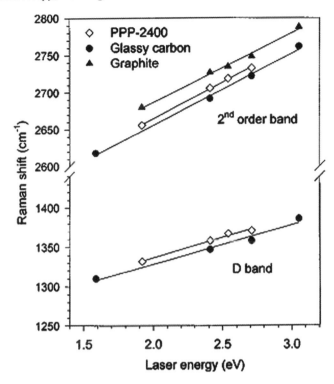

Figure 2.36 Dependence of the D band and its overtone G' band positions on the excitation laser energy (E_L) for different types of carbon materials. (Reproduced with kind permission from Marucci et al., ref. 48. Copyright Materials Research Society 1999.)

band) positions on the excitation laser energy (E_L) for different types of carbon materials as reported by Marucci et al.[48]

2.16 Instrumentation

While full description of Raman instrumentations is beyond the scope of this section, it is important for the reader to have basic idea on the essential elements of the Raman instrument. The most important thing to know is that Raman instrumentation is a very dynamic field. It enjoys major leaps with every advance made in electronics, optical devices, control software, or data capturing and analysis fields. Several book chapters devoted to Raman instrumentation have been published. The reader is encouraged to refer to these resources for details regarding such an important aspect of the Raman technique.[49–51] Figure 2.37 shows a schematic of a Raman micro-probe. Generally speaking, the instrument consists of six main parts: an excitation light source, an optical guiding system, an optical microscope, a spectrometer, a detector, and a data acquisition unit.

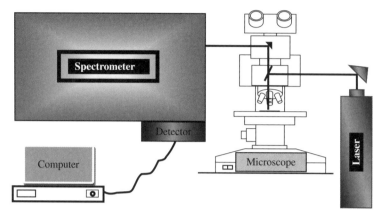

Figure 2.37 Schematic illustration of a typical Raman microscope.

The excitation light source is always a laser. It can be a monochromatic laser or a tunable dye laser. The most commonly used lasers line in Raman spectroscopy is the argon ion line at 514.5 nm. Other laser lines are very widely used as well. Solid-state lasers with an excitation wavelength in the 780 nm range are preferred for biological samples to avoid, or reduce, the detrimental effect of florescence. Recently, UV lasers in the 200–300 nm wavelength range have become preferred for carbon nanotube investigations.

Regular optical microscopes are used to focus the laser light on the sample, to collect the backscattered light and send it to the spectrometer, and to define the Cartesian axes for the experiment for polarization directions definition purposes. Moving microscope stages (with a micron and submicron step size) are widely used to conduct Raman mapping experiments. With a typical optical microscope, the spatial resolution of the technique will be limited to the size of the laser spot interaction volume with the sample. Confocal optical arrangements have been utilized to enhance the spatial resolution. A spatial resolution of 1 micron was successfully reached. This allowed good resolution depth profiling of interfacial regions in layered materials systems.[52] More recently, high-resolution near-field microscopes were employed to further enhance the spatial resolution and capabilities of the Raman technique.[53]

The purpose of the spectrometer is twofold: First to separate the Rayleigh scattered light from the Raman signal (by Rayleigh light rejection), which can be done by a holographic notch filter, an edge filter, or a monochromator. Notch filters will enable the collection of both Stokes and anti-Stokes lines, while edge filters will allow collection of the Stokes Raman lines only. In addition to being, currently, less expensive than notch filters, edge filters have the advantage of allowing observation of Raman lines closer to the Rayleigh line.

The second purpose of the spectrometer is to analyze the collected optical signal, which is carried out by dispersing the incoming light according to its

Raman Spectroscopy; the Diagnostic Tool

wavelength and then reimaging the output spectrum at the exit slit. The spectrometer consists of a group of gratings and mirrors that can be arranged in different ways to give one of the following spectrometer subclasses:

1. Monochromator, which presents one wavelength at a time from the exit slit, and can be tuned to select the required wavelength.
2. Scanning monochromator: in this case the monochromator is motorized to scan a range of wavelengths sequentially.
3. Polychromator: in this arrangement, selected wavelengths are presented at various exit slits.
4. Spectrograph: in this arrangement, a range of wavelengths is imaged at an exit plane, and there is no exit slit.
5. Imaging spectrograph: in this case, special corrective optics are used to maintain better image quality along the two axes of the imaging plane.

Monochromators usually have high resolution, high stray light rejection, but low throughput. Spectrographs, on the other hand, have lower resolution, lower stray light rejection, but higher throughput. A spectrometer can contain one or more of the aforementioned subclasses arranged in a way to give single, double, triple, or a single plus double monochromator systems. Spectrographs can also be combined with monochromators in the same system. Notably, the

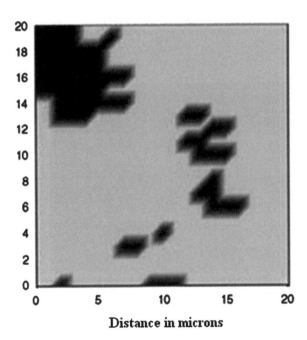

Figure 2.38 Raman map of the distribution of silicon carbide particulates in a silicon/silicon carbide composite. (Reproduced with kind permission from Amer *et al.*, ref. 55. Copyright Elsevier 2006.)

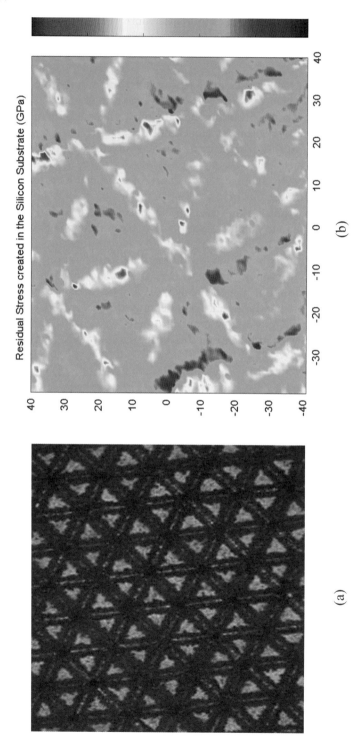

Figure 2.39 (a) Optical micrograph of a laser machined silicon wafer and (b) a 80×80 micron Raman map for stress distribution measured based on the shift in silicon Raman peak position. (Reproduced with kind permission from Amer *et al.*, ref. 56. Copyright Elsevier 2002.)

more sophisticated the spectrometer system is, the higher its resolution, the higher its stray light rejection capability, and the lower its throughput will be. Higher stray light rejection allows observation of Raman bands closer to the laser line.

Detectors can be either a single channel or multi-channel. Single-channel detectors (photomultiplier tubes, PMTs) read one wavelength at a time, and have long been used in Raman studies due to their high sensitivity, low background count, wide covered range, reliability, and relatively low cost. Their main disadvantage is the low sensitivity, requiring a long time to record a spectrum. Multi-channel detectors can be an intensified diode array or a charge coupled device (CCD). They are both an array of light sensitive units and they record a wide range of wavelengths at the same time. CCDs are the state-of-the-art detectors. They have the ability to record almost every striking photon of collected scattered light. It is hard to catch up with the rapid developments in the field of CCD cameras. Time-gated CCDs and time-gated confocal microscopy have recently shown strong potential for another lead in the field.[54]

The data acquisition unit is simply a PC interfaced with the detector to store, display, and manipulate the data. Advanced data manipulation software has been developed to enable better data presentation, especially in the Raman imaging branch of experimental Raman spectroscopy.

The optical guiding system consists of a group of mirrors to direct the laser beam from the source to the sample and back to the spectrometer, holographic filters to remove the plasma lines from the laser beam, and a set of polarizing units to control the polarization of the incident and the scattered beams. It is important to add that recently instruments have been developed with the ability to obtain microscopic images known as "Raman images". Two approaches have been described to obtain a Raman image. The first involves the illumination of the whole field-of-view of the microscope uniformly with the laser light. The microscope then transfers the sample image to a monochromator that selects predetermined wavelengths to be imaged on a TV camera at the exit focal plane. If the predetermined wavelength represents the Raman band of an element or a compound, the image will show the distribution of that element or compound within the selected area of the sample. The second approach to Raman imaging involves using a computer-controlled stepping motor driven sample stage. An area of the sample is mapped by recording spectra from a series of spatial points. The produced map can be either one- or two-dimensional profiles that show the intensity of the Raman band of a certain element or compound. Figure 2.38 shows a Raman map for the distribution of silicon carbide in a silicon/silicon carbide composite. The map was produced based upon the finger-printing SiC Raman peak intensity.[55]

Figure 2.39 shows an optical micrograph of a laser machined silicon wafer and 80×80 micron Raman map of the induced stresses as measured from the shift in the silicon peak position.[56]

Recommended General Reading

L. Tarasov, *This Amazing Symmetrical World*, translated from Russian by: Alexander Repyev, Mir Publishers, Moscow, 1986.

M. du Sautoy, *Symmetry; a Journey into the Patterns of Nature*, Harper Perrenial, London, 2008.

I. Stewart, *Why Beauty is Truth: The History of Symmetry*, Basic Books, New York, 2008.

R. Mirman, *Point Groups, Space Groups, Crystals, Molecules*, World Scientific, Singapore, 1999.

A. Vincent, *Molecular Symmetry and Group Theory*, John Wiley and Sons, New York, 1977.

R. Loudon, *Adv. Phys.*, 1964, **13**, 52.

D. A. Long, *The Raman Effect: A Unified Treatment of the Theory of Raman Scattering by Molecules*, John Wiley & Sons, London, 2002.

M. Cardona (ed.), *Light Scattering in Solids I*, Springer, Berlin, 1983.

H. A. Szymanski (ed.), *Raman Spectroscopy; Theory and Practice*, Plenum Press, New York, 1967.

References

1. M. Zeldin, *J. Chem. Educ.*, 1956, **43**, 17.
2. M. Hammermesh, *Group Theory and Its Applications to Physical Problems*, Addison Wesley, Reading, MA, 1972.
3. E. L. Muetterties, R. E. Merrifield, H. C. Miller, W. H. Knoth and J. R. Downing, *J. Am. Chem. Soc.*, 1962, **84**, 2506.
4. L. A. Paquette, R. L. Taransky, D. W. Balough and G. Kentgen, *J. Am. Chem. Soc.*, 1983, **105**, 5446.
5. H. W. Kroto, A. W. Allaf and S. P. Balm, *Chem. Rev.*, 1991, **91**, 1213.
6. J. Hollas, *Modern Spectroscopy*, John Wiley & Sons, Ltd., Chichester, 1987.
7. A. Weber, *Raman Spectroscopy of Gases and Liquids*, Springer Verlag, New York, 1979.
8. G. Herzberg, *Molecular Spectra and Molecular Structure. II Infrared and Raman Spectra of Polyatomic Molecules*, D. Van Nostrand, New York, 1945.
9. H. Wilhelm, M. Lelaurain, E. McRae and B. Humbert, *J. Appl. Phys.*, 1998, **84**, 6552.
10. (a) J. B. Cui, K. Amtmann, J. Ristein, L. Ley, *J. Appl. Phys.*, 1998, **83**, 7929; (b) E. S. Zouboulis and M. Grimsditch, *Phys. Rev. B.*, 1991, **43**, 490; (c) H. Herchen and M. A. Cappelli, *Phys. Rev. B.*, 1991, **43**, 740; (d) S. A. Solin and A. K. Ramdas, *Phys. Rev. B.*, 1970, **1**, 1687.
11. H. Mueller, *Phys. Rev.*, 1940, **58**, 805.
12. E. Anastassakis, F. H. Pollak and G. W. Rubloff, *Phys. Rev. B*, 1974, **9**, 551.

13. I. De Wolf, *Semiconductor Sci. Technol.*, 1996, **11**, 139.
14. M. S. Amer, in *Semiconductor Machining at the Micro-Nano Scale*, ed. J. Yan and J. Patten, Transworld Research Network, Kerala, India, 2007, pp. 333–360.
15. R. J. Young and S. Eichhorn, in *Raman Applications in Synthetic and Natural Polymer Fibers and Their Composites*, ed. M. Amer, John Wiley & Sons, Inc., Hoboken, 2009, p. 100.
16. C. Galiotis, *Compos. Sci. Technol.*, 1991, **42**, 125.
17. M. S. Amer, M. J. Koczak and L. S. Schadler, *Compos. Part A: Appl. Sci. Manufacturing*, 1996, **27**, 861.
18. M. S. Amer, *Int. J. Solids Structures*, 2005, **42**, 751.
19. R. Pucci and G. Piccitto, *The II Archimedes Workshop on Molecular Solids under Pressure*, Amsterdam, 1991.
20. H. Shimizu, Y. Nakamichi and S. Sasaki, *J. Chem. Phys.*, 1991, **95**, 2036.
21. H. Nakayama, H. Yamaguchi, S. Sasaki and H. Shimizu, *Physica B (Amsterdam)*, 1996, **219**, 523.
22. F. D. Hardcastle and I. E. Wachs, *J. Phys. Chem.*, 1991, **95**, 5031.
23. G. Chen, R. G. Haire and J. R. Peterson, *Appl. Spectrosc.*, 1992, **46**, 1495.
24. M. W. Urban and B. C. Cornilsen, in *Measurement of Non-stoichiometry in BaTiO3 using Raman Spectroscopy*, ed. R. L. Snyder, R. A. Condrate and P. F. Johnson, Plenum Press, New York, 1985, p. 89.
25. B. C. Cornilsen, E. F. Funkenbusch, C. P. Clarke, P. Singh and V. Lorprayoon, in *Advances in Materials Characterization*, ed. D. R. Rossington, R. A. Condrate and R. L. Snyder, Plenum Press, New York, 1983, **vol. 15**, p. 239.
26. F. Tuinstra and J. L. Koenig, *J. Chem. Phys.*, 1970, **53**, 1126.
27. D. S. Knight and W. B. White, *J. Mater. Res.*, 1989, **4**, 385.
28. L. Nikiel and P. W. Jagodzinski, *Carbon*, 1993, **31**, 1313.
29. M. S. Dresselhaus, M. A. Pimenta, P. C. Eklund and G. D. Dresselhaus, in *Raman Scattering in Fullerenes and Related Carbon Based Materials*, ed. W. H. Weber and R. Merlin, Springer, Berlin, 2000, p. 323.
30. J. Israelachivili, *Intermolecular & Surface Forces*, Academic Press, New York, 1992.
31. F. Franks, D. Eagland and R. Lumry, *Crit. Rev. Biochem. Mol. Biol.*, 1975, **3**, 165.
32. M. Mathlouthi, C. Luu, A. Meffroy-Biget and D. Luu, *Carbohydr. Res.*, 1980, **81**, 3.
33. P. Palmer, Y. Chen and M. Topp, *Chem. Phys. Lett.*, 2000, **321**, 62.
34. S. Dixit, W. Poon and J. Crain, *J. Phys.: Condens. Matter*, 2000, **12**, 323.
35. M. S. Amer and M. M. El-ashry, *Chem. Phys. Lett.*, 2006, **430**, 323.
36. J. F. Maguire, M. S. Amer and J. Busbee, *Appl. Phys. Lett.*, 2003, **82**, 2592.
37. S. Dixit, J. Crain, W. Poon, J. Finney and A. Soper, *Nature*, 2002, **416**, 829.
38. A. Laaksonen, P. G. Kusalik and I. M. Svischev, *J. Phys. Chem. A*, 1997, **101**, 5910.
39. L. Dougan, S. P. Bates, R. Hargreaves, J. P. Fox, J. Crain, J. L. Finney, V. Reat and A. K. Soper, *J. Chem. Phys.*, 2004, **121**, 6456.

40. T. O. Shegai and G. Haran, *J. Phys. Chem. B*, 2006, **110**, 2459.
41. M. Tsuboi and G. Thomas, *Appl. Spectrosc. Rev.*, 1997, **32**, 263.
42. J. F. Nye, *Physical Properties of Crystals*, Oxford Science Publications, Oxford, 1985.
43. M. S. Amer, J. Maguire, L. Cai, R. Biggers, J. Busbee and S. R. LeClair, *J. Appl. Phys.*, 2001, **89**, 8030.
44. M. Amer, J. F. Maguire, R. Biggers and S. R. LeClair, *Philos. Mag. Lett.*, 2002, **82**, 241.
45. N. Hayashi, in *Raman Applications in Liquid Crystals*, ed. M. Amer, John Wiley & Sons, Inc., Hoboken, 2009, p. 135.
46. M. J. Mathews, M. A. Pimenta, G. Dresselhous, M. S. Dresselhaus and M. Endo, *Phys. Rev. B.*, 1999, **59**, R6585.
47. R. P. Vidano, D. B. Fishbach, L. J. Willis and T. M. Loehr, *Solid State Commun.*, 1981, **39**, 341.
48. A. Marucci, S. D. M. Brown, M. A. Pimenta, M. J. Mathews, M. S. Dresselhaus, K. Nishimura and M. Endo, *J. Mater. Res.*, 1999, **14**, 1124.
49. M. Delhaye, J. Barbillat, J. Aubard, M. Bridoux and E. Da Silva, in *Raman Instrumentation*, ed. G. Turrell and J. Corset, Academic Press, London, 1996, p. 51.
50. J. Barbillat, in *Raman Imaging*, ed. G. Turrell and J. Corset, Academic Press, London, 1996, p. 175.
51. F. Adar, in *Evolution and Revolution of Raman Instrumentation*, ed. I. R. Lewis and H. G. Edwards, Marcel Dekker, New York, 2001, p. 11.
52. N. Everall, *Appl. Spectrosc.*, 2000, **54**, 1515.
53. A. Hartschuh, E. Sánchez, X. Xie and L. Novotny, *Phys. Rev. Lett.*, 2003, **90**, 95503.
54. V. Yakovlev, *Spectroscopy*, 2007, **22**, 34.
55. M. S. Amer, L. Durgam and M. M. El-Ashry, *Mater. Chem. Phys.*, 2006, **98**, 410.
56. M. S. Amer, L. Dosser, S. LeClair and J. F. Maguire, *Appl. Surf. Sci.*, 2002, **187**, 291.

CHAPTER 3
Fullerenes, the Building Blocks

3.1 Overview

In this chapter, fullerenes as nano-material building blocks will be discussed. The beginning of fullerenes and their current state will be covered. Their structures and production techniques will be examined. From a structural dimensionality viewpoint, fullerenes can be classified into zero-, one-, or two-dimensional structures. Zero-dimensional fullerenes include spheroidal cage-like nano-carbon species that were traditionally referred to as fullerenes; they include C_{60}, C_{70}, C_{80}, *etc*. One-dimensional fullerenes include tubular-like fullerenes, as well as single-, double-, and multi-walled carbon nanotubes. By two-dimensional fullerenes we refer to the recently isolated graphene sheets, including single- and multi-layered graphene sheets. Chapter 4 deals with the properties of these building blocks. It is important to realize that comparing the properties of graphite to those of graphene is an excellent way to demonstrate the size/performance dependence of a material system, and to demonstrate the change in behavior upon entering the nano-domain.

3.2 Introduction

As discussed in Chapter 1, for a concept to develop into a technology, the availability of suitable building blocks is a must. Material building blocks enable experimental verification of the theory behind the concept and provide the essential ingredient enabling devices and products to be built, bringing the new technology into reality. In our case, nanotechnology is based upon nano-building blocks that are essentially small thermodynamic systems (Chapter 1). Hence, for nanotechnology, any thermodynamic small system is, indeed, a building block. Consequently, any cluster of matter, regardless of its physical size, which would be on the order of any of the scale lengths discussed in Chapter 1, should satisfy the definition and can be considered as a building block. To this end, many systems can be defined as building blocks. Some of these building blocks are, themselves, made of smaller building blocks such as biological cells, or even biological species. If we consider the subject from such a viewpoint, the perimeter of nanotechnology building blocks will be too large to consider in detail. To keep the subject focused and of practical value, we will concentrate on some of the recently discovered and investigated building blocks which are essentially man-made. The word "some" in the previous statement was used intentionally since, in fact, recently developed and produced man-made nano-building blocks themselves are too many to discuss in detail. The list is already long, and new building blocks are produced, investigated, and reported often. The reported list in the literature includes nanowire, nano-rods, nano-cones, nano-horns, nanospheres, nanoshells, *etc.*, each of which has been produced from several different materials.

Based on our previous discussion of matter in the nano-domain, it can be inferred that nature is the best designer for nanosystems. Interestingly, nature used one particular element most frequently in its designs, especially for living systems: carbon. Hence, as we discuss the building blocks of nature's preferred technology, carbon-based building blocks should be a good choice for discussion. At the end of Chapter 1, we examined nanophenomena that were observed in systems based mainly on metallic clusters in the nano-domain. In this chapter we focus on carbon-based building blocks, namely, fullerenes. Fullerenes are a recently discovered form of crystalline carbon. They come in different geometrical shapes. Spherical, or generally speaking, balloon-like, shapes are usually referred to as fullerenes or buckyballs. Cylindrical shapes are more popularly known as single-walled carbon nanotubes (SWCNT), and, more recently, the single-sheet form referred to as graphene has also been produced and utilized as a building block for nanotechnology. Figure 3.1 shows the different types of carbon nanospecies[1,2] discussed in this chapter, namely, fullerenes, single- and multi-walled carbon nanotubes, carbon nano-onions, and graphene sheets. In addition, examples of recently created nano-hybrid structures[3-12] will be demonstrated and discussed from a materials science viewpoint. Figure 3.2 shows examples of such nano-hybrid structures (sometimes referred to as mesostructures). In these structures different nano-building blocks, are incorporated into new structures. For example, Figure 3.2(a) shows a mesostructure known as a *peapod*. In this structure fullerene molecules are inserted inside single-walled carbon nanotubes (C_{60}@SWCNT). Figure 3.2(b) shows a different type of structure, usually referred to as *nano-buds*. In nano-buds, fullerene molecules are attached to the surface of single-walled carbon nanotubes. Figure 3.2(c) shows an example of more complex mesostructures in which several fullerene molecules are attached at different sites on the surface of single-walled carbon nanotubes.

3.3 Fullerenes, the Beginnings and Current State

Historically, the possibility of creating *graphite balloons* similar to geodesic cages was first discussed in 1966 by David Jones who was writing under the pseudonym "Daedalus" in the journal the *New Scientist*.[13,14] The most famous

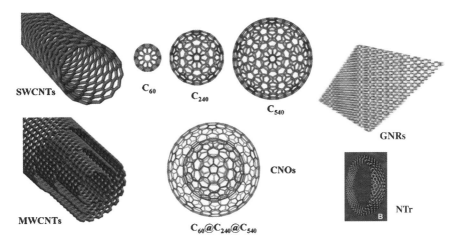

Figure 3.1 Different types of carbon nanospecies. Single-walled carbon nanotubes (SWCNTs), multi-walled carbon nanotubes (MWCNTs), [60], [240], and [540] fullerene, carbon nano-onions (CNOs), graphene nano-ribbons (GNRs), and nanotorus (NTr). (Adapted with kind permission from Delgado *et al.*, ref. 1. Copyright The Royal Society 2008; and from Liu *et al.* ref. 2, Copyright Nature Publishing Group 1997).

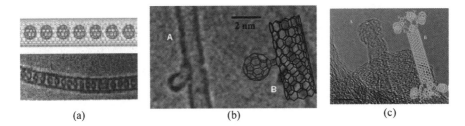

Figure 3.2 Transmission electron microscopy photographs and computer generated images of (a) peapods (C_{60}@SWCNT), (b) nano-bud, and (c) a mesostructure incorporating nanotubes and several [60] fullerenes. [(a) Reproduced with kind permission from Bandow *et al.*, ref. 11. Copyright Elsevier 2001. (b) and (c) Reproduced with kind permission from Nasibulin *et al.*, ref. 3. Copyright Nature Publishing Group 2007.]

form of fullerene molecules, however, is the 60 fullerene or C_{60}, commonly known as buckyball, made of 60 carbon atoms in a spherical shape that resembles a soccer ball. The earliest record of a fullerene molecule, however, was in an article (in Japanese) by Eiji Osawa in 1970.[15] In this article, Osawa speculated that such a molecule would be stable. In 1971, Osawa and Yoshida described the speculated molecule in more detail in a book – also in Japanese –

Figure 3.3 Buckminster Fuller and his famous geodesic dome design. (© Courtesy of Buckminster Fuller Institute).

on aromatic chemistry.[16] Two years later, in 1973, Bochavar and Gal'pern[17] used Hückel calculations to determine the energy levels and molecular orbitals in the C_{60} molecule. Later, in 1981, Davidson[18] applied general group theory techniques to a range of highly symmetrical molecules, one of which was the C_{60}. Hence, it is clear that by 1981 the idea of stable fullerene molecules did, in fact, exist and early studies had been carried out to explore the energy levels and molecular orbitals as well as symmetry properties of such molecule. The molecule itself was not yet experimentally observed.

In September 1985, such observation took place. While trying to simulate stellar nucleation conditions[19] of cyanopolyyenes[i] using the, then recently, developed laser vaporization cluster technique by Richard Smalley and his co-workers[20] at Rice University, USA, Curl, Kroto, Smalley, and their co-workers vaporized graphite and a serendipitous discovery was made. The C_{60} molecule was observed and found to be remarkably stable.[21–25] The molecule was named Buckminsterfullerene (later fullerene for short) after the famous architect Buckminster Fuller (1895–1983) who first created the geodesic cage or dome design that the fullerene molecule resembles.[26] Figure 3.3 shows a photograph of Buckminster Fuller and his famous geodesic dome design.

This interesting and long awaited discovery triggered a scientific race to investigate each and every aspect of the new molecule.[27–37] At first, however, progress was slow due to the fact that the amount of C_{60} produced by Smalley's method was minute. The real race of development started in 1990 with the findings of Krätschmer of the Max Planck Institute at Heidelberg, Huffman of the University of Arizona, and their co-workers, who could produce C_{60} molecules in macroscopic amounts using a simpler, more accessible technique

---

[i]Some carbon chain molecules known as cyanopolyyenes were earlier discovered in interstellar regions streaming out of red giant carbon stars. Cyanopolyyenes has the form H–C≡C–C≡C–C≡N with 5, 7, 11, and up to 33 carbon atoms.

than that used by Smalley.[30] The new technique vaporized graphite using a simple carbon arc in helium atmosphere. The soot deposited on the walls of the vessel, once dispersed in benzene, produced a reddish solution. Once dried, the solution produced beautiful crystals of "fullerite", which turned to be made of 90% C_{60} and 10% C_{70}. By using the method of Krätschmer and Huffman, C_{60} and other allotropes of fullerenes could be produced in reasonable amounts in a way accessible to many laboratories. This accelerated the fullerene investigation race and started what Curl described as "the Dawn of Fullerenes".[38] By 1991 fullerenes were the subject matter of 90% of the most cited papers, and the subject is still of current scientific interest.

In 1991, Sumio Iijima of the NEC laboratories in Japan reported:

"the preparation of a new type of finite carbon structure consisting of needle-like tubes. Produced using an arc-discharge evaporation method similar to that used for fullerene synthesis."[39]

Electron microscopy revealed that the needles consist of co-axial tubes of several graphitic sheets (between 2 and 50). These new molecular cylinders of graphitic sheets were called carbon nanotubes. In fact, Iijima's report on carbon nanotubes was not the first in the literature. As early as 1952, Radushkevich and Lukyanovich[ii] reported, in the *Journal of Physical Chemistry of Russia*, the first TEM evidence for the tubular nature of some nano-sized carbon filaments. In 1974, Oberlin, Endo, and Koyama working on benzene-derived carbon fibers reported that:

"These fibres have various external shapes and contain a hollow tube with a diameter ranging from 20 to more than 500 Å along the fibre axis."[40,41]

Figure 3.4 shows the transmission electron micrographs first produced by Endo and Iijima for what we currently refer to as *carbon nanotubes*. The history of carbon nanotubes and the question of who should be credited for their discovery was recently discussed by Monthioux and Kuznetsov.[42]

In 1993, Iijima and Ichihashi reported the observation of a more interesting carbon nanospecies, the single-walled carbon nanotubes.[43] They found the single-shell tubes in carbon soot formed in a carbon arc chamber similar to that used for fullerene production. Figure 3.5 shows the original transmission electron micrograph of single-walled carbon nanotubes.

The discovery of fullerene molecules in 1985 and the later discovery of carbon nanotubes in early 1990s established a new field of carbon nanosciences and triggered an active international scientific race to investigate the structure and properties of such fascinating molecules and to discover other allotropes of

---

[ii] L. V. Radushkevich and V. M. Lukyanovich, O strukture ugleroda, obrazujucegosja pri termiceskom razlozenii okisi ugleroda na zeleznom kontakte, *Zurn. Fisic. Chim.* 1952, **26**, 88–95.

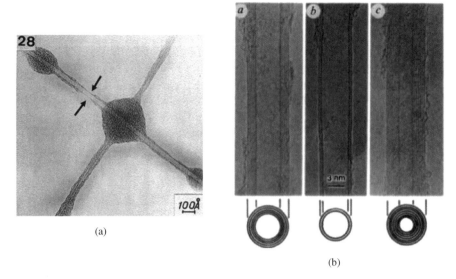

Figure 3.4 (a) Electron microscopy micrograph that first appeared in Endo's original dissertation and republished in 1974. (b) Transmission electron micrograph and cross sections of multi-walled carbon nanotubes. [(a) Reproduced with kind permission from Oberlin *et al.*, ref. 41. Copyright Elsevier 1976. (b) Reproduced with kind permission from Iijima, ref. 39. Copyright Nature Publishing Group 1991.]

this class of matter. The search for new allotropes of carbon was crowned in 2004 with the ability to isolate and manipulate single sheets of graphite currently referred to as *graphene* sheets or nano-ribbons.[44,45] Graphene sheets are essentially related to a much older form of graphite known as *exfoliated graphite*. In the exfoliated form, graphite is expanded by up to hundreds of times along its *c*-axis. Scientific and technological developments in the field of exfoliated graphite took place in the late 1960s when flexible graphite foils were made of exfoliated graphite and used for high-temperature gaskets and seals.[46] The ability to isolate single layers of graphene, however, at the age of nanotechnology spawned intensive research into the synthesis, properties, applications, and methods of the mass production of this new form of carbon.[47–56] Production methods of the graphene ribbons involve both traditional exfoliated graphite techniques and more sophisticated techniques based on unzipping of carbon nanotubes.[57] The importance of graphene sheets lies in the fact that they provide a unique opportunity for experimental investigation of truly two-dimensional systems – an opportunity that scientists never had before. In their profound recent article,[58] Geim and Novoselov described graphene as:

"the mother of all graphitic forms. It can be wrapped into a 0D buckyballs, rolled into 1D nanotubes, or stacked into 3D graphite."

Fullerenes, the Building Blocks

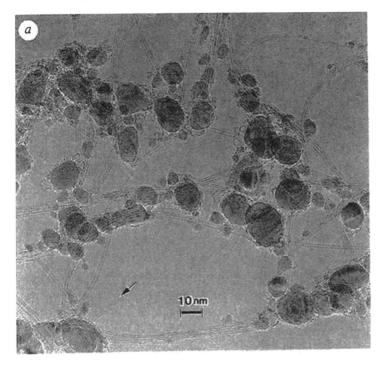

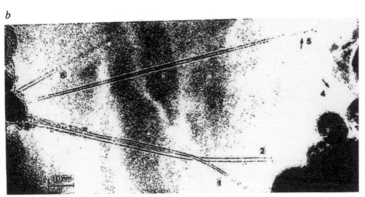

Figure 3.5 Original electron micrograph by Iijima and Ichihashi showing single-shell carbon nanotubes. The tube labeled 1 in (b) is 0.75 nm in diameter; the tube labeled 2 is 1.37 nm in diameter. Straight and terminated (4 and 5) tubes can be seen. (Reproduced with kind permission from Iijima and Ichihashi, ref. 43. Copyright Nature Publishing Group 1993.)

Figure 3.6 schematically represents the different branches of the graphitic family.

More recently, carbon aromatic chains – a unique flat one-dimensional form of nano-carbon – were derived from graphene sheets.[59] Despite that, since its inception, the carbon nanoworld was believed to be round (as in fullerene

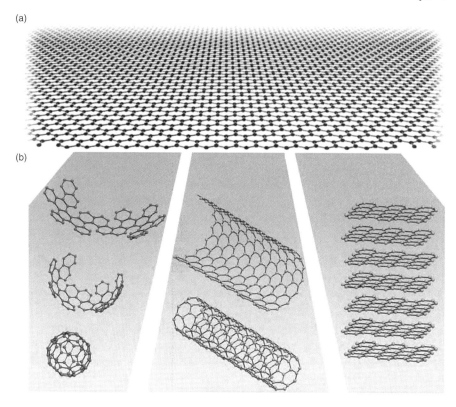

Figure 3.6 Different branches of the graphitic family; two-dimensional graphene that can be (a) wrapped into zero-dimensional fullerenes, (b) rolled into nanotubes, or (c) stacked into three-dimensional graphite. (Reproduced with kind permission from Geim and Novoselov, ref. 58. Copyright Nature Publishing Group 2007.)

molecules and cylinders) – the new flat forms of carbon nanospecies showed that even the carbon nano-world could be flat.

In the following sections we will examine the structure and properties of different types of fullerenes. We can discern three classes of fullerenes: zero-dimensional fullerenes, which include the different types of spheroidal fullerene molecules, one-dimensional fullerenes, which include cylindrical fullerenes such as nanotubes, and two-dimensional fullerenes that include flat graphene sheets and ribbons. Notably, while the zero-dimensionality description could be appropriate for small fullerene molecules with a limited number of carbon atoms (such as C_{60}) it cannot be claimed as accurate for giant fullerene molecules with a large number of carbon atoms (such as C_{540}). Also, based upon our discussion of the history and beginnings of fullerene science, the dimensionality classification, interestingly, follows the discovery trend of these fullerenes.

3.4 Zero-dimensional Fullerenes: The Structure

Fullerene molecules are closed hollow cage-like spheroidal molecules made of a number (n) of covalently bonded carbon atoms. The nomenclature most commonly used is either C_n or [n] fullerene. The structure of the carbon cage in fullerenes consists of several hexagons and pentagons. It is known that carbon in its sp^2 hybridization state creates planar sheets of connected hexagons. The introduction of pentagons in such a flat structure causes the sheet to wrinkle and curve due to slight changes in the C–C bond length. According to Euler's theorem for polyhedra, the number of faces (f), vertices (v), and edges (e) of a polyhedron should be correlated by:

$$f + v = e + 2 \tag{3.1}$$

Hence, if we consider a polyhedron consisting of h hexagons and p pentagons, the following three relations should be satisfied:

$$f = p + h, \quad 2e = 5p + 6h, \quad 3v = 5p + 6h \tag{3.2}$$

This yields that:

$$6(f + v - e) = p = 12 \tag{3.3}$$

Then, for any fullerene C_n having a cluster of n atoms (with n always even and larger than 20), and consisting of only pentagons and hexagons, closed cage-like structures consisting of 12 pentagons and (n–20)/2 hexagons can be constructed. This rule is important in predicting fullerene structures. The original discovery, as mentioned above, was for $n = 60$. Then, according to Euler's rule, the structure should be a cage-like molecule made of 60 covalently bonded carbon atoms arranged in 12 pentagons and 20 hexagons.

The smallest possible fullerene molecule should be the C_{20}. In this case the cage structure should consist of only 12 pentagons. While, theoretical studies[60] have noted that abutting pentagons raise the π-energy and cause structural destabilization, experimental observation of the C_{20} molecule was reported in 2000.[61,62] Prinzbach et al.[61] showed that the cage-structured C_{20} can be produced from its per-hydrogenated form (dodecahedrane $C_{20}H_{20}$) by replacing the hydrogen atoms with relatively weakly bound bromine atoms, followed by gas-phase debromination. The C_{20} molecules that were produced, however, were rather unstable, but their fleeting existence was confirmed using mass-selective anion photoelectron spectroscopy. Molecular simulations based on density functional theory calculations confirmed the experimental observation of a [20] fullerene molecule.[63,64] Other isomers of the 20-carbon cluster taking the shapes of a bowl containing both hexagons and pentagons and a ring were also reported. Figure 3.7 shows the three structural forms of a 20-carbon cluster: a caged fullerene, a bowl, and a ring.

Despite the "pentagon rule" stating that abutting pentagons are destabilizing, C_{20} molecules were produced and experimentally observed. The stability of

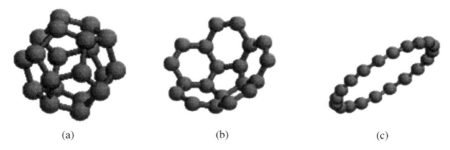

Figure 3.7 The three structures of a 20 carbon atom cluster: (a) [20] fullerene, (b) a bowl shaped graphene, and (c) a ring. (Reproduced with kind permission from Prinzbach *et al.*, ref. 61. Copyright Nature Publishing Group 2000.)

other forms of fullerene molecules – C_n, with $n = 24, 28, 32, 36$, and 50 – were also investigated and the structures were shown to be stable as well.[22]

On the other extreme of the size scale for fullerene molecules we usually find giant fullerene containing hundreds and even thousands of carbon atoms.[24,65–70] The largest fullerene molecule reported in theoretical studies is C_{4860}.[71] Notably, however, the exact shape and structure of large fullerenes is still not fully resolved. Fullerenes usually form isomers. As the number of carbon atoms in the fullerene molecule increases, the molecule can assume different geometrical shapes or structures, all of which can still satisfy both the Euler and the pentagon rules mentioned above.[72–75] Hence, it would be logical to realize that as the fullerene size increases the number of possible structures or isomers the molecule may assume also increases. These different possible structures of the different isomers would belong to different symmetry groups. Therefore, making definitive assignment of higher fullerene structures and symmetry becomes difficult.[76–78] For example, while C_{60} can be formed in only one structure that belongs to the icosahedral symmetry (I_h), C_{80} can be formed in seven different structures belonging to any of the six different symmetry groups (I_h), (D_{5d}), (D_2), (D_3), (D_{5h}), or (C_{2v}).[60,76] Figure 3.8 shows two of the seven C_{80} isomers belonging to the D_{5d} and the I_h symmetry point groups. Table 3.1 shows the number of isomers (N_i) obeying the isolated pentagon rule and symmetry point groups of common fullerene molecules.

Theoretical studies based on density functional theory (DFT) calculations shows that large fullerenes of icosahedral symmetry prefer faceted over spherical shapes.[79] Figure 3.10 shows the faceted as well as the hypothetical spherical shapes of large fullerenes. Very few of such molecules have been experimentally observed in individual forms. Many have been observed as one of the shells in a nano-onion structure. Nano-onion structures are discussed later Section 3.8.

Out of the huge possible number of fullerene molecules, merely five have been produced in significant quantities, rendering them most accessible and making their experimental investigation attractive. These fullerene molecules are C_{60}, C_{70}, C_{76}, C_{78}, and C_{84}. In terms of structural symmetry, their structures

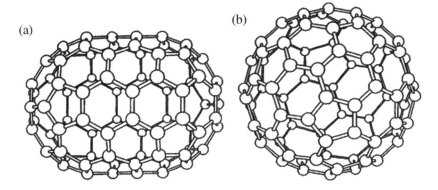

Figure 3.8 Two isomers of the C_{80} fullerene molecule belonging to (a) the D_{5d} point group and (b) the I_h point group. (Reproduced with kind permission from Schmalz et al., ref. 76. Copyright Elsevier 1986.)

Table 3.1 Number of isomers (N_i) obeying the isolated pentagon rule and symmetry point groups of common fullerene molecules (based on data in refs 76, 77 and 86)

Fullerene	N_i	Point group symmetry
C_{60}	1	I_h
C_{70}	1	D_{5h}
C_{76}	1	D_2
C_{78}	5	D_3, $2D_{3h}$, $2C_{2v}$
C_{80}	7	D_2, $2D_{3h}$, $2C_{2v}$, D_{5h}, D_{5d}, I_h
C_{82}	9	$3C_2$, C_{2v}, $2C_{3v}$, $3C_s$
C_{84}	24	See Figure 3.9 for illustration
C_{88}	35	See ref. 73 for the full list

have been identified to belong to eight different symmetry groups: C_{60} belongs to the I_h symmetry; C_{70} belongs to the D_{5h} symmetry; C_{76} belongs to the D_2 symmetry; C_{78} has five isomers that can belong to any of the three symmetry groups D_3, D_{3h}, or C_{2v}; C_{84}, however, has 24 isomers that can belong to different symmetry groups.[80–85] Figure 3.11 shows the eight possible lowest energy structures assumed by the five most accessible fullerenes. Atom numbering is based on the IUPAC system. The roman numbers used to distinguish fullerenes belonging to the same symmetry group are those given by Fowler and Manolopoulos in their atlas of fullerenes.[85]

The possibility of isomerization in higher fullerenes while still satisfying the isolated pentagon rule can be understood by the notion of the formation of a Stone–Wales (SW) rearrangement in the fullerene cage structure. The Stone–Wales rearrangement involves a 90° rotation of one of the carbon bonds in the structure.[86] Figure 3.12 shows the interconversion of the C_{78} molecule between the C_{2v} and D_{3h} isomers via a Stone–Wales rearrangement.[73,87]

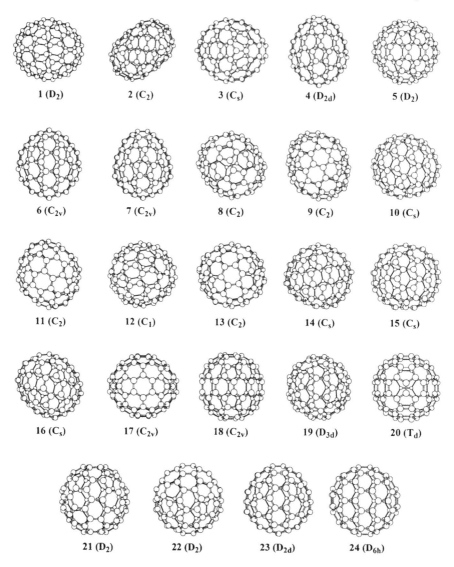

Figure 3.9 The 24 isomers of C_{84} that satisfy the isolated pentagon rule.

The stability of some isomers of higher order fullerenes is, indeed, an issue. For example, the C_{2v}(II) isomer of 78 fullerene (Figure 3.11f) was reported to be stable for only 5 months, after which it is completely degraded.[88] No issues regarding the stability of C_{60}, or many of the other produced fullerenes, have been reported so far in inert environments. While the fact that fullerenes are typically produced in electric arcs with temperatures well above 4000 °C in inert atmospheres supports the notion that once produced fullerenes are inherently stable, some studies showed that [60] fullerene is stable in a low pressure of inert atmospheres only up

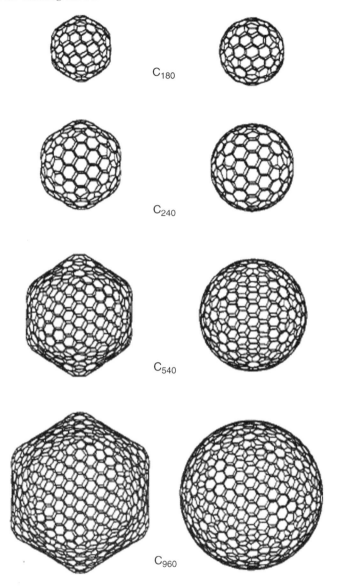

Figure 3.10 Facetted and spherical structures of large I_h fullerenes. (Reproduced with kind permission from Bakowies et al., ref. 79. Copyright American Chemical Society 1995.)

to 950 °C. Issues have also been raised regarding the stability of C_{60} in the presence of oxygen. While some recent studies based on infrared spectroscopy showed that the molecule is stable up to 600–650 °C, earlier studies based upon high-performance liquid chromatography suggested the degradation of C_{60} into $C_{120}O$ even at room temperature in the presence of oxygen.[89,90]

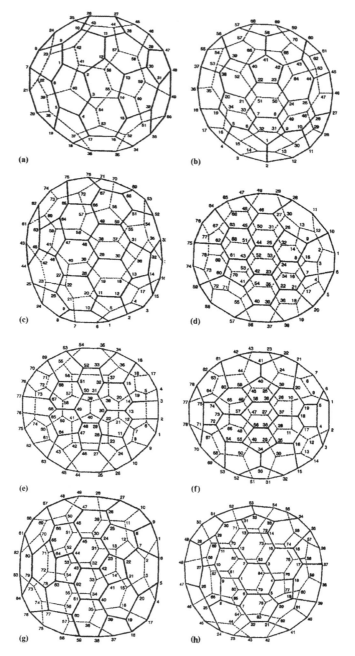

Figure 3.11 The eight possible lowest energy structures assumed by the five most accessible fullerenes. Structures of (a) C_{60}-I_h, (b) C_{70}-D_{5h}, (c) C_{76}-D_2, (d) C_{78}-D_3, (e) C_{78}-C_{2v}(I), (f) C_{78}-C_{2v}(II), (g) C_{84}-D_2(IV), and (h) C_{84}-D_{2d}(II). (Reproduced with kind permission from Taylor and Burley, ref. 78. Copyright Royal Society of Chemistry 2007.)

Fullerenes, the Building Blocks

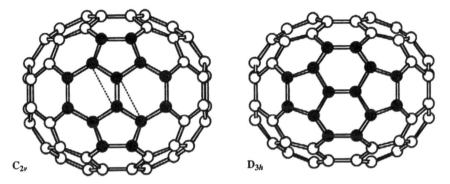

Figure 3.12 Interconversion of the C_{78} molecule between the C_{2v} and D_{3h} isomers *via* a Stone–Wales rearrangement of a single carbon bond. (Reproduced with kind permission from Diederich and Whetten, ref. 73. Copyright American Chemical Society 1992.)

Of the most accessible aforementioned fullerenes, C_{60} and C_{70} are the most investigated members. This most probably is due to their availability since the two isomers of fullerene are usually obtained in 75% and 24% yield, respectively, through the most common production method – the arc-discharge method of Krätschmer and Huffman.[30] The reminder of the process product (1%) is usually a mixture of other forms of carbon, including higher fullerenes, reaching to higher than C_{100}.[78,84] In the following section we examine the structures of these two most investigated fullerene molecules.

3.4.1 Structure of the [60] Fullerene Molecule

Once described as the most beautiful molecule,[91] C_{60} or [60] fullerene has a unique structure. The fullerene cage for C_{60} molecules has 12 pentagons and 20 hexagons (Figure 3.13). Inspection of the structure shows that each and every pentagon is surrounded by five hexagons. This makes C_{60} the smallest fullerene having no butting pentagons.

Two types of C–C bonds have been distinguished[iii] in the [60] fullerene structure (Figure 3.13). These are the C–C bond bordering a hexagon and a pentagon, known in the literature as "a_5", and the C–C bond bordering two hexagons, known in the literature as "a_6". The lengths of these bonds have been measured experimentally by NMR and neutron diffraction methods.[80,83,92,93] They were reported to range between 1.455 and 1.46 Å for the a_5 bond, and between 1.4 and 1.391Å for the a_6 bond. The neutron diffraction method always yielded the lower value of the length range.[94] Recently, it was shown that such bond length distribution matches the molecular simulation results for

---

[iii] In some texts, these bonds are further distinguished and classified as single and double bonds. The author prefers to avoid such distinction due to the well-known fact that such sharp distinction is hard to make in aromatic structures.

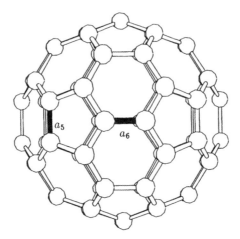

Figure 3.13 Structure of [60] fullerene molecules showing the two different types of C–C bonds; the "a_5" bordering a pentagon and a hexagon, and "a_6" bordering two hexagons.

an unperturbed C_{60} molecule. Any perturbation of the molecule by an external thermodynamic field, however, was shown to dramatically alter the C–C bond length distribution in the fullerene molecule.[95] Figure 3.14 shows the C–C bond distribution in a C_{60} fullerene molecule interacting with different numbers of water molecules, as determined by semi-empirical molecular simulation. Clearly, the chemical potential of the interacting water molecules tremendously alters the bond length distribution within the fullerene cage. The results show that as the number of interacting molecules increases, the C–C bond length distribution within the fullerene molecule broadens and shifts to lower values. This would definitely have an impact on fullerene properties, as we will show in the following sections.

The C–C bonds in a [60] fullerene molecule (the vertices of the molecule) are usually considered in the literature to form a *regular* truncated icosahedron. In fact, this is an accepted approximation since the length difference between the a_5 and the a_6 bonds causes the molecule to slightly deviate from being a regular truncated icosahedron. The diameter of the [60] fullerene molecule is an important parameter. Generally speaking, the diameter of an icosahedral fullerene (d_i) can be calculated based upon the number of carbon atoms (n) and the average C–C bond length (a_{C-C}) according to the equation:

$$d_i = \frac{5\sqrt{3}a_{c-c}}{\pi}\sqrt{\frac{n}{20}} \qquad (3.4)$$

Using an approximated average value of 1.44 Å yields the diameter of the C_{60} icosahedron as 6.88 Å. However, importantly, the effective diameter of the C_{60} molecule has to include the thickness of the π-electron shell associated

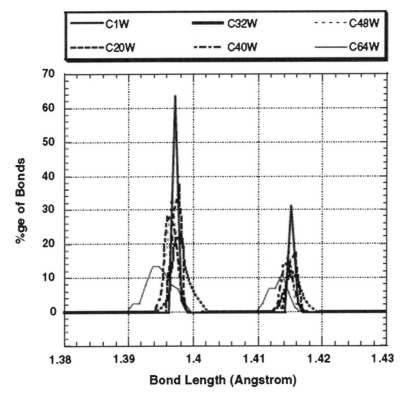

Figure 3.14 C–C bond distribution in a [60] fullerene molecule interacting with different numbers of water molecules. (Reproduced with kind permission from Amer et al., ref. 95. Copyright Elsevier 2005.)

with the sp^2 hybridization status of the carbon atoms forming the molecule (Figure 3.15). Estimating the π-electron shell thickness as 3.37 Å, based on the very well-characterized interplanar distance in graphite, gives an effective diameter of a C_{60} molecule very close to 1 nm. Recalling (from Chapter 1) that the correlation length (ξ) is typically on the order of 1 nm makes the fullerene molecule a perfect example of a nano-system where the system size is on the same order as one of the system's characteristic lengths scales discussed in Chapter 1.

The internal cavity of a [60] fullerene molecule has attracted much attention. From our previous discussion, the internal cavity for C_{60} would be on the order of a 4 Å diameter cavity surrounded by a 3 Å-thick shell of electrons (Figure 3.15). This triggered a line of investigation involving fullerenes (especially C_{60}) with incarcerated atoms. This class of fullerene is referred to as incar-fullerenes, endo-fullerenes, or endohedrals.[96–125] Different endo-fullerenes have been investigated theoretically and prepared experimentally. Incarceration of different elements in several types of fullerenes was also investigated. The different elements investigated included nitrogen, hydrogen,

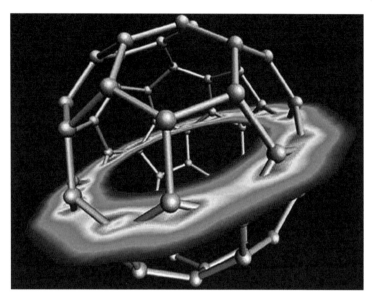

Figure 3.15 The 1 nm effective diameter of the [60] fullerene molecule, including the thickness of the π-electron shell associated with sp^2 hybridization of the carbon atoms. (Reproduced with kind permission from Tycko *et al.*, ref. 93. Copyright American Chemical Society 1991.)

noble gases (helium, neon, argon, krypton, and xenon), different metal nitrides, and different types of metallic, non-metallic, and even radioactive atoms. Table 3.2 shows different endo-fullerenes investigated and reported in the literature.

Endo-fullerenes, in general, and the easily produced *incartrimetalnitridofullerenes*,[126] in particular, have opened a new gate to another unchartered territory. The technological and biomedical applications of such class of fullerenes are very promising and still to be explored.[97,101,123,127] Figure 3.16 shows different types of endo-fullerene molecules in which an atom, multiple atoms, or a molecule are incarcerated in a fullerene cage of variable sizes or shapes.

3.4.2 Structure of the [70] Fullerene Molecule

The structure of next most investigated fullerene, [70] fullerene, is slightly different from that of [60] fullerene. According to the Euler construction rule for polyhedra, five extra hexagons exist in the structure of the [70] fullerene molecule compared to that of C_{60}. Figure 3.17 shows the structure of the [70] fullerene molecule with the extra five hexagons clearly observed. These five extra hexagons extend the cage structure in one direction, rendering a shape resembling that of a rugby ball. As mentioned above, the C_{70} molecule comes in only one structure (Figure 3.17) and belongs to the D_{5h} symmetry point group.

Unlike the C_{60} molecule, where all the carbon atom sites are equivalent, there are five different carbon atom sites in the C_{70} molecule.[128] This leads to eight

Table 3.2 Various endo-fullerenes investigated and reported in the literature.

Fullerene	Metal	Reference
Fullerenes incarcerating one atom		
C_{28}	Hf, Ti, U, Zr	109
C_{36}	U	109
C_{44}	K, La, U	109,112
C_{48}	Cs	112
C_{50}	U	109
C_{60}	Li, K, Ca, Co, Y, Cs, Ba, Rb, La, Ce, Pr, Nd, Sm, Eu, Gd, Tb, Dy, Ho, Er, Lu, U	110,114
C_{70}	Li, Ca, Y, Ba, La, Ce, Gd, Lu, U	109,111,114
C_{72}	U	109
C_{74}	Sc, La, Gd, Lu	112,113,115
C_{76}	La	109,115
C_{80}	Ca, Sr, Ba	116
C_{82}	Ca, Sc, Sr, Ba, Y, La, Ce, Pr, Nd, Sm, Eu, Tm, Lu	109,113,115
C_{84}	Ca, Sc, Sr, Ba, La	115,119,120
Fullerenes incarcerating two atoms		
C_{28}	U_2	109
C_{56}	U_2	109
C_{60}	Y_2, La_2, U_2	109
C_{74}	Sc_2	118
C_{76}	La_2	115
C_{80}	La_2, Ce_2, Pr_2	117,110
C_{82}	Sc_2, Y_2, La_2	117,115
C_{84}	Sc_2, La_2	113,115
Fullerenes incarcerating three atoms		
C_{82}	Sc_3	115
C_{84}	Sc_3	121
Fullerenes incarcerating four atoms		
C_{82}	Sc_4	116
Fullerenes incarcerating derivatives of ammonia NR_3, NR_2R', or $NRR'R''$		
C_{78}	Sc_3N	123
C_{76}	Dy_3N	122
C_{68}	Sc_3N, $Sc_2(Tm/Er/Gd/Ho/La)_2N$	124
C_{80}	Sc_3N, $ErSc_2N$, Er_2ScN, Lu_3N, $Lu(Gd/Ho)_2$, Y_3N, Ho_3N, Tb_3N, Dy_3N	122,125,126
C_{82-98}	Dy_3N	122

different types of C–C bonds in the molecular cage. Figure 3.18 shows the structure of the C_{70} molecule, elucidating the five nonequivalent carbon sites and the eight different C–C bonds connecting them. Theoretical calculations[129,130] as well as experimental investigations[131,132] were conducted to determine the C–C bond lengths in C_{70}. Table 3.3 shows examples of modeling and experimental results for the eight different C–C bonds in [70] fullerene as reported in the literature.

As we will discuss later, the five extra hexagons in the [70] fullerene molecule cause measurable differences in its properties and behavior compared to the properties and behavior of the [60] fullerene molecule. This is also true for the

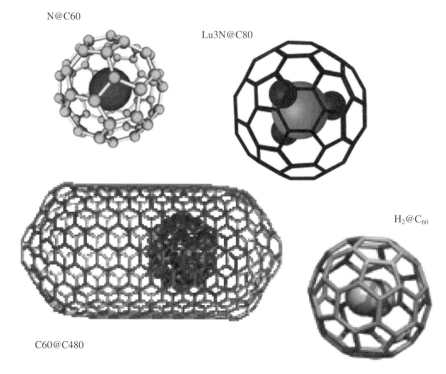

Figure 3.16 Different types of endo-fullerene molecules.

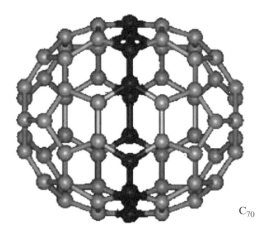

Figure 3.17 Structure of [70] fullerene showing the extra ten carbon atoms resulting in five extra hexagons.

properties and behavior of other fullerenes as well. While the chemical composition of different fullerenes is essentially the same, the size, shape, and chirality of different types of molecules manifest themselves strongly in the properties and behavior of these molecules.

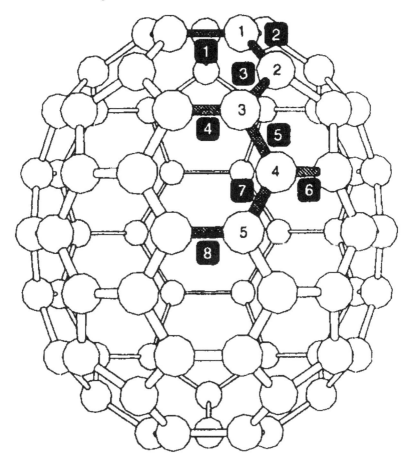

Figure 3.18 Structure of [70] fullerene molecule elucidating the five nonequivalent carbon sites and the eight different C–C bonds. (Reproduced with kind permission from Saito and Oshiyama, ref. 128. Copyright American Physical Society 1991.)

3.5 Production Methods of Fullerenes

Fullerenes can be generally produced in laboratory facilities in different ways involving the generation of a carbon-rich vapor or plasma. There are essentially three methods – with many modifications – to produce zero-dimensional fullerenes: the Huffman–Krätschmer,[30] or arc-discharge process,[133] benzene combustion in an oxygen-deficient environment,[134] and the condensation of polycyclic aromatic hydrocarbons.[135]

3.5.1 Huffman–Krätschmer Method

As mentioned above, this was the first method to produce fullerenes in significant amounts and its development, actually, marked the beginning of

Table 3.3 Examples of modeling and experimental results for the lengths (Å) of eight different C–C bonds in [70] fullerene as reported in the literature (see Figure 3.18 for bond locations).

Bond	Model		Experiments	
	TBMD[a]	LDA[b]	Electron diffraction[c]	X-Ray[d]
C1–C1	1.457	1.448	1.46	1.434
C1–C2	1.397	1.393	1.37	1.377
C2–C3	1.454	1.444	1.47	1.443
C3–C3	1.389	1.386	1.37	1.369
C3–C4	1.456	1.442	1.46	1.442
C4–C4	1.443	1.434	1.47	1.396
C4–C5	1.418	1.415	1.39	1.418
C5–C5	1.452	1.467	1.41	1.457

[a]Ref. 130 (TBMD = tight binding molecular dynamic method).
[b]Ref. 131 (LDA = local density analysis).
[c]Ref. 132.
[d]Ref. 133.

fullerene science. The Huffman–Krätschmer method involves arc-discharge between highly pure carbon rods in a helium or argon atmosphere at 100–200 Torr. A mechanical mechanism is needed to translate the electrodes together to maintain the electrode gap as the electrodes are consumed. Controlling such a gap was reported to be essential for the process and to prevent temperature drop.[78] At an estimated electrode tip temperature of 2000 °C the yield of fullerene in the produced soot is about 4%. However, at an estimated electrode tip temperature of ~4700 °C the yield of fullerene in the produced soot was reported to increase into the 7–10% range.[136] Notably, the yield calculations tend to account for all types of formed fullerenes. C_{60} and C_{70}, however, constitute the majority in the produced fullerene.

A simple bench-top reactor was developed to produce fullerene with 4% yield as well.[137] The reactor utilizes an inexpensive AC arc welding power supply to initiate and maintain a contact arc between two graphite electrodes in a helium atmosphere. Since this technique does not require a mechanism to translate the electrodes together as they become consumed, it is termed the "contact arc method". Instead, the technique utilizes flexible support for the upper electrode and relies on gravity to maintain contact between the two vertical electrodes. Figure 3.19 shows a schematic diagram of the bench-top contact arc apparatus.

In a leap forward, Parker *et al.*[138,139] used a plasma arc in a fixed gap between two horizontal electrodes. The developed apparatus is known as *fullerene generator* due to the high yield it generates, reaching ~40% in high boiling point solvents. The exceptionally high yield was attributed to the fine control of the arc gap combined with proper convection of the atmosphere in the apparatus and carful extraction. Figure 3.20 shows a schematic diagram of the plasma arc fullerene generator.

Several processing parameters are presently known to affect the fullerene yield in an arc-discharge process. The optimum atmosphere to be used was

Fullerenes, the Building Blocks

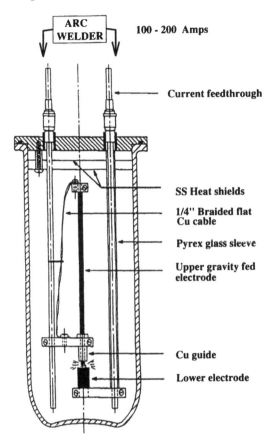

Figure 3.19 Schematic diagram of the bench-top contact arc apparatus. (Reproduced with kind permission from Koch *et al.*, ref. 137. Copyright American Chemical Society 1991.)

found to be highly pure helium. While most reactors are operated in the 100–200 Torr region, the optimum operation pressure was reported to be highly sensitive to the actual chamber design and should be determined for each specific reactor. Although the purity of the carbon electrodes was found not to affect the soot production rate, smaller-diameter electrodes gave higher yields of soot. In addition, small contaminations of hydrogen or moisture in the generation chamber seriously suppress fullerene generation.

3.5.2 Benzene Combustion Method

Evidence that fullerenes can be formed in flames was at first elusive, but progress was eventually made. In 1991 significant quantities of C_{60} and C_{70} were found in samples collected from low-pressure premixed benzene/oxygen flames.[134] Further investigations showed that fullerenes can be produced in

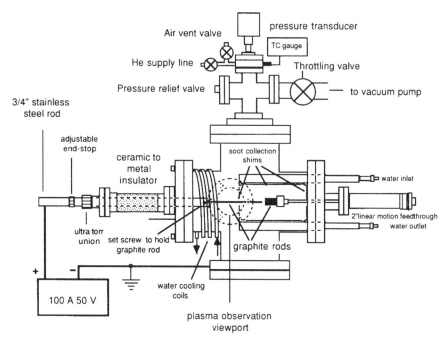

Figure 3.20 Schematic diagram of the plasma arc fullerene generator. (Reproduced with kind permission from Parker et al., ref. 138. Copyright Pergamon Press 1993.)

substantial quantities by sub-atmospheric pressure, laminar, premixed flames of benzene in an oxygen-deficient atmosphere with or without the presence of an inert gas. The largest yield of soot into fullerene was reported to be 20% at a pressure of 37.5 Torr. Figure 3.21 shows a schematic diagram of the burner and associated equipments used in the process. It was also reported that fullerene formation in the flame can take place with the presence of hydrogen and oxygen.[140] The promise of the combustion method encouraged the Frontier Carbon Corporation, a subsidiary of Mitsubishi Chemical Corporation, to construct a large-scale fullerene factory in Japan in 2003. The factory has the capacity of producing 5000 ton of fullerene annually.

3.5.3 Condensation Method

The condensation method is based upon condensation of polycyclic aromatic hydrocarbons through pyrolytic dehydrogenation or dehydrohalogenation processes. While the method does not produce fullerenes in sufficient quantities for practical applications, it provides a means of deducing the mechanism of fullerene formation. The method was used to produce only C_{60} fullerene from a molecular polycyclic aromatic precursor bearing chlorine substituents at key positions subjected to flash vacuum pyrolysis at 1100 °C

Fullerenes, the Building Blocks 133

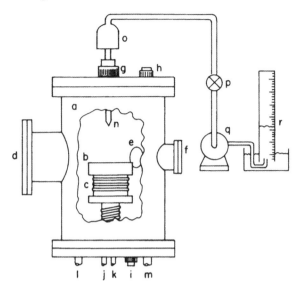

Figure 3.21 Schematic diagram of the burner and associated equipment used in the combustion process to produce fullerenes: a – low pressure chamber; b – copper-burner plate; c – water cooling coil; d to f – windows; g to i – feedthroughs; j – annular flame feed tube; k – core-flame feed tube; l and m – exhaust tubes; n – sampling probe; o – filter; p – valve; q – vacuum pump; r – gas meter. (Reproduced with kind permission from Howard *et al.*, ref. 134. Copyright Nature Publishing Group 1991.)

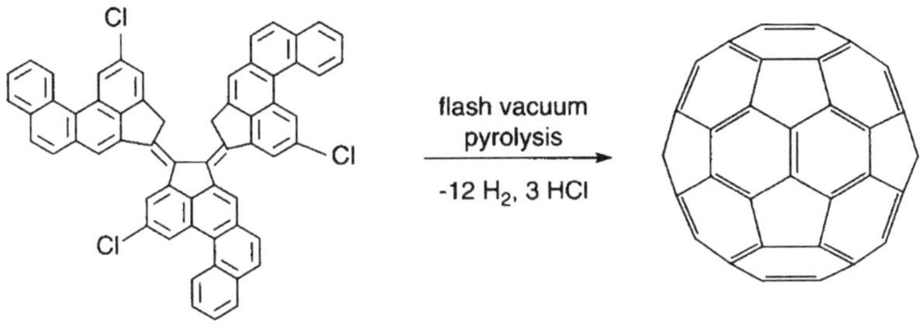

Figure 3.22 Formation of [60] fullerene from a chloroaromatic precursor through the condensation process. (Reproduced with kind permission from Scott *et al.*, ref. 141. Copyright AAAS 2002.)

through a 12-step reaction.[141] Figure 3.22 shows a schematic for the precursor and the formed C_{60}.

3.6 Extraction Methods of Fullerenes

The main method used to extract fullerene from the produced soot is the traditional Soxhlet extraction method. The Soxhlet method is traditionally used to

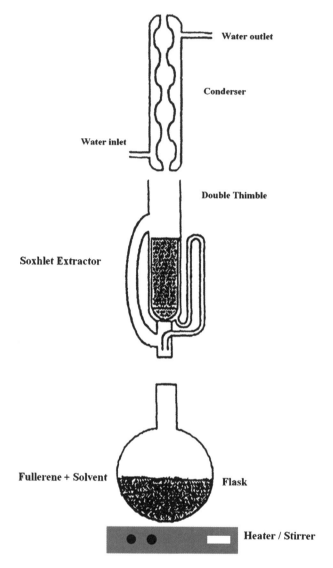

Figure 3.23 Schematic illustration of a Soxhlet extraction setup.

extract molecular moieties from solid phases using organic solvents capable of dissolving the molecular moieties. Figure 3.23 schematically illustrates a Soxhlet extraction unit. The solid sample containing the molecular moiety to be extracted is loaded into a thimble in the Soxhlet extractor. As the solvent is boiled in the flask at the bottom, solvent vapors rise through the side channel on the left of the extractor and condense near the bottom of the condenser unit, resulting in dripping hot distilled solvent into the thimble through the solid sample. The hot distilled solvent extracts the molecular moiety on its way down through the solid sample, and the solution, then, makes its way back to the flask *via* the tube to the right. This closed-loop system is usually operated for several hours, during which all extractable molecular moieties are collected in the flask. The non-extractable portion of the solid sample remains in the thimble.

Fullerenes are extracted from the produced soot using the Soxhlet apparatus with any of different types of solvents. Many solvents have been used in the extraction process, such as chloroform, toluene, benzene, *n*-hexane, 1,2-dichlorobenzene, *etc*. The type of the solvent controls the speed of the extraction process and dictates subsequent processes. For example, chloroform results in a very slow process. The use of 1,2-dichlorobenzene results in a very fast extraction process but requires a high vacuum process for removal of solvent traces. In addition, if carbon disulfide is used, it has to be vigorously removed under vacuum to avoid fullerene contamination with sulfur.[142] Selective extraction of various molecular weight fullerenes by varying the extraction solvent has also been reported.[138] Higher mass fullerenes are better extracted with more polar and higher boiling point solvents. Figure 3.24 shows a fullerene extraction and separation scheme. The laser desorption time of flight (TOF) mass spectra of soot extracts are also shown in the figure. The results shown are for soot produced using the combustion method described earlier.

The figure highlights two important points. First, the total fullerene yield is sharply dependent on the extraction scheme and solvents used. Secondly, the extracted fullerene molecular weight depends on the type of solvent used in the extraction process. For example, while benzene mainly extracts C_{60} and C_{70} with smaller amounts of higher fullerenes up to C_{96}, 1,2,3,5-tetramethylbenzene (TMB) extracts more of the higher fullerenes. In addition, the figure clearly shows that while hexane mainly extracts C_{60} and C_{70}, heptane additionally extracts C_{80} and C_{84}.

An efficient alternative method has also been proposed in the literature.[143] In this method, the produced carbon soot is dispersed in tetrahydrofuran (THF) at a concentration of $0.1\,\text{g}\,\text{mL}^{-1}$ at room temperature and then sonicated for 20 min. After filtration to remove insoluble remains, the THF is removed using a rotary evaporator operated at 50 °C, leaving fullerenes and other soluble impurities in the flask. In this case, the impurities contain many polyaromatic hydrocarbons that can be removed by washing the extract in diethyl ether before it is further purified.

A non-solvent based method, the *sublimation* method, was also utilized for fullerene extraction from the produced soot. C_{60} and C_{70} powders are known to sublime in vacuum at relatively low temperatures (*i.e.*, 350 and 460 °C, respectively).[144,145] The advantage of the sublimation extraction method is that it produces fullerene samples that are solvent free. This is crucial for experiments were any traces of solvent could greatly affect the investigation results. In the sublimation method, the soot sample is placed in one end of an evacuated quartz tube placed in a furnace with a temperature gradient. The tube end containing the soot sample is kept at the highest temperature zone of the furnace (600–700 °C). Fullerenes will sublime and drift down the temperature gradient to condense on the walls of the quartz tube. The higher the fullerene mass, the closer to the soot it will condense.

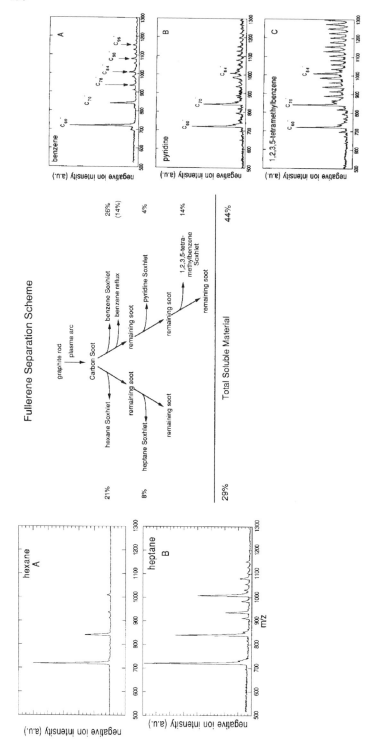

Figure 3.24 Fullerene extraction and separation scheme along with laser TOF mass spectra of extract prepared by sequential solvent extraction. (Adopted with kind permission from Parker *et al.*, ref. 138. Copyright Pergamon Press 1993.)

3.7 Purification Methods of Fullerene

Purification of fullerenes is intended to separate fullerene molecules from any impurities such as polyaromatic hydrocarbons and other types of carbon-based impurities. In addition, a purification process is used to separate fullerenes from each other based on their molecular mass and size. Mainly, solvent methods based on liquid chromatography (LC) and sublimation methods based on sublimation under temperature gradients are used for fullerene purification. The sublimation method for purification is essentially the same as the sublimation method for extraction discussed earlier. The effectiveness of the purification process is usually verified by fullerene sensitive characterizations techniques such as mass spectroscopy, nuclear magnetic resonance, and infrared, Raman, or optical absorption spectroscopy.

Liquid chromatography is the main technique for fullerene purification. In this technique, a solution containing a mixture of fullerenes to be separated (purified) (usually referred to as the mobile phase) is forced through a column filled with a high-surface area solid (usually referred to as the stationary phase). As the mixed solution passes through the column, different molecules in the solution experience different levels of interaction with the high-surface area solid in the column due to various physical and chemical mechanisms. Stronger interactions increase the retention time of the molecule in the column, or, in other words, decrease the migration rate of that particular molecule through the column. Hence, separated molecules should elute from the column in the order of decreasing retention time. Weakly interacting molecules will have shorter retention times and will elute first. Strongly interacting molecules will have longer retention times and will elute later.

With fullerenes, the identity of the separated fullerene is verified qualitatively by color and more quantitatively by other characterization techniques mentioned above. Regarding the qualitative identification, different fullerenes are identified by the color of their solution in certain solvents. For example, in toluene, C_{60} yields a magenta or purple solution, while C_{70} yields a reddish to orange solution. LC generally enables the separation of fullerenes based upon their size (or mass). The method, however, can still enable isolation of different symmetries of the same fullerene such as the C_{2v} and D_3 symmetries of the C_{78} fullerene.[84]

Effective separation (or purification) occurs when a sufficient difference in the retention time is achieved. This depends on the nature of the "stationary phase" used. Different types of stationary phases have been employed. The most widely used were based on silica gels and alumina. Carbon-based stationary phases (Elorite carbon) were proved efficient in firmly holding higher fullerenes while allowing [60], and [70] fullerenes to elute.[146] Recently, many solid phases have been developed for high-performance liquid chromatography of fullerenes.[147–149]

Fullerene purification procedures combining the extraction and purification steps were also reported in the literature.[150] The procedure is based on a modified Soxhlet technique discussed earlier. Figure 3.25 shows a schematic

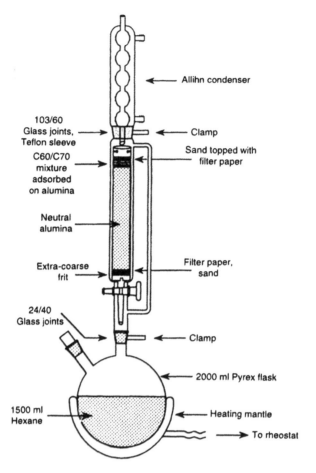

Figure 3.25 Schematic illustration of modified Soxhlet apparatus that combines the separation and purification steps. (Reproduced with kind permission from Khemani *et al.*, ref. 150. Copyright American Chemical Society 1992.)

illustration of the modified Soxhlet apparatus, which is reported to be capable of extracting 1 g of C_{60} and 0.1 g of C_{70} a day. As shown in the figure, the thimble in a regular Soxhlet apparatus has been replaced with a liquid chromatography column filled with neutral alumina. During operation, the distilled solvent (pure hexane) is condensed at the top of the column to pass down the column, extracting C_{60} molecules into the distillation flask. After about 20–30 hours of operation, the flask containing C_{60} in hexane is replaced with a second pure hexane flask and C_{70}, remaining in the column, is collected in the same way.

Nowadays, fullerenes are commercially available with claimed purity of 99.9%. Doubts have been raised regarding the ability of solvent-based processes to produce such a level of purity due to the reported presence of

Fullerenes, the Building Blocks 139

significant amounts of $C_{120}O$ as detected by high-performance liquid chromatography (HPLC).[90,151] An acceptable value for purity, however, is around 98%.[78]

A close look at the laser TOF mass spectroscopy (LTOF-MS) results presented in Figure 3.24 reveals that experimental extraction, and hence observation, of fullerenes with more than ~110 carbon atoms is not common. In fact, the largest experimentally extracted fullerene reported contained about 212–266 carbon atoms according to LTOF-MS results.[84,139] In an attempt to answer the question "Are giant fullerenes spherical or tubular?", Scuseria[152,153] of Rice University conducted a theoretical investigation using *ab initio* calculations to calculate the energy of higher fullerenes (ranging from C_{80} up to C_{240}) in both spherical (icosahedral) and tubular (cylindrical) geometries (Figure 3.26). His results showed that in all investigated fullerenes the

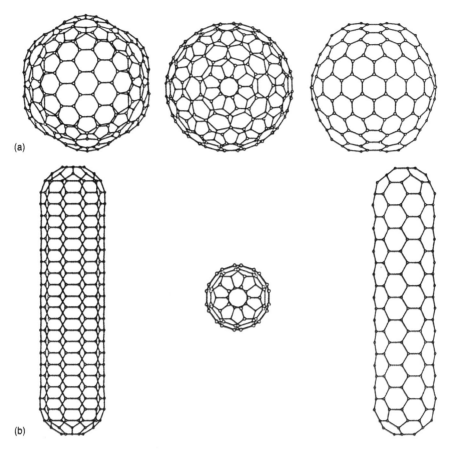

Figure 3.26 Equilibrium geometries of C_{240} as determined by *ab initio* calculations: (a) spherical I_h geometries and (b) D_{5d} cylindrical geometries. (Reproduced with kind permission from G. Scuseria, ref. 153. Copyright Elsevier 1995.)

icosahedral geometries are much more stable than the cylindrical "buckytube" geometries. In addition, scanning tunneling microscopy images of fullerenes with ∼300 carbon atoms according to LTOF-MS results showed that these giant fullerenes have roughly spherical shapes with diameters ranging between 1 and 2 nm with no evidence of tube-like geometries.[154]

These results and the fact that theoretical investigations predicted the possible stability of fullerenes with up to 4860 carbon atoms,[71] as we mentioned before, raises a very interesting question regarding the ability to extract and observe such gigantic fullerenes. The answer to such a question has been proposed[87] along the lines that such giant fullerenes could be either formed in amounts too small to observe experimentally or have a solubility, in the typically used extraction solvents, too low to enable their extraction. Another plausible reason for the discrepancy is the fact that all simulations are performed in an absolute vacuum environment. Processing, however, is not and the effect of such an environment could be why giant fullerenes, while shown to be more stable than tubular in the calculations, are not observed experimentally.

The other two possibilities for geometries of fullerenes with giant masses are either tubular or in the form of fullerene onions. We will start with the fullerene onion geometry and then discuss the tubular or cylindrical geometry, which is considered a one-dimensional fullerene.

3.8 Fullerene Onions

Fullerene nano-onions are quasi-spherical particles consisting of concentric graphitic shells. They were first observed when carbon soot particles and tubular graphitic structures were exposed to intense electron beam irradiation in transmission electron microscopy experiments.[155] Apparently, the extremely high local temperature within the electron beam allowed structural fluidity leading to the formation of the new form of carbon nanospecies.[156] Nano-onions were also observed as the result of nano-diamond particle annealing in vacuum at temperatures (1000–1500 °C),[157,158] and also upon generating an arc discharge between carbon electrodes submerged in water.[159,160] Figure 3.27 shows a high-resolution transmission electron micrograph of a carbon nano-onion. Clearly, the carbon nano-onions consist from several concentric shells. Up to 15 shells can be counted in a nano-onion, with roughly a 12 nm diameter, produced by the submerged carbon arc method.

More recently, carbon onions as large as 2 μm in diameter were observed after high shear and hydrostatic loading of graphite single crystals in a diamond anvil cell (DAC) at room temperature.[161] The number of layers and, hence, the size of the carbon onion was found to increase with increasing loading pressure. Small onions with 4–6 layers were formed at lower pressure ranges (46 GPa after shear) and large onions with up to 60 layers were formed under higher pressures (71 GPa after shear). Figure 3.28 shows high-resolution transmission electron micrographs of carbon onions observed after high shear and

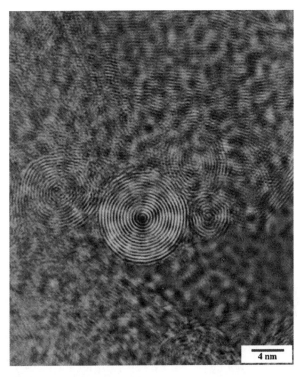

Figure 3.27 High-resolution transmission electron micrograph showing carbon nano-onions produced by arc discharge between carbon electrodes submerged in water. (Reproduced with kind permission from Roy et al., ref. 160. Copyright Elsevier 2003.)

hydrostatic pressure treatment at room temperature. Interestingly, the shell to shell distance was found to range between 0.317 and 0.36 nm for different size onions, with the smaller distance characteristic of the larger onions. Figure 3.29 shows a scanning electron micrograph of the giant carbon onions observed as a result of this treatment of single graphite crystals.

Carbon onions have also been produced and observed by many other means such as carbon soot annealing,[162] regular arc discharge,[163] carbon ion beam implantation on silver and copper substrates,[164–167] plasma spraying of nano-diamond,[168] catalytic decomposition of methane over an Ni/Al catalyst,[169] as well as after C_{60} thermobaric treatment.[170] In addition, carbon onions having an amorphous-like silicon carbide (SiC) core have been produced from polycrystalline SiC powder subjected to laser shock compression.[171,172] Carbon onions have been also shown to exist in the interstellar dust and are believed to be the reason behind the 217.5 nm ultraviolet radiation absorption band.[173,174]

Our state of knowledge regarding carbon nano-onions can be safely described as being at the start of a steep learning curve. Despite several

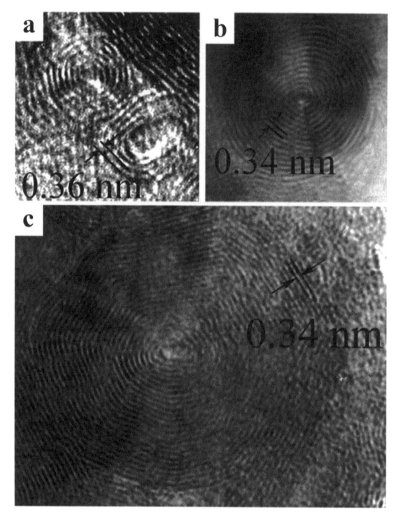

Figure 3.28 Carbon onions formed in graphite single crystals under pressure and shear deformation: (a) an onion with 6–7 layers (24 GPa pressure before shear and 46 GPa after shear, 15 shear cycles), (b) an onion with 14–15 layers (48 GPa pressure before shear and 63 GPa after shear, 20 shear cycles), and (c) an onion with about 30 layers (57 GPa pressure before shear and 71 GPa after shear, 5 shear cycles) (Reproduced with kind permission from Blank *et al.*, ref. 161. Copyright Institute of Physics 2007.)

investions[175–183] into the production, formation mechanism, and properties of carbon nano-onions, the field is virtually unexplored and there is still plenty of space to fill regarding the properties and applications of such a unique class of carbon nano-building blocks.

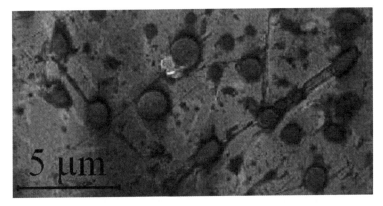

Figure 3.29 Scanning electron micrograph of giant carbon onions about 2 μm in diameter. (Reproduced with kind permission from Blank *et al.*, ref. 161. Copyright Institute of Physics 2007.)

3.9 One-dimensional Fullerene: The Structure

3.9.1 Single-walled Carbon Nanotubes (SWCNTs)

One-dimensional fullerenes are typically in the form of tubes or cylinders with diameters ranging between a fraction of a nanometer and few tens of nanometers. They can be in the form of a single tube, which we refer to as single-walled carbon nanotubes (SWCNTs), or in the form of a group of concentric tubes positioned one inside the other, in which case we refer to them as multi-walled carbon nanotubes (MWCNTs). The tubes can also be in the form of only two concentric tubes, which are referred to as double-walled carbon nanotubes (DWCNTs). Carbon nanotubes can be formed by rolling graphene sheets (a single sheet, double sheets, or multiple sheets) into a seamless cylinder. Figure 3.30 shows a schematic of how graphene sheets can be rolled to form carbon nanotubes.[184] From the science and application viewpoint, SWCNTs are more interesting and exciting. Hence we will start by describing their structure first.

A SWCNT can be formed by rolling a single sheet of graphite (a graphene sheet). Figure 3.31 shows the typical hexagonal (honeycomb) lattice structure of a graphene sheet made of covalently bonded sp^2 hybridized carbon atoms. The two principal axes directions and the unit vectors (a_1 and a_2) of the primitive hexagonal cell are shown in the figure. If we connect any two carbon atoms on the flat sheet geometry by a straight line, we can always roll the sheet around an axis normal to that line such that the selected line becomes a circumference of the tube. In this case, the two atoms at the start point and end point of the selected line have to coincide and become the same atom on the tube surface (Figure 3.31). If we assign any random atom on the sheet the coordinates (0,0) and consider it the origin of a two-dimensional space, the coordinates of any other atom on the flat sheet can be easily described in terms

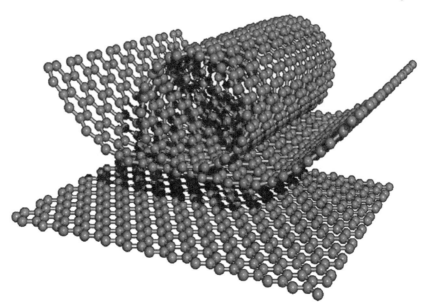

Figure 3.30 Schematic of how graphene sheets can be rolled to form carbon nanotubes. (Reproduced with kind permission from Endo *et al.* ref. 184. Copyright The Royal Society 2004.)

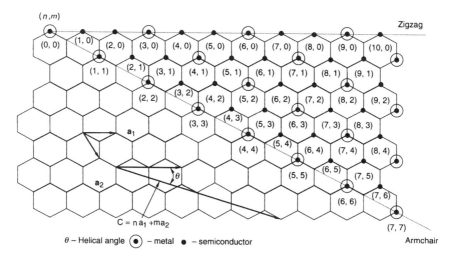

Figure 3.31 Typical hexagonal structure of a graphene sheet made of covalently bonded sp^2 hybridized carbon atoms. (Adapted with kind permission from Dresselhaus *et al.*, ref. 187. Copyright Elsevier, 1995.)

Fullerenes, the Building Blocks

of the unit vectors (a_1 and a_2) as shown in Figure 3.31. Then, the line connecting any two carbon atoms on the flat sheet geometry can always be expressed as a vector (C_h) in terms of multiples of the unit vectors (a_1 and a_2). Assigning the vector start point the origin coordinates (0,0) would simplify the representation and the circumference of a carbon nanotube can be expressed in vector notation as:

$$\vec{C_h} = n\vec{a_1} + m\vec{a_2} \equiv (n,m) \qquad (3.5)$$

The vector representing the circumference (p) of a tube is the chiral vector (C_h). The angle between the chiral vector and the a_1 vector is known as the chiral angle (θ) (Figure 3.31). A SWCNT can be uniquely defined by its chiral vector. The word uniquely, in this context, means both geometrically and, as we will realize later, physical and chemical property wise. Hence nanotubes are usually referred to by the two vector multiples (n,m) defining their chiral vector.

Once the sheet is rolled to form a tube, the carbon atoms will form a helix of graphite lattice points. For each tube with a chiral angle (θ) between 0° and 30° there is an equivalent tube with a chiral angle $\theta' = \theta + 30°$. The two equivalent tubes will have two equivalent right- and left-hand helices. Owing to the six-fold rotation symmetry of the graphene lattice, chiral vectors that are a multiple of 60° apart are equivalent and the tubes having such chiral vectors are essentially the same [a (n,m) tube and a (m,n) tube are equivalent]. For these two reasons, chirality discussions of nanotubes in the literature are restricted to chiral angles $0 \leq \theta \leq 30°$. Figure 3.32 shows a scanning tunneling microscopy image of two SWCNTs, depicting the helix and the chiral angle of the tube.[185,186]

Depending on the chiral vector of the tube, or in other words depending on the chiral angle of the tube, the architectural configuration of C–C bonds in SWCNTs can be classified into three different categories: armchair tubes (the term *serpentine* tubes might also be encountered in the literature), zigzag tubes (the term *sawtooth* tubes might also be encountered in the literature), and chiral tubes. As shown in Figure 3.31, armchair tubes are formed when the chiral vector of the tube results in a chiral angle that equals 30°. In this case, the two tube vector multiples are equal to each other, i.e., $n = m$. All zigzag tubes will have a chiral angle equal to zero ($\theta = 0$) and their m vector multiple will always equal zero (Figure 3.31). Any other tube with a chiral angle $0 < \theta < 30°$ is classified as a chiral tube. Figure 3.33 shows the C–C bond arrangements in both zigzag and armchair tubes.

Hence a (10,10) tube is an arm chair tube, and a (8,0) tube is a zigzag tube. Any tube with mixed vector multiples, For example (7,8) is a chiral tube. To describe these three different arrangements in SWCNTs we can say that; in armchair tubes, the hexagons are pointing normal to the tube axis, in zigzag tubes, the hexagons are pointing along the tube axis, and in chiral tubes, the hexagons are pointing at an angle to the tube axis (Figure 3.33).

From Figure 3.31, it is clear that the unit vectors a_1 and a_2 have equal magnitude (length). This magnitude can be correlated to the C–C bond length

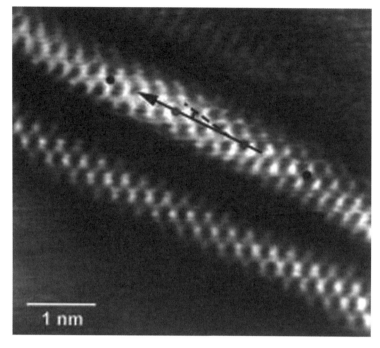

Figure 3.32 STM images of SWCNTs. The solid black arrow highlights the tube axis, and the dashed line indicates the zigzag direction. (Reproduced with kind permission from Hu *et al.*, ref. 186. Copyright American Chemical Society 1999.)

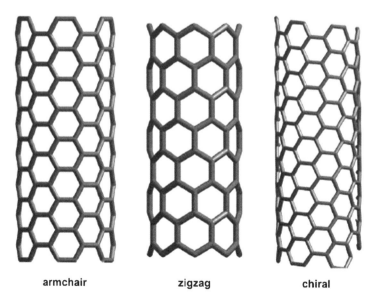

Figure 3.33 Configurations of C–C bonds in armchair, zigzag, and chiral single-walled carbon nanotubes.

in the graphene sheet. Taking the C–C bond length in the graphene sheet as 1.421 Å[187] leads to:

$$|a_1| = |a_2| \equiv a = 1.421\sqrt{3}\text{Å} = 2.461\text{Å} \qquad (3.6)$$

In addition, the length of the chiral vector (C_h) and, hence, the tube circumference (p) can be calculated using the unit vector magnitude (a) and the tube vector multiples as:

$$p = |C_h| = a\sqrt{n^2 + nm + m^2} \qquad (3.7)$$

Hence, the diameter of a SWCNT (d) can be calculated using its chirality or vector multiples as:

$$d = \frac{a}{\pi}\sqrt{n^2 + nm + m^2} \qquad (3.8)$$

Notably, the calculated tube diameter according to Equation (3.8) is, actually, a theoretical diameter calculated based upon pure geometrical considerations and an assumed length of the C–C bond. As mentioned during our discussion of zero-dimensional fullerene structures, the C–C bond length is known to change due to surface curvature as well as due to the presence of chemical potentials close to the fullerene molecules (Figures 3.13, 3.14, and 3.18). This, however, does not make Equation (3.8) a less valuable tool in obtaining acceptable *estimations* for the diameters of SWCNTs.

According to Equation (3.8), the diameter of SWCNTs can be readily estimated based on their vector multiples. For example, a (10,10) armchair nanotube would have a theoretical diameter of 1.37 Å. Also, a (9,0) zigzag nanotube should have a theoretical diameter of 7.15 Å. These values should also be increased roughly by 3.37 Å if the thickness of π-electron shells is to be accounted for. This, again, puts SWCNTs in the center of interest as nanosystems by themselves and makes them ideal as building blocks for nanostructured systems as well.

The chiral angle (θ) can also be defined using the tube chirality multiples as:[187–190]

$$\sin\theta = \frac{\sqrt{3}m}{2\sqrt{n^2 + nm + m^2}}; \quad \cos\theta = \frac{2n + m}{2\sqrt{n^2 + nm + m^2}} \qquad (3.9)$$

Note that the chirality of a SWCNT not only determines the tube's geometrical properties (such as diameter) but mostly its thermal, optical, and electronic properties as well.[127,185,187,188,191–199] Based on their chirality, the electronic band structure of nanotubes has been found to exhibit either metallic or semiconducting behavior. It has been shown that metallic conduction occurs when:

$$n - m = 3q \qquad (3.10)$$

where q is an integer (0, 1, 2, 3, ...). This means that all armchair single-walled carbon tubes should be metallic, while one-third of zigzag and chiral tubes are expected to be metallic. The reminder of SWCNTs is expected to be semiconducting. Metallic and semiconducting tubes are identified in Figure 3.31 according to their chirality.

The ends of SWCNTs are typically capped with one half of a fullerene molecule. The nature of the fullerene end cap depends on the diameter and chirality of the nanotube. Figure 3.34 shows different SWCNTs capped with different half-fullerene molecules.[200,201] As shown in the figure, one half of a C_{60} molecule can correctly cap each end of a (5,5) armchair tube. One half of a C_{70} molecule can correctly cap the ends of a (9,0) zigzag tube, and one half of a C_{80} fullerene molecule can correctly cap the ends of a (10,5) chiral nanotube. Owing to the large aspect ratio of SWCNTs (>1000), the end caps can be neglected in tube analysis without losing the generality of the treatment.[202] In addition, the same fullerene molecule can cap different types of SWCNTs. For example, the C_{60} molecule can cap both the (5,5) armchair and the (9,0) zigzag nanotubes.[203] Figure 3.35 shows the two types of SWCNTs that can be capped by a C_{60} molecule. Notably, the two different types of carbon nanotubes result essentially from the way the [60] fullerene molecule is bisected.

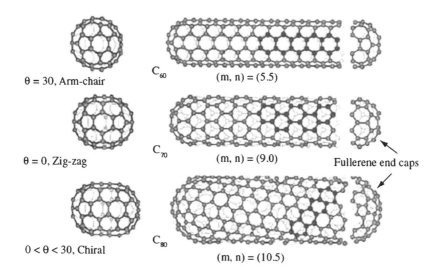

Figure 3.34 Fullerene molecule end caps compatible with different types of single-walled carbon nanotubes. (Reproduced with kind permission from Kalamkarov *et al.* ref. 200. Copyright Elsevier 2006.)

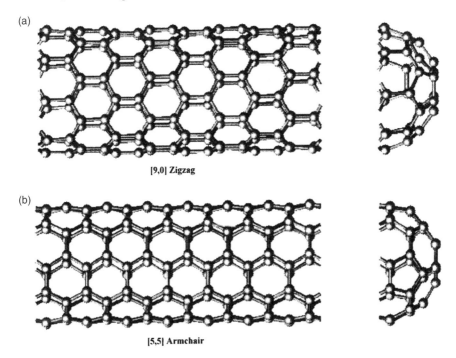

Figure 3.35 Single-walled carbon nanotubes of different chirality can be end capped by one half of a C_{60} molecule: (a) the zigzag (9,0) tube and (b) the armchair (5,5) tube.

If the fullerene molecule is bisected normal to a five-fold axis, the armchair configuration is formed. However, if the fullerene molecule is bisected normal to a three-fold axis, the zigzag configuration is formed.

Considering the SWCNTs as a one-dimensional fullerene, or a one-dimensional crystal, enables the definition of a unit cell that can be translated along the tube axis. For all tube types, such a translational unit cell is cylindrical in shape. The two-dimensional graphene sheet lattice model shown in Figure 3.36 is an excellent method to illustrate how the unit cell of a SWCNT can be constructed.[187-188,204-205]

As shown in Figure 3.36, the method involves drawing a straight line through the carbon atom selected as the origin (0,0), normal to the chiral vector (C_h), and extending this line until it intersects an exactly equivalent lattice point. The length of the tube transitional unit cell is the magnitude of the vector (**T**) as shown in Figure 3.36 for the case of a (6,3) nanotube. Figure 3.37 shows the transitional unit cell for (a) a (5,5) armchair tube and (b) a (9,0) zigzag tube. From Figures 3.36 and 3.37, the length of the translational unit cell clearly depends not only on the diameter of the tube, but also on the tube chirality. As we have shown before, while both the (5,5) and the (9,0) tubes are equal in diameter (both can be capped by half of a C_{60} molecule), the unit cell is longer

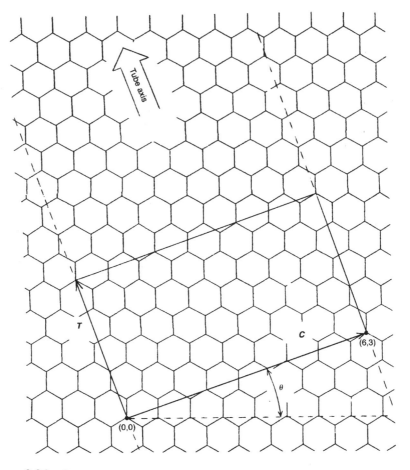

Figure 3.36 Construction of a nanotube unit cell; see text for details.

for the zigzag tube. While the length of the unit cell in the (5,5) armchair tube is equal to a [2.461 Å according to Equation (3.6)], the unit cell length equals $a\sqrt{3}$ (4.263 Å) in the case of the (9,0) zigzag tube (Figure 3.37).

Expressions have been derived[187,202,204,205] for the length of the unit cell (T) in terms of the magnitude of the chiral vector (C_h) and the highest common divisor (d_H) of the coordinate multiples n and m. It has been shown that:

$$T = \frac{C_h}{d_H}\sqrt{3}, \quad \text{if } n - m \neq 3qd_H \tag{3.11}$$

while:

$$T = \frac{C_h}{d_H}\frac{1}{\sqrt{3}}, \quad \text{if } n - m = 3qd_H \tag{3.12}$$

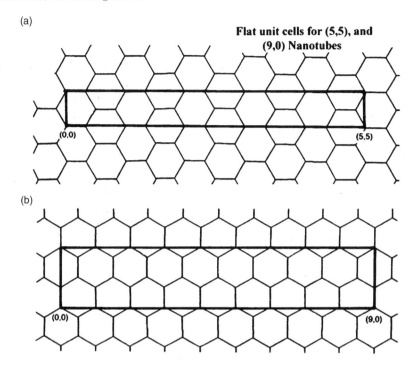

Figure 3.37   Unit cells for a (5,5) armchair nanotube (a) and a (9,0) zigzag nanotube (b) shown on a flat graphene sheet.

It has also been shown that for a tube specified by (n,m) the number of carbon atoms per unit cell of a tube is $2N$ such that:

$$N = \frac{2(n^2 + nm + m^2)}{d_H} \quad \text{if } n - m \neq 3qd_H \tag{3.13}$$

and:

$$N = \frac{2(n^2 + nm + m^2)}{3d_H} \quad \text{if } n - m = 3qd_H \tag{3.14}$$

where q is, again, an integer.

These expressions enable the determination of the unit cell length and number of carbon atoms therein. For typical experimentally observed SWCNTs, the diameters range between 1 and 30 nm. This makes unit cells very large. For example, a 10 nm diameter tube would have a unit cell length in the range of 50–55 nm, containing around 65 000 carbon atoms. Such large unit cells render the molecular simulation of large nanotubes a very computationally expensive task. While the details of simulation investigation of different types of fullerenes is beyond the scope of this book, it is important to mention that the computationally expensive nature of molecular simulation, especially of carbon nanotubes, has led to the utilization of many other methods of simulations such as *molecular structural mechanics*. In the molecular structural mechanics approach, a SWCNT is simulated as a space frame structure, with the covalent bonds and the carbon atoms as connecting beams and joint nodes, respectively.[190,200] While such an approach could yield approximate estimates for some physical constants and properties of carbon nanotubes, it can be seriously misleading if used to predict the physics of a nanosystem based on our discussion of these systems presented in Chapter 1.

We have mentioned earlier that the smallest zero-dimensional fullerene is the C_{20} molecule. The question regarding the smallest one-dimensional fullerene or carbon nanotube has also been raised. In 2000, Qin *et al.*[206,207] reported the discovery of the smallest carbon nanotube (0.4 nm in diameter) confined inside an 18-shell carbon nanotube. Figure 3.38 shows the transmission electron micrograph and a superimposed atomic model of the nanotube.

Figure 3.38 High-resolution transmission electron micrograph and a superimposed atomic model of a 0.4 nm diameter nanotube confined in an 18 shell MWCNT. (Reproduced with kind permission from Qin *et al.* ref. 206. Copyright Nature Publishing Group 2000.)

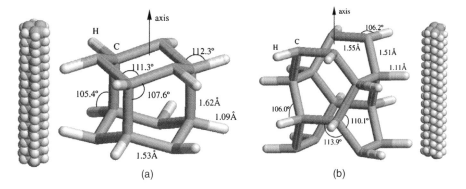

Figure 3.39 Calculated, relaxed structures of (a) (3,0) and (b) (2,2) sp^3 nanotubes, showing both a single unit cell and a space-filling model of the tubular structures. (Reproduced with kind permission from Stojkovic et al., ref. 208. Copyright American Physical Society 2001.)

In 2001, Stojkovic and co-workers described how sp^2 carbon in regular nanotubes can be replaced by sp^3 carbon to produce extremely small-diameter (in the range of 0.4 nm) carbon nanotubes with only minimal bond-angle distortion. These tubes were shown to have chiral multiples of either a zigzag (3,0) or an armchair (2,2), and were predicted to have a stiffness higher than that of traditional sp^2-bonded carbon nanotubes, therefore forming the stiffest one-dimensional systems known.[208] Figure 3.39 shows the calculated, relaxed structures of (a) the (3,0) and (b) the (2,2) sp^3 nanotubes – both a single unit cell and a space-filling model of the tubular structures. More recently, however, Zhao et al. argued, based on experimental and simulation results, that the (2,2) armchair nanotube is in fact 3 Å in diameter.[209]

The symmetry of SWCNTs has also been thoroughly investigated and reported.[187,205,210–217] While the symmetry of all-carbon nanotubes is described by the space groups in one-dimensional space groups[218,219] (line groups, see Section 2.7.3.1), SWCNTs can still be classified into two distinct groups from a symmetry viewpoint. One group, which includes armchair and zigzag nanotubes, can be represented by *symmorphic* space groups, meaning that none of the symmetry operations requires rotation and translation operations to be combined. This group is referred to as *achiral* tubes. The second group, which includes chiral tubes, can be represented by *nonsymmorphic* space groups in which some of the symmetry operations involve both rotation and translation operations. Figure 3.40 shows the symmetry elements for chiral, zigzag, and armchair carbon nanotubes. As shown in the figure, all carbon nanotubes possess rotation axes normal to their surface passing through the hexagon center and the midpoint of C–C bonds. In addition, only achiral (zigzag and armchair) nanotubes possess reflection, glide, and rotation–reflection symmetry elements.

Regarding the symmetry groups for achiral SWCNTs (*n,n*) or (*n,0*), it was shown that since all such tubes possess rotation symmetry axes in addition to mirror and/or glide planes (Figure 3.40), they belong to either D_{nh} or D_{nd} groups. Assuming that all achiral tubes have an inversion center, it follows that achiral tubes with n even belong to the D_{nh} group, and that achiral tubes with n odd belong to the D_{nd} group. Hence, a (5,5) armchair or a (5,0) zigzag tube should belong to the D_{5d} symmetry group. Also, a (6,6) armchair or a (6,0) zigzag tube belong to the D_{6h} symmetry group. Tables 3.4 and 3.5 show the character tables for n odd and even achiral SWCNTs, respectively.

Regarding the symmetry of chiral (n,m) nanotubes the determination of the symmetry group is not as straightforward as in the case of achiral tubes. Here we will only outline the general method used to determine the symmetry group. The reader is referred to more elaborated work on the subject[210,213,215,217,220] if more details are needed. Since chiral tubes do not possess any mirror planes, their symmetry should belong to the C symmetry groups. In addition, due to their nonsymmorphic symmetry, their basic symmetry operation $R = (\psi|\tau)$ involves a rotation by an angle ψ followed by a translation τ. This operation corresponds to the vector $\vec{R}$ such that:

$$\vec{R} = p\vec{a}_1 + q\vec{a}_2 \qquad (3.15)$$

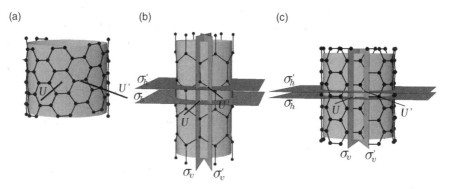

Figure 3.40 Symmetry elements for chiral (8,6) nanotube (a), zigzag (6,0) (b), and armchair (6,6) (c) nanotubes. U and U' denote rotation axes, σ denotes mirror planes, σ'_v is a glide plane, and σ'_h is a rotation–reflection plane. (Reproduced with kind permission from Damnjanovic *et al.* ref. 220. Copyright American Physical Society 1999.)

Table 3.4 Character table for D_{2j+1} point group (odd n achiral) single-walled nanotubes.

Mode	E	$2C^1_{\phi j}$	$2C^2_{\phi j}$	...	$2C^j_{\phi j}$	$(2j+1)C'_2$
A_1	1	1	1	...	1	1
A_2	1	1	1	...	1	−1
E_1	2	$2\cos|\phi_j$	$2\cos 2|\phi_j$	...	$2\cos|j\phi_j$	0
E_2	2	$2\cos|2\phi_j$	$2\cos|4\phi_j$	...	$2\cos|2j\phi_j$	0
⋮	⋮	⋮	⋮	⋮	⋮	⋮
E_j	2	$2\cos|j\phi_j$	$2\cos|2j\phi_j$	...	$2\cos|j^2\phi_j$	0

Table 3.5 Character table for D_{2j} point group (even n) achiral single-walled nanotubes.

Mode	E	C_2	$2C_{\phi j}^1$	$2C_{\phi j}^{12}$	...	$2C_{\phi j}^{j-1}$	$(2j)\ C_2'$	$(2j)\ C_2''$
A_1	1	1	1	1	...	1	1	1
A_2	1	1	1	1	...	1	−1	−1
B_1	1	−1	1	1	...	1	1	−1
B_2	1	−1	1	1	...	1	−1	1
E_1	2	−2	$2\cos|\phi_j|$	$2\cos 2|\phi_j|$	...	$2\cos|2(j-1)\phi_j|$	0	0
E_2	2	2	$2\cos|2\phi_j|$	$2\cos|4\phi_j|$	...	$2\cos|2(j-1)\phi_j|$	0	0
⋮	⋮	⋮	⋮	⋮	⋮	⋮	⋮	⋮
E_{j-1}	2	$(-1)^{j-1}2$	$2\cos|(j-1)\phi_j|$	$2\cos|2(j-1)\phi_j|$	...	$2\cos|(j-1)^2\phi_j|$	0	0

here, (p,q) are the coordinates reached when the symmetry operation $(\psi|\tau)$ acts on an atom at the coordinates $(0,0)$. The values of p and q are given by:[187,217]

$$mp - nq = d_H \quad (3.16)$$

with the condition that: $q < m/d_H$ and $p < n/d_H$ it can be shown that: $\psi = 2\pi \frac{\Omega}{Nd_H}$ and $\tau = \frac{Td_H}{N}$ where Ω is defined as:

$$\Omega = p(9m + 2n) + q(n + 2m)/((d_H/d_R)) \quad (3.19)$$

The value of d_R depends on the tube chirality. For tubes where $n - m$ is not a multiple of $3d_H$, $d_R = d_H$; for tubes where $n - m$ is a multiple of $3d_H$, $d_R = 3d_H$. For chiral tubes with $d_H = 1$, the symmetry group belongs to the $C_{N/\Omega}$ group. For chiral tubes with $d_H \neq 1$, the symmetry belongs to the group expressed by the direct product $C_{dH} \otimes C_{N/\Omega}$. Tables 3.4 and 3.5 show the character tables for achiral SWCNTs with odd and even n, respectively.[187,188]

3.9.2 Multi-walled Carbon Nanotubes (MWCNTs)

As we discussed earlier, the first carbon nanotubes to be discovered were multi-walled nanotubes (MWNTs). As shown in Figures 3.4 and 3.38, multi-walled carbon nanotubes MWCNTs consist of multiple concentric single-walled nanotubes one inside the other. Typically, inner diameters of readily available MWCNTs range between 1 and 3 nm while outer diameters range between 2 and 20 nm.[221,222] Multi-walled carbon nanotubes with larger outer diameters reaching hundreds of nanometers also exist. The important question regarding the structural correlation between the successive layers in a MWCNT has been addressed and investigated.[223,224] For MWCNTs, the interlayer distance between concentric tubes was measured using high-resolution transmission electron microscopy and was reported to be around 3.4 Å.[39] This means that the circumference of successive tubes should increase by about 21 Å. It can be easily seen that this is not possible for zigzag tubes that require the successive

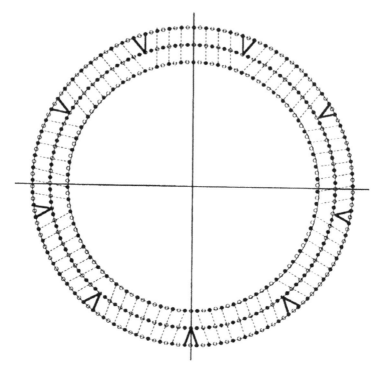

Figure 3.41 Schematic of a three-layer zigzag MWCNT as proposed by Zhang et al., illustrating how interfacial dislocations (bold lines) can be introduced to accommodate hoop stresses. (Reproduced with kind permission from Zhang et al., ref. 223. Copyright Elsevier 1993.)

tubes circumference to increase by precise multiples of 0.246 nm, the width of a single hexagon. For the zigzag MWCNTs, the closest approximation for the measured interlayer separation is when successive tubes circumferences differ in nine hexagons, producing an inter-tube separation of 3.52 Å. In this case, the traditional ABAB stacking of perfect graphite can only be maintained over short distances in the circumferential direction. Figure 3.41 schematically illustrates a three-layer zigzag MWCNT as proposed by Zhang.[223] The figure illustrates how interfacial dislocations can be introduced to accommodate hoop stresses.

In the case of armchair tubes, however, the length of the repeating unit is 0.42 nm, and five-times that would be very close to the required increase in the successive tube circumference (2.1 nm). Hence, multi-walled armchair structures can be assembled in which the ABAB graphitic stacking sequence is maintained, keeping the interlayer close to 3.4 Å. For chiral tubes, the restrictions on multi-walled tubes are more complicated and, generally speaking, it was concluded that it is not possible to have two consecutive layers having exactly the same chiral angle and separated by the exact interlayer

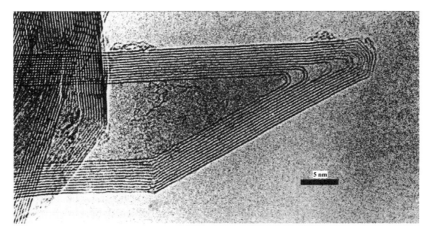

Figure 3.42 High-resolution transmission electron micrograph of a typical MCNT end cap. (Reproduced with kind permission from Harris et al., ref. 156. Copyright Royal Society of Chemistry 1994.)

spacing of 3.4 Å.[156,202,225] End caps in MWCNTs are not as simple or well understood as those in SWCNTs.[226,227] Figure 3.42 shows a typical MWCNT end cap.

A special case of MWCNTs that has captured a lot of attention is the double-walled carbon nanotubes (DWCNTs). While DWCNTs were among the first nanotube to be reported (Figure 3.4b), attention was really paid to their properties later as the field of nanotechnology started to be better understood. With their mean outer diameter ranging between 9 and 18 Å and with only two tubes, DWCNTs provide an excellent model system and an opportunity to investigate the physics of nanotubes on both experimental and theoretical levels.[228–235] They are also expected to exhibit unique properties compared to those of single- and multi-walled carbon nanotubes.[233,236] Such properties include mechanical, thermal, electrical, and chemical properties, leading to a unique ability to store hydrogen for energy applications and a unique ability to be functionalized with certain amino acids.[234,237–252]

The fact that DWCNTs are attracting a lot of attention due to their unique properties compared to these of single- and multi-walled carbon nanotubes should, indeed, elucidate the core idea of nanosystems and nanotechnology as we discussed in Chapter 1. From the surface nature viewpoint, single-, double-, and multi-walled carbon nanotubes should be essentially identical and similar to other forms of fullerenes. However, the physics of that particular cylindrical fullerene with only two layers, and outer diameter ranging between 0.9 and 1.8 nm (the double-walled nanotubes) is still unique. It must be that the uniqueness of DWCNTs is because the system scale is resonating with the important length scales defining nanosystems as we discussed before. This point will be further discussed as we examine the properties of fullerene building blocks and fullerene-based nanostructured systems in Chapter 4.

3.9.3 Production of Carbon Nanotubes

The methods of carbon nanotube production can mainly be classified into two groups. One group is based upon sublimation/condensation of graphite and the other is based upon pyrolysis of hydrocarbons. In both categories the processes involved are catalytic in nature. In the sublimation/condensation based methods, the energy required for the sublimation process is provided by different sources such as arc-discharge, laser radiation, resistive heating, or by concentrated solar radiation. The major disadvantages of all methods of this group include large energy consumption, relatively low yield of nanotubes, high cost due to the high cost of graphite, complications due to the required high vacuum and controlled atmosphere, and the difficulties of automation and up-scaling. Pyrolysis-based methods, in contrast, are relatively much less complicated. This group of methods does not require high temperatures, can proceed under atmospheric pressure, can be run continuously, and provides a comparatively high yield. In addition, pyrolysis-based methods enable better control of the nanotube diameter and produce longer nanotubes than those produced by sublimation/condensation methods. The major disadvantage of pyrolysis methods is that they produce lower quality nanotubes having structural imperfections.

In the following sections, we first discuss different production methods belonging to the sublimation/condensation group as well as those belonging to the pyrolysis group.

3.9.3.1 Arc-discharge Method

Original production[39,253] of carbon nanotubes was carried out in the same *arc-discharge* method used to produce fullerenes (Figure 3.20) where a DC arc is used to evaporate carbon electrodes in vacuum or inert atmosphere. This method is one of the sublimation/condensation production methods. The only difference is that in nanotube production the carbon electrodes were kept apart at a distance instead of being in continuous contact as in the case for fullerene production. The initial carbon nanotube yield using this method was rather poor. Subsequent studies[222,254–257] showed that several factors are important in producing a high yield and good quality of carbon nanotubes. These factors included the pressure of the inert atmosphere inside the evaporation chamber, the stability of the plasma field, quality and composition of graphite electrodes, and efficient cooling of the electrodes and chamber. The most important factor of these is the pressure of helium inside the production chamber, which appears to be optimized at 500 Torr.[254] Interestingly, in this method, if the helium pressure was kept under 100 Torr the process mainly produces [60] fullerene molecules. This point elucidates the extremely important role played by the production atmosphere in determining which form of fullerene is most stable, and hence is the main product of the process. The fact that simulation studies (usually conducted in a pure vacuum) showed that spheroidal geometries of giant fullerenes are more favorable than tubular geometries (Section 3.6) might be clouded by the eliminated atmosphere factor.

Regarding the arc current, it is now known that maintaining the current as low as possible enables maintaining stable plasma and increases the nanotube production yield. In addition, a graphite rod containing metal catalyst (Ni, Fe, Co, Pt, Pd, *etc.*) used as the anode with a pure graphite cathode was found to increase the nanotube yield.[43,117,198]

The maximum yield is achieved with Ni-Co. It was also observed that the addition of sulfur increases the catalytic effect of other metal catalysts.[258,259] Efficient cooling has also been shown to enable avoiding excessive sintering of the soot – an essential condition for good quality nanotubes. SWCNTs have also been obtained in high yield in the arc discharge method by using a bimetallic Ni-Y catalyst in a helium atmosphere.[260] The efficiency by this method was sharply increased by positioning the two graphite electrodes at 30° angle instead of the conventional 180° alignment. This new arrangement is referred to as the *arc-plasma-jet method* and it typically yields SWCNTs at a rate of 1 g min^{-1}.[261]

The morphology of the produced nanotube was also found to depend on the nature of the chamber atmosphere. MWCNTs were primarily formed in a methane atmosphere.[262] The important role of hydrogen presence in the formation and production of MWCNTs was confirmed by multiple studies.[263–266] Interestingly, round fullerenes will not form in an atmosphere of a gas including hydrogen atoms.[267] Water as an atmosphere for the arc discharge method was also used to produce well graphitized MWCNTs.[268] In addition, a new morphology of carbon nanostructures in the shape of a petal was found to form in such an atmosphere.[264] The production of double-walled carbon nanotubes (DWCNTs) by arc discharge method in a hydrogen atmosphere under similar operating conditions has also been reported.[269,270] Production of high quality double- and single-walled nanotubes was achieved by a *high temperature pulsed arc discharge* method that utilizes a DC pulsed arc discharge located inside a furnace to maintain homogeneous production condition.[271,272]

Carbon nanotubes grown by the arc discharge method are, in general, not aligned but are rather produced in the form of a web. Partially aligned SWCNTs can be grown by convection[273] or a directed arc plasma method.[274] Because of the high production temperature of the arc discharge method, the produced carbon nanotubes are expected to be of high quality.

3.9.3.2 Other Condensation Methods

Carbon nanotubes could also be produced through the condensation of carbon vapor in the absence of an electric field. The process starts with the evaporation of a graphite target and then the carbon vapor would condense into carbon nanotubes, mainly single-walled, on a cooled substrate under the appropriate conditions. Earliest reports involving such a method for carbon nanotubes production came from a group at the Russian Academy of Science who used an electron beam to evaporate graphite in high vacuum (10^{-6} Torr).[275,276] The carbon vapor condensed on a quartz substrate producing, among other forms, what was later described as imperfect MWCNTs.[277] The condensation of carbon vapor was also used to produce carbon nanotubes after evaporating carbon by resistively heating carbon foils and condensing the vapor on freshly cleaved highly oriented pyrolytic graphite (HOPG) substrate under high vacuum (10^{-8} Torr). The so-produced nanotubes were essentially SWCNTs with diameters ranging between 1 and 7 nm.[278,279] Laser beams (typically a YAG or a CO_2 laser) were also used to evaporate carbon in an inert gas atmosphere to produce carbon nanotubes by the same method.[280–282] Concentrated sunlight was also reported as an applicable graphite sublimation method to produce carbon nanotubes.[283–285]

3.9.3.3 HiPco Process and Other Pyrolytic Methods

For over a century, it has been known that filament-like carbon species can be produced by catalytic decomposition of carbon-containing gas on a hot surface. This phenomenon was first discovered by the Schultzenbergers in 1890 while experimenting with the passage of cyanogens over red-hot porcelain.[202,286] The phenomenon was re-investigated in the 1950s and it was established that carbonous filaments can be produced by the interaction of a wide range of hydrocarbon and other carbon-containing gases (such as CO) with metal surfaces such as iron, nickel, platinum, and cobalt.[287–289] In the case of carbon monoxide source, the technique mainly depends on the CO disproportionation reaction that can be described as follows:

$$2CO_{(g)} \Leftrightarrow C_{(s)} + CO_{2(g)} \qquad (3.20)$$

The very early, and unrealized, production of carbon nanotubes by Endo and co-workers (Figure 3.4a) from benzene vapor was based on this process. In 1999, Mukhopadhyay et al. reported and described the optimization details of a large-scale production method of quasi-aligned carbon nanotube bundles.[290] The method includes catalytic decomposition of acetylene over well-dispersed metal particles embedded in commercially available zeolite at a lower temperature (in the range of 700 °C). The produced tubes were mainly homogeneous, perfectly graphitized multi-walled tubes with inner diameter ranging from 2.5 to 4 nm, and an outer diameter ranging from 10 to 12 nm.

In 2001 Smalley's group at Rice University rediscovered the technique and developed a new method that they called "HiPco" that utilized high-pressure carbon monoxide as a precursor for SWCNT mass production.[291] In this process SWCNTs are grown in a high-pressure (in the range 30–50 atm), high-temperature (in the range 900–1100 °C), flowing CO atmosphere. Iron is added to the gas flow in the form of iron pentacarbonyl [$Fe(CO)_5$]. Upon heating, the $Fe(CO)_5$ decomposes and the iron atoms condense into clusters. These clusters serve as catalytic particles upon which SWCNTs nucleate and grow in the gas phase via the CO disproportionation reaction. The technique produces highly pure SWCNTs at rates of up to 450 mg h^{-1}.

Since then, many other pyrolytic methods have been described in which carbon nanotubes can be produced. For example, the modified, self-regulated arc discharge method has also been used to decompose liquid hydrocarbons in carbon nanotube structures.[292] An electric furnace has also been used to commercially produce carbon nanotubes from toluene.[293] More recently, production methods based upon decomposition of ethanol (with ferrocene as a catalyst) in a vertical furnace,[294] and over silica coated cobalt catalyst[295] have been reported. Acetylene diluted with argon has also been reported to be a good precursor for pure SWCNT production.[296] The current status of carbon nanotube production by pyrolytic methods was recently reviewed by E. Rakov.[297]

Purification of carbon nanotubes[78,298–303] is, typically, based upon oxidative treatment using oxygen or other oxidants such as nitric acid or hydrogen peroxide to remove carbonaceous impurities as well as metal catalyst impurities. To reduce carbon nanotube damage due to oxidative purification treatments, other acidic treatments were suggested using hydrochloric acid (HCl). Such alternative acidic treatments, while shown to be less damaging to the nanotubes, are considered less effective in removing the impurities. More recently, a room temperature, liquid bromine based treatment was reported to be more effective in removing impurities and less damaging to the nanotubes than other treatments. Table 3.6 shows different carbon nanotube purification treatments reported in the literature.[299,304–308] Figure 3.43 shows an electron micrograph of carbon nanotubes before and after purification.

3.10 Two-dimensional Fullerenes – Graphene

Two-dimensional fullerenes are simple single or few layers of graphene sheets. This type of fullerene was the latest to be prepared and investigated. The physics of single or few layers of graphene sheets have been shown to be very unique and vastly different from the physics of bulk layered graphite.[58] Despite Landau[309] and Peierls[310] argument, more than half a century ago, stating that strictly two-dimensional (2D) crystals are thermodynamically unstable and cannot exist, and despite Mermin's[311] confirmation, in 1968, of the argument and ample experimental support, single and few layers of graphene were experimentally observed and reported less than 5 years ago.[44–45,58,312] Thermodynamically, the melting temperature of free standing thin films should rapidly decrease with decreasing film thickness. Once the film thickness reaches a limit of, typically, a few tens of atomic distances, the film should become unstable, *i.e.*, decompose or transform into particles (3D structures) unless it is part of a three-dimensional system, *i.e.*, grown on a substrate.[313–315] A scientific debate is currently active regarding the exact thermodynamic bases of stability for experimentally observed graphene ribbons. Furthermore, several other questions regarding the reasons behind the experimentally observed wrinkles, and the exact chirality of the film edges (zigzag or armchair), are still being investigated. The superior electrical and thermal properties of graphene

Table 3.6 Carbon nanotube purification methods used in the literature.

Purification procedure	Reference	Weight % iron remaining after purification (lit., %)	Weight % iron remaining after purification[a] (Ref. 299)	NIR luminescence intensity[b] (Ref. 299)
HCl (35%), 4 h at 60 °C	308	<1	10.2–14.4	Good
Microwave radiation for 2 min, then HCl (35%), 4 h at 60 °C	304	9	10.8–12.6	Good
Microwave radiation for 20 min, then HCl (35%), 4 h at 60 °C	304	7	10.5–12.5	Good
H_2SO_4 (98%) + HNO_3 (70%), 4 h at 60 °C	306	N/A	9.7–14	Poor
H_2SO_4 (25%), 10 min at 20 °C[c]	306	N/A	13.9–14.8	Good
HNO_3 (10%), 4 h at 60 °C	307	<1	0.6–0.8	Poor
$O_2 + SF_6$(g) at 200–400 °C for 3–7 days, then HCl (35%), 12 h at 60 °C	305	1.5	3.0	Good
Br_2 (l), room temp, 4 h	299	N/A	2.8–3.6[d]	Good
			1.6–1.8[e]	Poor

[a] Iron analysis by ICP-AE as described in Ref. 299.
[b] Comparison with the NIR luminescence intensity for the original unpurified, raw SWCNTs (100%); an intensity greater than 25% of the original is considered as "good," below 5% as "poor"; excitation laser at 660 nm with luminescence observed in the 900–1600 nm range.
[c] H_2SO_4 mixed with H_2O_2 (30%) at 0 °C then diluted with water.
[d] After a first purification cycle.
[e] After a second purification cycle.

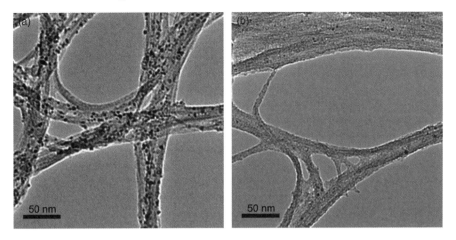

Figure 3.43 Transmission electron micrograph of (a) as-produced SWCNTs and (b) bromine-purified SWCNTs purified to a residual iron content of *ca.* 3% by weight. (Reproduced with kind permission from Mackeyev *et al.*, ref. 299. Copyright Elsevier 2007.)

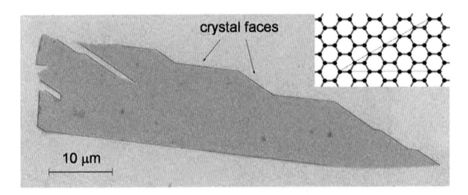

Figure 3.44 Scanning-electron micrograph of a relatively large graphene crystal, which shows that most of the crystal's faces are zigzag and armchair edges, as indicated by blue and red lines and illustrated in the inset. (Reproduced by kind permission from Geim and Novoselov, ref. 58. Copyright Nature Publishing Group 2007.)

triggered a "graphene rush".[312,316–342,47,49–54,56,59] In fact, publication statistics show that over 1500 papers have been published with the term "graphene" in their title since 2004, which is roughly one paper per day, reflecting the potential of such building blocks in future nanotechnology. Figure 3.44 shows a scanning-electron micrograph of a relatively large graphene crystal, which shows that most of the crystal's faces are zigzag or armchair edges as indicated by blue and red lines and illustrated in the inset.

The atomic structure of single-layered graphene sheets on insulating SiO_2 substrates (the most commonly used device configurations) has been investigated using atomic resolution scanning tunneling microscopy (STM).[320] While the hexagonal lattice symmetry of graphene was confirmed, atomic resolution STM images reveal the presence of triangular symmetries, which break the hexagonal lattice symmetry of the graphitic lattice. In addition, structural corrugations of the graphene sheet partially conform to the underlying silicon oxide substrate, emphasizing the important of substrate topography on the configuration, and hence the performance, of graphene sheets. Figure 3.45 shows a scanning tunneling microscopy micrograph depicting both hexagonal and triangular lattices experimentally observed in a single-layered graphene sheet on SiO_2. Other point defects and deviations from hexagonal symmetry can also be observed.

It is widely known that point defects (or imperfections) such as vacancies, substitutional, or interstitial imperfections and topological defects are usually associated with lattice perturbation. In a nanosystem, like graphene, such perturbations can be significant enough to change the electronic and the vibrational properties of the system, and therefore can modify the thermodynamic

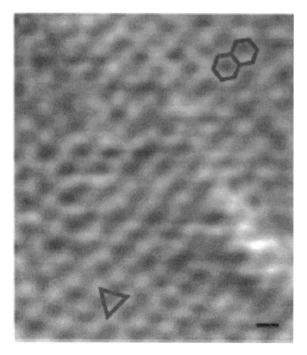

Figure 3.45 Scanning tunneling microscopy micrograph depicting both hexagonal and triangular lattices experimentally observed in a single-layered graphene sheet on SiO_2. Other point defects and deviations from hexagonal symmetry can also be observed. (Reproduced with kind permission from Ishigami *et al.*, ref. 320. Copyright American Chemical Society 2007.)

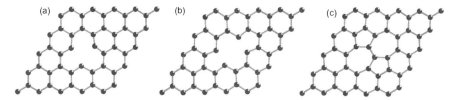

Figure 3.46 Optimized atomic structure of graphene with three different point defects: (a) a mono-vacancy (MV) defect, (b) a di-vacancy (DV) defect, and (c) a Stone–Wales (SW) defect. The latter two structures are almost flat, while for the MV one the unpaired atom has a large out-of-plane displacement of 0.49 Å. (Reproduced with kind permission from Popov et al., ref. 343. Copyright Elsevier 2009.)

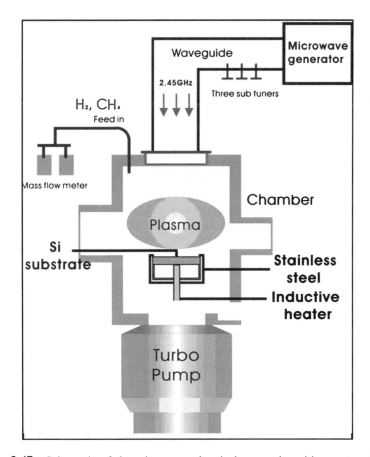

Figure 3.47 Schematic of the microwave chemical vapor deposition system for the growth of graphene sheets. (Reproduced with kind permission from Yuan et al., ref. 47. Copyright Elsevier 2009.)

166 Chapter 3

properties. In 2009, Popov et al.[343] investigated, using density functional theory (DFT) calculations, three different possible types of point defects and their effect on electronic and optical performance of single graphene sheets. The three different types of defects investigated were a mono-vacancy (MV), a di-vacancy (DV), and a Stone–Wales (SW) defect (described in Section 3.4). Figure 3.46 shows the optimized atomic structure of graphene with three different point defects: (a) a MV defect, (b) a DV defect, and (c) a SW defect. All three

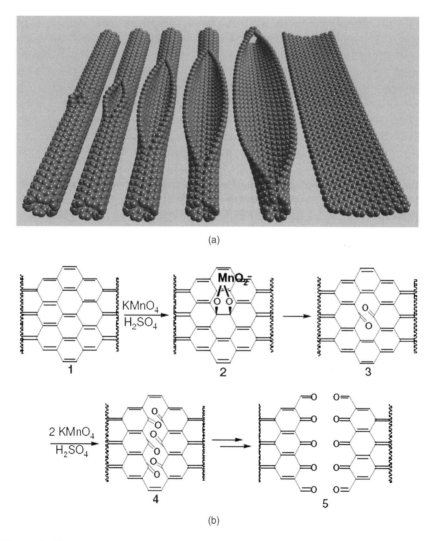

Figure 3.48 (a) Schematic of the longitudinal unzipping of a SWCNT into a single graphene sheet. (b) Proposed chemical mechanism of nanotube unzipping into graphene sheets. (Reproduced with kind permission from Kosynkin et al., ref. 335. Copyright Nature Publishing Group 2009.)

Fullerenes, the Building Blocks

structures are almost flat, except for the MV one, where the unpaired atom has a large out-of-plane displacement of 0.49 Å.

The defects were found to modify the electronic structure and the phonons of graphene, giving rise to new optical transitions and defect-related phonons. In addition, DFT and quantum Monte Carlo simulations of Stone–Wales defects in graphene reported by Ma *et al.*[336] reveal that the structure of the Stone–Wales defect in graphene is more complex than thus far appreciated. In their report, Ma *et al* pointed out that rather than being a simple in-plane transformation of two carbon atoms, Stone–Wales defects result in out-of-plane wave-like defect structures that extend over several nanometers.[336] To this end, graphene is clearly another frontier to be explored.

Graphene has been recently prepared by several approaches.[51,58,321,344] Some of these approaches, including several lithographic,[318,319] chemical,[317,329] and other synthetic procedures,[332] are known to produce microscopic samples of graphene. Macroscopic quantities of graphene were also produced using conventional[326] as well as microwave plasma chemical vapor deposition,[47] in addition to traditional liquid-phase exfoliation of graphite.[53] Figure 3.47 shows a schematic of the recently (2009) developed microwave plasma chemical vapor deposition instrument used to grow macroscopic quantities of graphene sheets.

Another unique approach to preparing graphene is by longitudinal unzipping of single- and multi-walled carbon nanotubes using oxidation treatment.[335] This approach is very well suited for research purposes since it enables the preparation of very well controlled samples for scientific investigations. The oxidation treatment involves suspending carbon nanotubes in concentrated sulfuric acid followed by treatment with 500 wt% KMnO$_4$ for 1 h at room temperature (22 °C). Figure 3.48(a) shows a schematic of the unzipping process, while Figure 3.48(b) shows the oxidation chemical reaction leading to the longitudinal unzipping of the nanotube into a graphene sheet. Figure 3.49 shows transmission electron micrographs depicting the transformation of MWCNTs into oxidized graphene sheets. The graphene sheet yield using

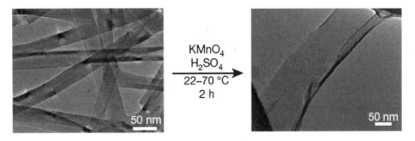

Figure 3.49 Transmission electron image depicting the transformation of MWCNTs (left) into oxidized graphene sheets (right). (Reproduced with kind permission from Kosynkin *et al.*, ref. 335. Copyright Nature Publishing Group 2009.)

this method is almost 100% and, hence, no further purification process is required.

We have discussed the structure, production methods, and purification methods mainly of the fullerene building blocks. In the next chapter, we discuss the properties of such fullerene building blocks individually and the unique properties of nanostructured systems based on such fullerene building blocks.

References

1. J. Delgado, M. Herranz and N. Martín, *J. Mater. Chem.*, 2008, **18**, 1417.
2. J. Liu, H. Dai, J. H. Hafner, D. T. Colbert, R. E. Smalley, S. J. Tans and C. Dekker, *Nature*, 1997, **385**, 780.
3. A. G. Nasibulin, P. V. Pikhitsa, H. Jiang, D. P. Brown, A. V. Krasheninnikov, A. S. Anisimov, P. Queipo, A. Moisala, D. Gonzalez and G. Lientschnig, *Nat. Nanotechnol.*, 2007, **2**, 156.
4. A. G. Nasibulin, A. S. Anisimov, P. V. Pikhitsa, H. Jiang, D. P. Brown, M. Choi and E. I. Kauppinen, *Chem. Phys. Lett.*, 2007, **446**, 109.
5. M. Remskar, M. Virsek and A. Jesih, *Nano Lett.*, 2007, **8**, 76.
6. Y. Tian, D. Chassaing, A. G. Nasibulin, P. Ayala, H. Jiang, A. S. Anisimov and E. I. Kauppinen, *J. Am. Chem. Soc.*, 2008, **130**, 7188.
7. X. Wu and X. C. Zeng, *ACS Nano*, 2008, **2**, 1459.
8. S. Berber, Y.-K. Kwon and D. Tománek, *Phys. Rev. Lett.*, 2002, **88**, 185502.
9. A. Gloter, K. Suenaga, H. Kataura, R. Fujii, T. Kodama, H. Nishikawa, I. Ikemoto, K. Kikuchi, S. Suzuki and Y. Achiba, *Chem. Phys. Lett.*, 2004, **390**, 462.
10. R. Almairac, J. Cambedouzou, S. Rols and J. L. Sauvajol, *Eur. Phys. J. B*, 2006, **49**, 147.
11. S. Bandow, M. Takizawa, H. Kato, T. Okazaki, H. Shinohara and S. Iijima, *Chem. Phys. Lett.*, 2001, **347**, 23.
12. I. V. Krive, R. I. Shekhter and M. Jonson, *Low Temperature Phys.*, 2006, **32**, 887.
13. D. E. Jones, *New Sci.*, 1966, **32**, 245.
14. D. E. H. Jones, *The Inventions of Daedalus: A Compendium of Plausible Schemes*, Freeman, Oxford, 1982.
15. E. Osawa, (in Japanese) *Kagaku* (*Kyoto*) 25 854-863 (1970); *Chem. Abstr.*, 1971, **74**, 698.
16. Z. Yoshida and E. Osawa, (in Japanese) *Aromaticity*, Kagakudojin, Kyoto, 1971, p. 174.
17. D. A. Bochavar and E. G. Gal'pern, *Dokl. Acad. Nauk SSSR, Chem., Engl.*, 1972, **209**, 239.
18. R. A. Davidson, *Theor. Chim. Acta*, 1981, **58**, 193.
19. H. Kroto, *Chem. Soc. Rev.*, 1982, **11**, 435.

20. T. G. Dietz, M. A. Duncan, D. E. Powers and R. E. Smalley, *J. Chem. Phys.*, 1981, **74**, 6511.
21. H. W. Kroto, J. R. Heath, S. C. O'Brien, R. F. Curl and R. E. Smalley, *Nature*, 1985, **318**, 162.
22. H. W. Kroto, *Nature*, 1987, **329**, 529.
23. H. Kroto, *Science*, 1988, **242**, 1139.
24. H. W. Kroto and K. McKay, *Nature*, 1988, **331**, 328.
25. H. W. Kroto, A. W. Allaf and S. P. Balm, *Chem. Rev.*, 1991, **91**, 1213.
26. H. Kroto, *Nature*, 1986, **322**, 766.
27. R. Ettl, I. Chao, F. Diederich and R. L. Whetten, *Nature*, 1991, **353**, 149.
28. Y. Quo, N. Karasawa and W. A. Goddard, *Nature*, 1991, **351**, 464.
29. R. J. Wilson, G. Meijer, D. S. Bethune, R. D. Johnson, D. D. Chambliss, M. S. de Vries, H. E. Hunziker and H. R. Wendt, *Nature*, 1990, **348**, 621.
30. W. Krätschmer, L. D. Lamb, K. Fostiropoulos and D. R. Huffman, *Nature*, 1990, **347**, 354.
31. M. Meier, G. Wang, R. Haddon, C. Brock, M. Lloyd and J. Selegue, *Nature*, 1991, **353**, 147.
32. C. Smart, B. Eldridge, W. Reuter, J. A. Zimmerman, W. R. Creasy, N. Rivera and R. S. Ruoff, *Chem. Phys. Lett.*, 1992, **188**, 171.
33. R. C. Haddon, *Science*, 1993, **261**, 1545.
34. L. Becker, J. Bada, R. Winans, J. Hunt, T. Bunch and B. French, *Science*, 1994, **265**, 642.
35. C. J. Hawker, P. M. Saville and J. W. White, *J. Org. Chem.*, 1994, **59**, 3503.
36. H. Kroto, *Fullerenes, Nanotubes, Carbon Nanostruct.*, 1994, **2**, 333.
37. H. P. Lang, V. Thommengeiser and H. J. Guntherodt, *Mol. Crystals Liquid Crystals Sci. Technol., Sect. A*, 1994, **244**, A283.
38. R. F. Curl and R. E. Smalley, *Science*, 1988, **242**, 1017.
39. S. Iijima, *Nature*, 1991, **354**, 56.
40. A. Oberlin, M. Endo and T. Koyama, *J. Cryst. Growth*, 1976, **32**, 335.
41. A. Oberlin, M. Endo and T. Koyama, *Carbon*, 1976, **14**, 133.
42. M. Monthioux and V. Kuznetsov, *Carbon*, 2006, **44**, 1621.
43. S. Iijima and T. Ichihashi, *Nature*, 1993, **363**, 603.
44. K. Novoselov, A. Geim, S. Morozov, D. Jiang, Y. Zhang, S. Dubonos, I. Grigorieva and A. Firsov, *Science*, 2004, **306**, 666.
45. K. S. Novoselov, D. Jiang, F. Schedin, T. J. Booth, V. V. Khotkevich, S. V. Morozov and A. K. Geim, *Proc. Natl. Acad. Sci. U.S.A.*, 2005, **102**, 10451.
46. D. D. L. Chung, *J. Mater. Sci.*, 1987, **22**, 4190.
47. G. D. Yuan, W. J. Zhang, Y. Yang, Y. B. Tang, Y. Q. Li, J. X. Wang, X. M. Meng, Z. B. He, C. M. L. Wu, I. Bello, C. S. Lee and S. T. Lee, *Chem. Phys. Lett.*, 2009, **467**, 361.
48. J. S. Park, A. Reina, R. Saito, J. Kong, G. Dresselhaus and M. S. Dresselhaus, *Carbon*, 2009, **47**, 1303.
49. L. Jiao, L. Zhang, X. Wang, G. Diankov and H. Dai, *Nature*, 2009, **458**, 877.

50. J. Campos-Delgado, Y. A. Kim, T. Hayashi, A. Morelos-Gómez, M. Hofmann, H. Muramatsu, M. Endo, H. Terrones, R. D. Shull, M. S. Dresselhaus and M. Terrones, *Chem. Phys. Lett.*, 2009, **469**, 177.
51. J. Zhu, *Nat. Nanotechnol.*, 2008, **3**, 528.
52. X. Li, G. Zhang, X. Bai, X. Sun, X. Wang, E. Wang and H. Dai, *Nat. Nanotechnol.*, 2008, **3**, 538.
53. Y. Hernandez, V. Nicolosi, M. Lotya, F. Blighe, Z. Sun, S. De, I. McGovern, B. Holland, M. Byrne and Y. Gun'Ko, *Nat. Nanotechnol.*, 2008, **3**, 563.
54. X. Du, I. Skachko, A. Barker and E. Andrei, *Nat. Nanotechnol.*, 2008, **3**, 491.
55. A. Das, S. Pisana, B. Chakraborty, S. Piscanec, S. Saha, U. Waghmare, K. Novoselov, H. Krishnamurthy, A. Geim and A. Ferrari, *Nat. Nanotechnol.*, 2008, **3**, 210.
56. A. Fasolino, J. Los and M. Katsnelson, *Nat. Mater.*, 2007, **6**, 858.
57. M. Terrones, *Nature*, 2009, **458**, 845.
58. A. Geim and K. Novoselov, *Nat. Mater.*, 2007, **6**, 183.
59. C. Jin, H. Lan, L. Peng, K. Suenaga and S. Iijima, *Phys. Rev. Lett.*, 2009, **102**, 205501.
60. D. J. Klein, W. A. Seitz and T. G. Schmalz, *Nature*, 1986, **323**, 703.
61. H. Prinzbach, A. Weiler, P. Landenberger, F. Wahl, J. Wörth, L. Scott, M. Gelmont, D. Olevano and B. Issendorff, *Nature*, 2000, **407**, 60.
62. M. F. Jarrold, *Nature*, 2000, **407**, 26.
63. M. Saito and Y. Miyamoto, *Phys. Rev. Lett.*, 2001, **87**, 035503.
64. Z. Chen, T. Heine, H. Jiao, A. Hirsch, W. Thiel and P. von Ragué Schleyer, *Chem. – Eur. J.*, 2004, **10**, 963.
65. A. Sandhu, *Nat. Nanotechnol.*, 2007.
66. B. Dunlap and R. Zope, *Chem. Phys. Lett.*, 2006, **422**, 451.
67. P. Calaminici, G. Geudtner and A. M. Köster, *J. Chem. Theory Comput.*, 2009, **5**, 29.
68. F. Diederich and M. Gomez-Lopez, *Chimia*, 1998, **52**, 551.
69. F. Diederich and R. L. Whetten, *Acc. Chem. Res.*, 1992, **25**, 119.
70. F. Diederich and M. Gómez-López, *Chem. Soc. Rev.*, 1999, **28**, 263.
71. B. Wang, H. Wang, J. Chang, H. Tso and Y. Chou, *J. Mol. Struct.: THEOCHEM*, 2001, **540**, 171.
72. B. Zhang, C. Wang and K. Ho, *Chem. Phys. Lett.*, 1992, **193**, 225.
73. F. Diederich and R. L. Whetten, *Acc. Chem. Res.*, 1992, **25**, 119.
74. K. Nakao, N. Kurita and M. Fujita, *Phys. Rev. B.*, 1994, **49**, 11415.
75. K. Kikuchi, N. Nakahara, T. Wakabayashi, S. Suzuki, H. Shiromaru, Y. Miyake, K. Saito, I. Ikemoto, M. Kainosho and Y. Achiba, *Nature*, 1992, **357**, 142.
76. T. G. Schmalz, W. A. Seitz, D. J. Klein and G. E. Hite, *Chem. Phys. Lett.*, 1986, **130**, 203.
77. M. Dresselhaus, G. Dresselhaus and P. Eklund, *Science of Fullerenes and Carbon Nanotubes*, Academic Press, San Diego, 1996.

78. R. Taylor and G. A. Burley, in *Fullerenes, Principles and Applications*, ed. F. Langa and J. F. Nierengarten, The Royal Society of Chemistry, Cambridge, 2007, p. 1.
79. D. Bakowies, M. Buehl and W. Thiel, *J. Am. Chem. Soc.*, 1995, **117**, 10113.
80. R. Taylor, J. P. Hare, A. K. Abdul-Sada and H. W. Kroto, *J. Chem. Soc., Chem. Commun.*, 1990, 1423.
81. A. Avent, P. Birkett, A. Darwish, H. Kroto, R. Taylor and D. Walton, *Tetrahedron*, 1996, **52**, 5235.
82. D. Manolopoulos, P. Fowler, R. Taylor, H. Kroto and D. Walton, *J. Chem. Soc., Faraday Trans.*, 1992, **88**, 3117.
83. R. Taylor, *J. Chem. Soc.*, 1993, **5**, 813.
84. F. Diederich, R. Ettl, Y. Rubin, R. L. Whetten, R. Beck, M. Alvarez, S. Anz, D. Sensharma, F. Wudl, K. C. Khemani and A. Koch, *Science*, 1991, **252**, 548.
85. P. Fowler and D. Manolopoulos, *An Atlas of Fullerenes*, Clarendon Press, Oxford, 1995.
86. A. Stone and D. Wales, *Phys. Lett.*, 1986, **128**, 501.
87. C. Thilgen, F. Diederich and R. L. Whetten, in *Buckminsterfullerenes*, ed. W. E. Billups and M. A. Ciufolini, VCH Publishers, New York, 1993, p. 59.
88. R. Taylor, *Fullerene Sci. Technol.*, 1999, **7**, 305.
89. A. M. Vassallo, L. S. K. Pang, P. A. Cole-Clarke and M. A. Wilson, *J. Am. Chem. Soc.*, 2002, **113**, 7820.
90. R. Taylor, M. P. Barrow and T. Drewello, *Chem. Commun.*, 1998, 2497.
91. H. Aldersey-Williams, *The Most Beautiful Molecule: The Discovery of the Buckyball*, John Wiley & Sons, Inc., New York, 1995.
92. R. Taylor, G. Langley, A. Avent, T. Dennis, H. Kroto and D. Walton, *J. Chem. Soc.*, 1993, **6**, 1029.
93. R. Tycko, R. C. Haddon, G. Dabbagh, S. H. Glarum, D. C. Douglass and A. M. Mujsce, *J. Phys. Chem.*, 1991, **95**, 518.
94. W. I. F. David, R. M. Ibberson, J. C. Matthewman, K. Prassides, T. J. S. Dennis, J. P. Hare, H. W. Kroto, R. Taylor and D. R. M. Walton, *Nature*, 1991, **353**, 147.
95. M. S. Amer, J. A. Elliott, J. F. Maguire and A. H. Windle, *Chem. Phys. Lett.*, 2005, **411**, 395.
96. H. Kato, Y. Kanazawa, M. Okumura, A. Taninaka, T. Yokawa and H. Shinohara, *J. Am. Chem. Soc.*, 2003, **125**, 4391.
97. T. Akasaka and S. Nagase, *Endofullerenes: A New Family of Carbon Clusters*, Kluwer Academic Publishers, Dordrecht, 2002.
98. P. Scharff, *Carbon*, 1998, **36**, 481.
99. N. N. Breslavskaya, A. A. Levin and A. L. Buchachenko, *Russ. Chem. Bull.*, 2004, **53**, 18.
100. A. Hirsch, H. Bestmann and C. Moll, *The Chemistry of the Fullerenes*, Thieme, Stuttgart, 1994.
101. A. Krestinin, M. Kislov and A. Ryabenko, *J. Nanosci. Nanotechnol.*, 2004, **4**, 390.

102. M. Takata, E. Nishibori, B. Umeda, M. Sakata, E. Yamamoto and H. Shinohara, *Phys. Rev. Lett.*, 1997, **78**, 3330.
103. S. Iida, Y. Kubozono, Y. Slovokhotov, Y. Takabayashi, T. Kanbara, T. Fukunaga, S. Fujiki, S. Emura and S. Kashino, *Chem. Phys. Lett.*, 2001, **338**, 21.
104. K. Laasonen, W. Andreoni and M. Parrinello, *Science*, 1992, **258**, 1916.
105. E. B. Iezzi, J. C. Duchamp, K. R. Fletcher, T. E. Glass and H. C. Dorn, *Nano Lett.*, 2002, **2**, 1187.
106. S. Stevenson, *Anal. Chem.*, 1994, **66**, 2675.
107. S. Stevenson, *J. Phys. Chem.*, 1998, **102**, 2833.
108. T. Guo, M. D. Diener, Y. Chai, M. J. Alford, R. E. Haufler, S. M. McClure, T. Ohno, J. H. Weaver, G. E. Scuseria and R. E. Smalley, *Science*, 1992, **257**, 1661.
109. J. Ding, N. Lin, L. Weng, N. Cue and S. Yang, *Chem. Phys. Lett.*, 1996, **261**, 92.
110. Y. Chai, T. Guo, C. Jin, R. E. Haufler, L. P. F. Chibante, J. Fure, L. Wang, J. M. Alford and R. E. Smalley, *J. Phys. Chem.*, 2002, **95**, 7564.
111. F. D. Weiss, J. L. Elkind, S. C. O'Brien, R. F. Curl and R. E. Smalley, *J. Am. Chem. Soc.*, 2002, **110**, 4464.
112. L. Moro, R. S. Ruoff, C. H. Becker, D. C. Lorents and R. Malhotra, *J. Phys. Chem.*, 1993, **97**, 6801.
113. A. Gromov, N. Krawez, A. Lassesson, D. Ostrovskii and E. Campbell, *Curr. Appl. Phys.*, 2002, **2**, 51.
114. H. Shinohara, H. Yamaguchi, N. Hayashi, H. Sato, M. Ohkohchi, Y. Ando and Y. Saito, *J. Phys. Chem.*, 1993, **97**, 4259.
115. T. J. S. Dennis and H. Shinohara, *Chem. Commun.*, 1998, 883.
116. R. E. Smalley, *Acc. Chem. Res.*, 1992, **25**, 38.
117. D. S. Bethune, C. H. Klang, M. S. de Vries, G. Gorman, R. Savoy, J. Vazquez and R. Beyers, *Nature*, 1993, **363**, 605.
118. Z. Xu, T. Nakane and H. Shinohara, *J. Am. Chem. Soc.*, 1996, **118**, 11309.
119. T. J. S. Dennis and H. Shinohara, *Chem. Phys. Lett.*, 1997, **278**, 107.
120. Y. Kubozono, H. Maeda, Y. Takabayashi, K. Hiraoka, T. Nakai, S. Kashino, S. Emura, S. Ukita and T. Sogabe, *J. Am. Chem. Soc.*, 1996, **118**, 6998.
121. S. Yang and L. Dunsch, *J. Phys. Chem. B*, 2005, **109**, 12320.
122. M. M. Olmstead, A. Bettancourt-Dias, J. C. Duchamp, S. Stevenson, D. Marciu, H. C. Dorn and A. L. Balch, *Angew. Chem. Int. Ed.*, 2001, **40**, 1223.
123. S. Stevenson, P. W. Fowler, T. Heine, J. C. Duchamp, G. Rice, T. Glass, K. Harich, E. Hajdu, R. Bible and H. C. Dorn, *Nature*, 2000, **408**, 427.
124. M. Krause, V. N. Popov, M. Inakuma, N. Tagmatarchis, H. Shinohara, P. Georgi, L. Dunsch and H. Kuzmany, *J. Chem. Phys.*, 2004, **120**, 1873.
125. M. Krause, H. Kuzmany, P. Georgi, L. Dunsch, K. Vietze and G. Seifert, *J. Chem. Phys.*, 2001, **115**, 6596.
126. S. Stevenson, G. Rice, T. Glass, K. Harich, F. Cromer, M. R. Jordan, J. Craft, E. Hadju, R. Bible, M. M. Olmstead, K. Maitra, A. J. Fisher, A. L. Balch and H. C. Dorn, *Nature*, 1999, **401**, 55.

127. J. Lee, H. Kim, S. J. Kahng, G. Kim, Y. W. Son, J. Ihm, H. Kato, Z. W. Wang, T. Okazaki, H. Shinohara and Y. Kuk, *Nature*, 2002, **415**, 1005.
128. S. Saito and A. Oshiyama, *Phys. Rev. B*, 1991, **44**, 11532.
129. C. Z. Wang, C. T. Chan and K. M. Ho, *Phys. Rev. B*, 1992, **46**, 9761.
130. W. Andreoni, F. Gygi and M. Parrinello, *Chem. Phys. Lett.*, 1992, **189**, 241.
131. D. R. McKenzie, C. A. Davis, D. J. H. Cockayne, D. A. Muller and A. M. Vassallo, *Nature*, 1992, **355**, 622.
132. S. van Smaalen, V. Petricek, J. de Boer, M. Dusek, M. Verheijen and G. Meijer, *Chem. Phys. Lett.*, 1994, **223**, 177.
133. R. E. Haufler, J. Conceicao, L. P. F. Chibante, Y. Chai, N. E. Byrne, S. Flanagan, M. M. Haley, S. C. O'Brien and C. Pan, *et al., J. Phys. Chem.*, 1990, **94**, 8634.
134. J. B. Howard, J. T. McKinnon, Y. Makarovsky, A. L. Lafleur and M. E. Johnson, *Nature*, 1991, **352**, 139.
135. R. Taylor, G. J. Langley, H. W. Kroto and D. R. M. Walton, *Nature*, 1993, **366**, 728.
136. A. Mittlebach, W. Hoenle, H. G. van Schnering, J. Carlsen, R. Janiak and H. Quast, *Angew. Chem. Int. Edn. Engl.*, 1992, **31**, 1640.
137. A. S. Koch, K. C. Khemani and F. Wudl, *J. Org. Chem.*, 1991, **56**, 4543.
138. D. H. Parker, K. Chatterjee, P. Wurz, K. R. Lykke, M. J. Pellin and L. M. Stock, in *The Fullerenes*, ed. H. Kroto, J. E. Fisher and D. E. Cox, Pergamon Press, Oxford, 1993, p. 29.
139. D. H. Parker, P. Wurz, K. Chatterjee, K. R. Lykke, J. E. Hunt, M. J. Pellin, J. C. Hemminger, D. M. Gruen and L. M. Stock, *J. Am. Chem. Soc.*, 1991, **113**, 7499.
140. J. B. Howard, A. L. Lafleur, Y. Makarovsky, S. Mitra, C. J. Pope and T. K. Yadav, in *The Fullerenes*, ed. H. Kroto, J. E. Fisher and D. E. Cox, Pergamon Press, Oxford, 1993, p. 45.
141. L. T. Scott, M. M. Boorum, B. J. McMahon, S. Hagen, J. Mack, J. Blank, H. Wegner and A. de Meijere, *Science*, 2002, **295**, 1500.
142. A. Darwish, H. Kroto, R. Taylor and D. R. M. Walton, *Fullerene Sci. Technol.*, 1993, **1**, 571.
143. M. Diack, R. L. Hettich, R. N. Compton and G. Guiochon, *Anal. Chem.*, 1992, **64**, 2143.
144. J. Abrefah, D. R. Olander, M. Balooch and W. J. Siekhaus, *Appl. Phys. Lett.*, 1992, **60**, 1313.
145. C. Pan, M. P. Sampson, Y. Chai, R. H. Hauge and J. L. Margrave, *J. Phys. Chem.*, 1991, **95**, 2944.
146. A. Darwish, H. Kroto, R. Taylor and D. R. M. Walton, *J. Chem. Soc., Chem. Commun.*, 1994, 15.
147. X. Xia, N. Monteiro-Riviere and J. Riviere, *J. Chromatogr. A*, 2006, **1129**, 216.
148. N. Casadei, M. Thomassin, Y. Guillaume and C. André, *Anal. Chim. Acta*, 2007, **588**, 268.
149. D. Bouchard and X. Ma, *J. Chromatogr. A*, 2008.

150. K. C. Khemani, M. Prato and F. Wudl, *J. Org. Chem.*, 1992, **57**, 3254.
151. M. Wohlers, A. Bauer, T. Rueleh, F. Neitzel, H. Werner and R. Schloegle, *Fullerene Sci. Technol.*, 1996, **4**, 781.
152. G. Scuseria, in *Buckminsterfullerenes*, ed. W. E. Billups and M. A. Ciufolini, VCH Publishers, New York, 1993, p. 103.
153. G. Scuseria, *Chem. Phys. Lett.*, 1995, **243**, 193.
154. L. D. Lamb, D. R. Huffman, R. K. Workman, S. Howells, T. Chen, D. Sarid and F. Ziolo, *Science*, 1992, **255**, 1413.
155. D. Ugarte, *Nature*, 1992, **359**, 707.
156. P. Harris, S. Tsang, J. Claridge and M. Green, *J. Chem. Soc., Faraday Trans.*, 1994, **90**, 2799.
157. E. Obraztsova, M. Fujii, S. Hayashi, V. Kuznetsov, Y. Butenko and A. Chuvilin, *Carbon*, 1998, **36**, 821.
158. O. O. Mykhaylyk, Y. M. Solonin, D. N. Batchelder and R. Brydson, *J. Appl. Phys.*, 2005, **97**, 074302.
159. N. Sano, H. Wang, I. Alexandrou, M. Chhowalla, K. Teo, G. Amaratunga and K. Iimura, *J. Appl. Phys.*, 2002, **92**, 2783.
160. D. Roy, M. Chhowalla, H. Wang, N. Sano, I. Alexandrou, T. Clyne and G. Amaratunga, *Chem. Phys. Lett.*, 2003, **373**, 52.
161. V. D. Blank, V. N. Denisov, A. N. Kirichenko, B. A. Kulnitskiy, S. Y. Martushov, B. N. Mavrin and I. A. Perezhogin, *Nanotechnology*, 2007, **18**.
162. W. Bacsa, W. De Heer, D. Ugarte and A. Châtelain, *Chem. Phys. Lett.*, 1993, **211**, 346.
163. X. Wang, B. Xu, X. Liu, H. Jia and I. Hideki, *Physica B (Amsterdam)*, 2005, **357**, 277.
164. T. Cabioc'h, A. Kharbach, A. Le Roy and J. Riviere, *Chem. Phys. Lett.*, 1998, **285**, 216.
165. T. Cabioc'h, E. Thune and M. Jaouen, *Chem. Phys. Lett.*, 2000, **320**, 202.
166. T. Cabioc'h, J. Girard, M. Jaouen, M. Denanot and G. Hug, *Europhys. Lett.*, 1997, **38**, 471.
167. T. Cabioc'h, J. Riviere and J. Delafond, *J. Mater. Sci.*, 1995, **30**, 4787.
168. A. Gubarevich, J. Kitamura, S. Usuba, H. Yokoi, Y. Kakudate and O. Odawara, *Carbon*, 2003, **41**, 2601.
169. C. He, N. Zhao, X. Du, C. Shi, J. Ding, J. Li and Y. Li, *Scr. Mater.*, 2006, **54**, 689.
170. V. D. Blank, B. A. Kulnitskiy, G. A. Dubitsky and I. Alexandrou, *Fullerenes, Nanotubes, Carbon Nanostruct.*, 2005, **13**, 167.
171. T. Kobayashi, T. Sekine and H. He, *Chem. Mater.*, 2003, **15**, 2681.
172. T. Gorelik, S. Urban, F. Falk, U. Kaiser and U. Glatzel, *Chem. Phys. Lett.*, 2003, **373**, 642.
173. L. Henrard, P. Lambin and A. Lucas, *Astrophys. J.*, 1997, **487**, 719.
174. M. Chhowalla, H. Wang, N. Sano, K. Teo, S. Lee and G. Amaratunga, *Phys. Rev. Lett.*, 2003, **90**, 155504.
175. L. Joly-Pottuz, N. Matsumoto, H. Kinoshita, B. Vacher, M. Belin, G. Montagnac, J. Martin and N. Ohmae, *Tribol. Int.*, 2008, **41**, 69.

176. S. Tomita, T. Sakurai, H. Ohta, M. Fujii and S. Hayashi, *J. Chem. Phys.*, 2001, **114**, 7477.
177. T. Cabioc'h, E. Thune, J. Riviere, S. Camelio, J. Girard, P. Guerin, M. Jaouen, L. Henrard and P. Lambin, *J. Appl. Phys.*, 2002, **91**, 1560.
178. A. Hirata, M. Igarashi and T. Kaito, *Tribol. Int.*, 2004, **37**, 899.
179. V. Kuznetsov, Y. Butenko, A. Chuvilin, A. Romanenko and A. Okotrub, *Chem. Phys. Lett.*, 2001, **336**, 397.
180. S. Tomita, M. Fujii and S. Hayashi, *Phys. Rev. B*, 2002, **66**, 245424.
181. O. Mykhaylyk, Y. Solonin, D. Batchelder and R. Brydson, *J. Appl. Phys.*, 2005, **97**, 074302.
182. V. Pol, M. Motiei, A. Gedanken, J. Calderon-Moreno and M. Yoshimura, *Carbon*, 2004, **42**, 111.
183. E. Koudoumas, O. Kokkinaki, M. Konstantaki, S. Couris, S. Korovin, P. Detkov, V. Kuznetsov, S. Pimenov and V. Pustovoi, *Chem. Phys. Lett.*, 2002, **357**, 336.
184. M. Endo, T. Hayashi, Y. Ahm Kim, M. Terrones and M. S. Dresselhaus, *Philos. Trans. R. Soc. London, Ser. A*, 2004, **362**, 2223.
185. T. W. Odom, J.-L. Huang, P. Kim and C. M. Lieber, *Nature*, 1998, **391**, 62.
186. J. Hu, T. Odom and C. Lieber, *Acc. Chem. Res.*, 1999, **32**, 435.
187. M. S. Dresselhaus, G. Dresselhaus and R. Saito, *Carbon*, 1995, **33**, 883.
188. M. S. Dresselhaus, *Carbon*, 1995, **33**, 871.
189. M. S. Dresselhaus and G. Dresselhaus, *Nanostruct. Mater.*, 1997, **9**, 33.
190. A. Ávila and G. Lacerda, *Mater. Res.*, 2008, **11**, 325.
191. C. Dekker, *Phys. Today*, 1999, **52**, 22.
192. Y. Zhang and S. Iijima, *Appl. Phys. Lett.*, 1999, **75**, 3087.
193. H. J. Choi, J. Ihm, S. G. Louie and M. L. Cohen, *Phys. Rev. Lett.*, 2000, **84**, 2917.
194. E. Thostenson, Z. Ren and T. Chou, *Compos. Sci. Technol.*, 2001, **61**, 1899.
195. P. Avouris, *Acc. Chem. Res.*, 2002, **35**, 1026.
196. R. Baughman, A. Zakhidov and W. de Heer, *Science*, 2002, **297**, 787.
197. P. Corio, P. S. Santos, M. A. Pimenta and M. S. Dresselhaus, *Chem. Phys. Lett.*, 2002, **360**, 557.
198. S. Iijima, *Physica B (Amsterdam)*, 2002, **323**, 1.
199. A. G. Souza Filho, A. Jorio, G. G. Samsonidze, G. Dresselhaus, M. S. Dresselhaus, A. K. Swan, M. S. Ünlü, B. B. Goldberg, R. Saito, J. H. Hafner, C. M. Lieber and M. A. Pimenta, *Chem. Phys. Lett.*, 2002, **354**, 62.
200. A. Kalamkarov, A. Georgiades, S. Rokkam, V. Veedu and M. Ghasemi-Nejhad, *Int. J. Solids Struct.*, 2006, **43**, 6832.
201. M. S. Dresselhaus, G. Dresselhaus and R. Saito, *Phys. Rev. B*, 1992, **45**, 6234.
202. P. J. F. Harris, *Carbon Nanotubes and Related Structures*, Cambridge Press, Cambridge, 1999.
203. C. T. White, J. W. Mintmire, R. C. Mowrey, D. Brenner, D. H. Robertson, J. A. Harrson and B. I. Dunlap, in *Buckminsterfullerenes*,

ed. W. E. Billups and M. A. Ciufolini, VCH Publishers, New York, 1993, p. 125.
204. M. S. Dresselhaus, G. Dresselhaus and P. C. Eklund, *Science of Fullerences and Carbon Nanotubes*, Academic Press, San Diego, 1996.
205. R. A. Jishi, D. Inomata, K. Nakao, M. Dresselhaus and G. Dresselhaus, *J. Phys. Soc. Jpn.*, 1994, **63**, 2252.
206. L. Qin, X. Zhao, K. Hirahara, Y. Miyamoto, Y. Ando and S. Iijima, *Nature*, 2000, **408**, 50.
207. L. Qin, X. Zhao, K. Hirahara, Y. Ando and S. Iijima, *Chem. Phys. Lett.*, 2001, **349**, 389.
208. D. Stojkovic, P. Zhang and V. Crespi, *Phys. Rev. Lett.*, 2001, **87**, 125502.
209. X. Zhao, Y. Liu, S. Inoue, T. Suzuki, R. Jones and Y. Ando, *Phys. Rev. Lett.*, 2004, **92**, 125502.
210. R. A. Jishi, L. Venkataraman, M. S. Dresselhaus and G. Dresselhaus, *Chem. Phys. Lett.*, 1993, **209**, 77.
211. E. Di Donato, M. Tommasini, C. Castiglioni and G. Zerbi, "The electronic structure of achiral nanotubes: a symmetry based treatment," in *Electronic Properties of Synthetic Nanostructures: XVIII International Winterschool/Euroconference on Electronic Properties of Novel Materials*, ed. H. Kuzmany, J. Fink and M. Mehring, American Institute of Physics, 2004, AIP Conference Proceedings 723, pp. 359–363.
212. M. Damnjanovi, I. Miloševi, T. Vukovi and R. Sredanovi, *Phys. Rev. B*, 1999, **60**, 2728.
213. S. Reich, C. Thomsen and J. Maultzsch, *Carbon Nanotubes, Basic Concepts and Physical Properties*, Wiley-VCH Verlag, Weinheim, 2004.
214. C. T. White, D. H. Robertson and J. W. Mintmire, *Phys. Rev. B*, 1993, **47**, 5485.
215. R. A. Jishi, L. Venkataraman, M. S. Dresselhaus and G. Dresselhaus, *Phys. Rev. B*, 1995, **51**, 11176.
216. N. Hamada, S. -i. Sawada and A. Oshiyama, *Phys. Rev. Lett.*, 1992, **68**, 1579.
217. P. Eklund, J. Holden and R. Jishi, *Carbon*, 1995, **33**, 959.
218. I. B. Bozovic, M. Vujicic and F. Herbut, *J. Phys. A: Math. Gen.*, 1978, **11**, 2133.
219. M. Damnjanovi, *Phys. Lett. A*, 1983, **94**, 337.
220. M. Damnjanovic, I. Miloševic, T. Vukovic and R. Sredanovic, *Phys. Rev. B*, 1999, **60**, 2728.
221. E. Dujardin, T. Ebbesen, H. Hiura and K. Tanigaki, *Science*, 1994, **265**, 1850.
222. T. Ebbesen, *Annu. Rev. Mater. Sci.*, 1994, **24**, 235.
223. X. F. Zhang, X. B. Zhang, G. V. Tendeloo, S. Amelinckx, M. O. d. Beeck and J. V. Landuyt, *J. Cryst. Growth*, 1993, **130**, 368.
224. D. Reznik, C. H. Olk, D. A. Neumann and J. R. D. Copley, *Phys. Rev. B*, 1995, **52**, 116.
225. D. Brenner, J. Harrison, C. White and R. Colton, *Thin Solid Films*, 1991, **206**, 220.

226. S. Lair, W. Herndon, L. Murr and S. Quinones, *Carbon*, 2006, **44**, 447.
227. W. Zhu, D. Miser, W. Chan and M. Hajaligol, *Mater. Chem. Phys.*, 2003, **82**, 638.
228. S. Bandow, M. Takizawa, K. Hirahara, M. Yudasaka and S. Iijima, *Chem. Phys. Lett.*, 2001, **337**, 48.
229. S. Bandow, G. Chen, G. Sumanasekera, R. Gupta, M. Yudasaka, S. Iijima and P. Eklund, *Phys. Rev. B*, 2002, **66**, 75416.
230. Y. A. Kim, H. Muramatsu, T. Hayashi, M. Endo, M. Terrones and M. S. Dresselhaus, *Chem. Phys. Lett.*, 2004, **398**, 87.
231. J. Arvanitidis, D. Christofilos, K. Papagelis, T. Takenobu, Y. Iwasa, H. Kataura, S. Ves and G. A. Kourouklis, *Phys. Rev. B*, 2005, **72**.
232. T. Shimada, T. Sugai, C. Fantini, M. Souza, L. G. Cançado, A. Jorio, M. A. Pimenta, R. Saito, A. Grüneis, G. Dresselhaus, M. S. Dresselhaus, Y. Ohno, T. Mizutani and H. Shinohara, *Carbon*, 2005, **43**, 1049.
233. C. Reinhold, *Nano Today*, 2006, **1**, 15.
234. V. Gadagkar, S. Saha, D. V. S. Muthu, P. K. Maiti, Y. Lansac, A. Jagota, A. Moravsky, R. O. Loutfy and A. K. Sood, *J. Nanosci. Nanotechnol.*, 2007, **7**, 1753.
235. A. Latgé and D. Grimm, *Carbon*, 2007, **45**, 1905.
236. Z. Gu, K. Wang, J. Wei, C. Li, Y. Jia, Z. Wang, J. Luo and D. Wu, *Carbon*, 2006, **44**, 3315.
237. E. Mamontov, C. Burnham, S. Chen, A. Moravsky, C. Loong, N. de Souza and A. Kolesnikov, *J. Chem. Phys.*, 2006, **124**, 194703.
238. D. Tang, L. Ci, W. Zhou and S. Xie, *Carbon*, 2006, **44**, 2155.
239. G. Marcolongo, G. Ruaro, M. Gobbo and M. Meneghetti, *Chem. Commun.*, 2007, 4925.
240. X. Yao and Q. Han, *Eur. J. Mechanics-A/Solids*, 2007, **26**, 298.
241. G. Chen, J. Qiu and H. Qiu, *Scr. Mater.*, 2008, **58**, 457.
242. G. M. do Nascimento, T. Hou, Y. A. Kim, H. Muramatsu, T. Hayashi, M. Endo, N. Akuzawa and M. S. Dresselhaus, *Nano Lett.*, 2008.
243. Y. Honda, M. Takeshige, H. Shiozaki, T. Kitamura, K. Yoshikawa, S. Chakrabarti, O. Suekane, L. Pan, Y. Nakayama, M. Yamagata and M. Ishikawa, *J. Power Sources*, 2008, **185**, 1580.
244. F. Villalpando-Paez, H. Son, D. Nezich, Y. P. Hsieh, J. Kong, Y. A. Kim, D. Shimamoto, H. Muramatsu, T. Hayashi, M. Endo, M. Terrones and M. S. Dresselhaus, *Nano Lett.*, 2008, **8**, 3879.
245. Z. Q. Zhang, H. W. Zhang, Y. G. Zheng, J. B. Wang and L. Wang, *Curr. Appl. Phys.*, 2008, **8**, 217.
246. I. Elishakoff and D. Pentaras, *J. Sound Vibration*, 2009, **322**, 652.
247. J. W. Kang, O. K. Kwon, J. H. Lee, Y. G. Choi and H. J. Hwang, *Solid State Commun.*, 2009, **149**, 1574.
248. C. Lamprecht, J. Danzberger, P. Lukanov, C. M. Tîlmaciu, A. M. Galibert, B. Soula, E. Flahaut, H. J. Gruber, P. Hinterdorfer, A. Ebner and F. Kienberger, *Ultramicroscopy*, 2009, **109**, 899.
249. C. H. Lee, W. S. Su, R. B. Chen and M. F. Lin, *Physica E (Amsterdam)*, 2009, **41**, 1226.

250. R. Marega, G. Accorsi, M. Meneghetti, A. Parisini, M. Prato and D. Bonifazi, *Carbon*, 2009, **47**, 675.
251. K. Yaghmaei and H. Rafii-Tabar, *Curr. Appl. Phys.*, 2009, **9**, 1411.
252. H. W. Zhang, Z. Q. Zhang and L. Wang, *Curr. Appl. Phys.*, 2009, **9**, 750.
253. S. Iijima, P. Ajayan and T. Ichihashi, *Phys. Rev. Lett.*, 1992, **69**, 3100.
254. T. W. Ebbesen and P. M. Ajayan, *Nature*, 1992, **358**, 220.
255. T. W. Ebbesen, *Annu. Rev. Mater. Sci.*, 1994, **24**, 235.
256. T. W. Ebbesen, *Carbon Nanotubes: Preparation and Properties*, CRC Press, Boca Raton, 1997.
257. S. Seraphin, D. Zhou, J. Jiao, J. Whiters and R. Loutfy, *Carbon*, 1993, **31**, 685.
258. M. Yudasaka, Y. Kasuya, F. Kokai, K. Takahashi, M. Takizawa, S. Bandow and S. Iijima, *Appl. Phys. A*, 2002, **74**, 377.
259. Y. Ando, X. Zhao, T. Sugai and M. Kumar, *Mater. Today*, 2004, **7**, 22.
260. C. Journet, W. K. Maser, P. Bernier, A. Loiseau, M. L. de la Chapelle, S. Lefrant, P. Deniard, R. Lee and J. E. Fischer, *Nature*, 1997, **388**, 756.
261. Y. Ando, X. Zhao, K. Hirahara, K. Suenaga, S. Bandow and S. Iijima, *Chem. Phys. Lett.*, 2000, **323**, 580.
262. Y. Ando, *Fullerene Sci. Technol.*, 1994, **2**, 173.
263. M. Wang, X. Zhao, M. Ohkohchi and Y. Ando, *Fullerene Sci. Technol.*, 1996, **4**, 1027.
264. Y. Ando, X. Zhao and M. Ohkohchi, *Carbon*, 1997, **35**, 153.
265. X. Zhao, M. Ohkohchi, M. Wang, S. Iijima, T. Ichihashi and Y. Ando, *Carbon*, 1997, **35**, 775.
266. X. K. Wang, X. W. Lin, V. P. Dravid, J. B. Ketterson and R. P. H. Chang, *Appl. Phys. Lett.*, 1995, **66**, 2430.
267. Y. Tai, K. Inukai, T. Osaki, M. Tazawa, J. Murakami, S. Tanemura and Y. Ando, *Chem. Phys. Lett.*, 1994, **224**, 118.
268. L. P. Biró, Z. E. Horváth, L. Szalmás, K. Kertész, F. Wéber, G. Juhász, G. Radnóczi and J. Gyulai, *Chem. Phys. Lett.*, 2003, **372**, 399.
269. Y. Saito, T. Nakahira and S. Uemura, *J. Phys. Chem. B*, 2003, **107**, 931.
270. J. Hutchison, N. Kiselev, E. Krinichnaya, A. Krestinin, R. Loutfy, A. Morawsky, V. Muradyan, E. Obraztsova, J. Sloan and S. Terekhov, *Carbon*, 2001, **39**, 761.
271. T. Sugai, H. Omote, S. Bandow, N. Tanaka and H. Shinohara, *J. Chem. Phys.*, 2000, **112**, 6000.
272. T. Shimada, T. Sugai, Y. Ohno, S. Kishimoto, T. Mizutani, H. Yoshida, T. Okazaki and H. Shinohara, *Appl. Phys. Lett.*, 2004, **84**, 2412.
273. H. Takikawa, M. Ikeda, K. Hirahara, Y. Hibi, Y. Tao, P. Ruiz, T. Sakakibara, S. Itoh and S. Iijima, *Physica B (Amsterdam)*, 2002, **323**, 277.
274. H. Huang, H. Kajiura, S. Tsutsui, Y. Hirano, M. Miyakoshi, A. Yamada and M. Ata, *Chem. Phys. Lett.*, 2001, **343**, 7.
275. L. Chernozatonskii, Z. Kosakovskaya, Y. Gulyaev, N. Sinitsyn, G. Torgashov and Y. Zakharchenko, *J. Vacuum Sci. Technol., B*, 1996, **14**, 2080.

276. Z. Kosakovskaya, L. A. Chernozatonskii and G. N. Fedorova, *JETP Lett.*, 1992, **56**, 26.
277. L. Chernozatonskii, Z. Kosakovskaja, A. Kiselev and N. Kiselev, *Chem. Phys. Lett.*, 1994, **228**, 94.
278. M. Ge and K. Sattler, *Science*, 1993, **260**, 515.
279. M. Ge and K. Sattler, *Appl. Phys. Lett.*, 1994, **65**, 2284.
280. A. Thess, R. Lee, P. Nikolaev, H. Dai, P. Petit, J. Robert, C. Xu, Y. H. Lee, S. G. Kim, A. G. Rinzler, D. T. Colbert, G. E. Scuseria, D. Tomanek, J. E. Fischer and R. E. Smalley, *Science*, 1996, **273**, 483.
281. T. Guo, P. Nikolaev, A. G. Rinzler, D. Tomanek, D. T. Colbert and R. E. Smalley, *J. Phys. Chem.*, 1995, **99**, 10694.
282. W. K. Maser, E. Muñoz, A. M. Benito, M. T. Martínez, G. F. de la Fuente, Y. Maniette, E. Anglaret and J. L. Sauvajol, *Chem. Phys. Lett.*, 1998, **292**, 587.
283. D. Laplaze, P. Bernier, G. Flamant, M. Lebrun, A. Brunelle and S. Della-Negra, *J. Phys. -London-B Atomic Mol. Opt. Phys.*, 1996, **29**, 4943.
284. D. Laplaze, P. Bernier, C. Journet, J. L. Sauvajol, D. Bormann, G. Flamant and M. Lebrun, *J. Phys. III*, 1997, **7**, 463.
285. T. Guillard, L. Alvarez, E. Anglaret, J. Sauvajol, P. Bernier, G. Flamant and D. Laplaze, *J. Phys. IV*, 1999, **9**, 399.
286. P. Schultzenberger and L. Schultzenberger, *C. R. Acad. Sci., Paris*, 1890, **111**, 774.
287. R. T. K. Baker and P. S. Harris, *Chem. Phys. Carbon*, 1978, **14**, 83.
288. A. Kock, P. De Bokx, E. Boellaard, W. Klop and J. Geus, *J. Catal*, 1985, **96**, 468.
289. P. De Bokx, A. Kock, E. Boellaard, W. Klop and J. Geus, *J. Catal.*, 1985, **96**, 454.
290. K. Mukhopadhyay, A. Koshio, T. Sugai, N. Tanaka, H. Shinohara, Z. Konya and J. B. Nagy, *Chem. Phys. Lett.*, 1999, **303**, 117.
291. M. J. Bronikowski, P. A. Willis, D. T. Colbert, K. A. Smith and R. E. Smalley, *J. Vac. Sci. Technol., A*, 2001, **19**, 1800.
292. V. A. Ryzhkov, *Physica B*, 2002, **323**, 324.
293. N. K. Shimbun, *Mater. Today*, 2002, **5**, 9.
294. R. Bhowmick, B. M. Clemens and B. A. Cruden, *Carbon*, 2008, **46**, 907.
295. T. Iguchi, S. Takenaka, K. Nakagawa, Y. Orita, H. Matsune and M. Kishida, *Top. Catal.*, 2009, **52**, 563.
296. K. Dasgupta, R. Venugopalan, G. K. Dey and D. Sathiyamoorthy, *J. Nanoparticle Res.*, 2008, **10**, 69.
297. E. G. Rakov, *Nanotechnol. Russ.*, 2008, **3**, 575.
298. I. W. Chiang, B. E. Brinson, A. Y. Huang, P. A. Willis, M. J. Bronikowski, J. L. Margrave, R. E. Smalley and R. H. Hauge, *J. Phys. Chem. B*, 2001, **105**, 8297.
299. Y. Mackeyev, S. Bachilo, K. B. Hartman and L. J. Wilson, *Carbon*, 2007, **45**, 1013.
300. E. Gregan, S. M. Keogh, A. Maguire, T. G. Hedderman, L. O. Neill, G. Chambers and H. J. Byrne, *Carbon*, 2004, **42**, 1031.

301. A. W. Musumeci, E. R. Waclawik and R. L. Frost, *Spectrochim. Acta, Part A*, 2008, **71**, 140.
302. K. Shen, S. Curran, H. Xu, S. Rogelj, Y. Jiang, J. Dewald and T. Pietrass, *J. Phys. Chem. B*, 2005, **109**, 4455.
303. N. Komatsu, T. Ohe and K. Matsushige, *Carbon*, 2004, **42**, 163.
304. E. Va'zquez, V. Georgakilas and M. Prato, *Chem. Commun.*, 2002, **20**, 2308.
305. Y.-Q. Xu, H. Peng, R. H. Hauge and R. E. Smalley, *Nano Lett.*, 2004, **5**, 163.
306. A. G. Rinzler, J. Liu, H. Dai, P. Nikolaev, C. B. Huffman and F. J. Rodriguez-Macias, *Appl. Phys. A*, 1998, **67**, 29.
307. H. Kajiura, S. Tsutsui, H. Huang and Y. Murakami, *Chem. Phys. Lett.*, 2002, **364**, 586.
308. J.-M. Moon, K. H. An, Y. H. Lee, Y. S. Park, D. J. Bae and G.-S. Park, *J. Phys. Chem. B*, 2001, **105**, 5677.
309. L. D. Landau, *Phys. Z. Sowjetunion*, 1937, **11**, 26.
310. R. E. Peierls, *Ann. I. H. Poincare*, 1935, **5**, 177.
311. N. D. Mermin, *Phys. Rev.*, 1968, **176**, 250.
312. K. S. Novoselov, A. K. Geim, S. V. Morozov, D. Jiang, M. I. Katsnelson, I. V. Grigorieva, S. V. Dubonos and A. A. Firsov, *Nature*, 2005, **438**, 197.
313. J. Venables, G. Spiller and M. Hanbucken, *Rep. Prog. Phys.*, 1984, **47**, 399.
314. J. Evans, P. Thiel and M. Bartelt, *Surf. Sci. Rep.*, 2006, **61**, 1.
315. M. Zinke-Allmang, L. Feldman and M. Grabow, *Surf. Sci. Rep.*, 1992, **16**, 377.
316. H. Yanagisawa, T. Tanaka, Y. Ishida, M. Matsue, E. Rokuta, S. Otani and C. Oshima, *Surf. Interface Anal.*, 2005, **37**, 133.
317. H. C. Schniepp, J.-L. Li, M. J. McAllister, H. Sai, M. Herrera-Alonso, D. H. Adamson, R. K. Prud'homme, R. Car, D. A. Saville and I. A. Aksay, *J. Phys. Chem. B*, 2006, **110**, 8535.
318. Z. Chen, Y. Lin, M. Rooks and P. Avouris, *Physica E (Amsterdam)*, 2007, **40**, 228.
319. M. Y. Han, B. Ozyilmaz, Y. Zhang and P. Kim, *Phys. Rev. Lett.*, 2007, **98**, 206805.
320. M. Ishigami, J. Chen, W. Cullen, M. Fuhrer and E. Williams, *Nano Lett.*, 2007, **7**, 1643.
321. M. Katsnelson, *Mater. Today*, 2007, **10**, 20.
322. M. Lemme, T. Echtermeyer, M. Baus, H. Kurz and A. Aachen, *IEEE Electron Device Lett.*, 2007, **28**, 282.
323. M. Trushin and J. Schliemann, *Phys. Rev. Lett.*, 2007, **99**, 216602.
324. A. A. Balandin, S. Ghosh, W. Bao, I. Calizo, D. Teweldebrhan, F. Miao and C. N. Lau, *Nano Lett.*, 2008, **8**, 902.
325. S. P. Berciaud, S. Ryu, L. E. Brus and T. F. Heinz, *Nano Lett.*, 2008, **9**, 346.
326. J. Campos-Delgado, J. Romo-Herrera, X. Jia, D. Cullen, H. Muramatsu, Y. Kim, T. Hayashi, Z. Ren, D. Smith and Y. Okuno, *Nano Lett.*, 2008, **8**, 2773.

327. S. Ghosh, I. Calizo, D. Teweldebrhan, E. P. Pokatilov, D. L. Nika, A. A. Balandin, W. Bao, F. Miao and C. N. Lau, *Appl. Phys. Lett.*, 2008, **92**, 151911.
328. F. Guinea, *J. Low Temp. Phys.*, 2008, **153**, 359.
329. X. Li, X. Wang, L. Zhang, S. Lee and H. Dai, *Science*, 2008, **319**, 1229.
330. Z. Wang, H. Lim, S. Ng, B. Özyilmaz and M. Kuok, *Carbon*, 2008, **46**, 2133.
331. Z. Wei, D. E. Barlow and P. E. Sheehan, *Nano Lett.*, 2008, **8**, 3141.
332. X. Yang, X. Dou, A. Rouhanipour, L. Zhi, H. J. Rader and K. Mullen, *J. Am. Chem. Soc.*, 2008, **130**, 4216.
333. S. Alwarappan, A. Erdem, C. Liu and C.-Z. Li, *J. Phys. Chem. C.*, 2009, **113**, 8853.
334. J. Hu, X. Ruan and Y. P. Chen, *Nano Lett.*, 2009, **9**, 2730.
335. D. V. Kosynkin, A. L. Higginbotham, A. Sinitskii, J. R. Lomeda, A. Dimiev, B. K. Price and J. M. Tour, *Nature*, 2009, **458**, 872.
336. J. Ma, D. Alfe, A. Michaelides and E. Wang, *Phys. Rev. B*, 2009, **80**, 033407.
337. Z. H. Ni, T. Yu, Z. Q. Luo, Y. Y. Wang, L. Liu, C. P. Wong, J. Miao, W. Huang and Z. X. Shen, *ACS Nano*, 2009, **3**, 569.
338. D. L. Nika, S. Ghosh, E. P. Pokatilov and A. A. Balandin, *Appl. Phys. Lett.*, 2009, **94**, 203103.
339. J. Park, Y. H. Ahn and C. Ruiz-Vargas, *Nano Lett.*, 2009, **9**, 1742.
340. S. Shivaraman, M. V. S. Chandrashekhar, J. J. Boeckl and M. G. Spencer, *J. Electron. Mater.*, 2009, **38**, 725.
341. M. L. Teague, A. P. Lai, J. Velasco, C. R. Hughes, A. D. Beyer, M. W. Bockrath, C. N. Lau and N. C. Yeh, *Nano Lett.*, 2009, **9**, 2542.
342. Z.-S. Wu, S. Pei, W. Ren, D. Tang, L. Gao, B. Liu, F. Li, C. Liu and H.-M. Cheng, *Adv. Mater.*, 2009, **21**, 1756.
343. V. N. Popov, L. Henrard and P. Lambin, *Carbon*, 2009, **47**, 2448.
344. J. Meyer, A. Geim, M. Katsnelson, K. Novoselov, T. Booth and S. Roth, *Nature*, 2007, **446**, 60.

CHAPTER 4
The Nano-frontier; Properties, Achievements, and Challenges

4.1 Introduction

It is well known that a major goal of materials science is to produce matter that is ordered on all length scales, from the molecular (1–100 Å) *via* the nano (10–100 nm) to the meso (1–100 μm). In fact, the identification and sophisticated use of distinct and common features of mesosystems may hold the key to the design of new materials with unique properties.[1] Nowadays, we know that this high degree of order at certain length-scales leads to unique physical and mechanical properties that are different from and, indeed, superior to those of the same material in a disordered form.[2–5] Highly ordered structures with unique physical and mechanical properties address and enable advanced applications in medical, biological, and technological fields.[6–8] Mesoscopic systems at the micro- and the nano-scale hold the potential for whole new forms of matter that have never existed on earth. As discussed in Chapter 1, conventional matter is governed ultimately by the range and strength of inter-atomic and intermolecular interactions and by the physics of bulk domains. By constructing "nanoparticle" based meso-structures with engineered or designed inter-particle interactions, based on the physics of the nanodomain, it will be possible to create useful new structures and forms of matter that will represent a breakthrough of almost unimaginable scientific frontiers. This is becoming increasingly recognized as evidenced by the numerous examples of unique nanophenomena (Section 1.8).

In previous chapters, we discussed the characteristics of nanotechnology, the details of Raman spectroscopy as a characterization technique, and the structures and production methods of fullerene as essential building blocks for the technology of the future. We are now in a position to examine, in detail, the properties of these building blocks and to discuss an overview of the physics of nanostructured systems created once such building blocks are put together.

In this chapter we will cover the behavior of individual fullerene building blocks and the observed behavior of mesosystems based on such building blocks. Over a decade ago, three-dimensional carbon-based structures with periodicity on the nanoscale were produced and shown to have unique properties.[8] Nowadays, the quest continues with more amazing findings.[5–7] We start with the Raman scattering behavior of such building blocks and their systems.

4.2 Raman Scattering of Fullerenes

With a dimensionality on the order of 1 nm, fullerenes – spheres, cylinders, and sheets – have demonstrated unique light scattering, in general, and Raman scattering, in particular, properties. In fact, Raman spectroscopy plays a crucial role in addressing fundamental questions regarding the physics of fullerenes. Critical issues such as the structure and properties of fullerenes and their molecular crystals have been addressed and largely understood based on Raman investigations. Raman spectroscopy has also been utilized to investigate thermodynamic equations-of-state and phase transitions in certain fullerene molecular solids. In addition, surface interactions and adsorption behavior of C_{60}, and other fullerene forms, both have been greatly elucidated based on Raman scattering studies.

4.2.1 Raman Scattering of C_{60} Molecules and Crystals

As we mentioned before, a [60] fullerene molecule (C_{60}) has 60 carbon atoms in a cage-like structure belonging to the icosahedral (I_h) point group symmetry. Owing to the symmetry of the C_{60} molecule, its 174 normal vibration modes ($3N - 6$) can be reduced to 46 distinct symmetry species according to the following symmetries:

$$\Gamma = 2A_g + 3F_{1g} + 4F_{2g} + 6G_g + 8H_g + A_u + 4F_{1u} + 5F_{2u} + 6G_u + 7H_u \quad (4.1)$$

Only ten of these vibrational modes, two belonging to the A_g ($2A_g$) and eight belonging to the H_g species ($8H_g$), are Raman active. The four (F_{1u}) modes are infrared active, while the remaining 32 modes are optically silent.[9–13] Three of the Raman active modes [H_g (7), A_g (2), and H_g (8)] are surface modes. They correspond to pentagon shear, pentagon pinch, and hexagon shear modes, respectively.[14–17] The A_g (1) is a "radial breathing mode" (RBM) where all atoms are radially displaced at equal magnitude and in phase. Figure 4.1 schematically shows the atomic displacement vectors corresponding to the ten Raman active modes of C_{60} molecules.

Such vibration modes are known as the intramolecular (or, simply, molecular) vibration modes since they are generated from atomic degrees of freedom within the molecule itself as we discussed in Chapter 2. Later, we will examine other types of vibration modes (referred to as intermolecular, or lattice, vibration modes) that have been observed in groups of C_{60} fullerene molecules

Figure 4.1 Schematic diagram depicting the atomic displacement vectors corresponding to Raman active vibration modes in C_{60}: (a) A_g (1), (b) A_g (2), (c) H_g (1), (d) H_g (2), (e) H_g (3), (f) H_g (4), (g) H_g (5), (h) H_g (6), (i) H_g (7), and (j) H_g (8). (Reproduced with kind permission from Schlüter et al., ref. 11. Copyright Elsevier 1992.)

interacting and forming crystals or cooperative structures. The energies and, hence, the frequencies of the Raman active intramolecular vibration modes of C_{60} were theoretically investigated and calculated[18] as early as 1987. Experimental measurements of C_{60} Raman spectrum, however, had to wait until the

Table 4.1 Symmetry assignment and theoretical and experimental frequencies of Raman active modes in a C_{60} molecule.

Mode	Frequency (cm^{-1})	Theoretical calculations	
		Giannozzi and Baroni[45]	Adams et al.[31]
H_g (1)	272	259 (−4.8%)	259 (−4.8%)
H_g (2)	433	425 (−1.8%)	427 (−1.4%)
A_g (1)	496	495 (−0.2%)	494 (−0.4%)
H_g (3)	709	711 (0.3%)	694 (−2.1%)
H_g (4)	772	783 (1.4%)	760 (−1.6%)
H_g (5)	1099	1020 (1.9%)	1103 (0.4%)
H_g (6)	1252	1281 (2.3%)	1328 (6.1%)
H_g (7)	1426	1452 (1.8%)	1535 (7.7%)
A_g (2)	1469	1504 (2.3%)	1607 (9.3%)
H_g (8)	1575	1578 (0.2%)	1628 (3.4%)

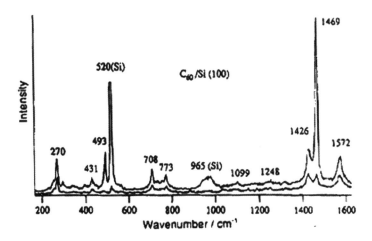

Figure 4.2 Polarized Raman spectra of C_{60} on a silicon substrate. Upper trace is for ($\parallel/\parallel$) polarization, which shows both A_g and H_g modes. Lower trace is for ($\parallel/\perp$) polarization, which only shows H_g modes. (Reproduced with kind permission from Eklund et al., ref. 48. Copyright Elsevier 1992.)

Krätschmer–Huffman method[19] was developed in 1990 (Section 3.3) due to sample availability issues. For the past 20 years, the Raman activity of C_{60} has been extensively investigated both theoretically[20–30] and experimentally.[31–46] Table 4.1 shows the theoretical *versus* the experimental frequencies of the Raman active modes for a C_{60} molecule. Phonon calculations and polarized Raman scattering studies[46–50] of C_{60} solid films showed that in the ($\parallel/\parallel$) polarization arrangement both A_g and H_g modes are observable. In the ($\parallel/\perp$) polarization setup, however, only the H_g modes can be experimentally observed. Figure 4.2 shows experimental Raman spectrum of C_{60} obtained from a solid C_{60} film on a silicon substrate.

Clearly, from Table 4.1, the experimentally measured frequencies always differ from their theoretically calculated values. While most theoretical calculations studies tend to judge the accuracy of their results by comparing them to experimental values, it should be emphasized that the unavoidable presence of perturbation fields around the C_{60} molecule under experimental investigations renders such comparison meaningless. Calculated Raman frequencies are performed on individual isolated molecules. In contrast, experimental measurements are performed on fullerenes in a condensed state in the form of solid crystals (fullerite) or thin sublimated films. Perturbation fields resulting from the chemical potentials of neighboring molecules, molecules of the testing environment, and testing temperatures should all have a measureable effect on the Raman spectrum of the tested molecules (Section 2.11).

In the solid state, C_{60} molecules assume a simple cube (SC) unit cell that exhibits a phase transition into a face-centered cubic (FCC) upon heating around a transition temperature $T_{01} \approx 260$ K.[51] The transition is also associated with restriction of rotation of the C_{60} molecules. Above the transition temperature fullerene molecules rotate almost freely in the FCC unit cell. Once the "orientational ordering temperature" is reached, rotation of the fullerene molecules is restricted to rotation about one of the {111} directions in the cell, as shown schematically in Figure 4.3.

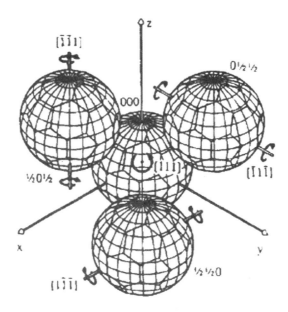

Figure 4.3 The four molecules in the FCC unit cell of molecular C_{60} solid. In this orientation, molecules at (0,0,0), (1/2,1/2,0), (1/2,0,1/2), and (0,1/2,1/2) rotate by the same angle about the local axes [111], [1,$\bar{1}$,$\bar{1}$], [$\bar{1}$,$\bar{1}$,1], and [$\bar{1}$,1,$\bar{1}$], respectively. The sense of rotation about each axis is shown. (Reproduced with kind permission from Copley et al., ref. 51. Copyright, Taylor and Francis Publishers 1993.)

In the solid state, C_{60} is nearly an ideal molecular solid with weak intermolecular van der Waals interactions. This fullerene–fullerene interaction and the associated symmetry change (due to crystal formation) induce a new set of intermolecular lattice vibration modes or phonons. Generally, lattice vibrational modes can be classified into acoustic, optical, or librational modes. Optical "lattice" modes are found in all solids with primitive unit cells containing more than two atoms or molecules. Hence, such optical modes are expected in the case of FCC C_{60} crystals. Librational modes, however, are specific for molecular crystals. They typically originate from the inertia of the individual molecule and are associated with a hindered rotation (rocking) of the molecules about their equilibrium lattice sites. For C_{60} molecular crystals, the frequencies of the librational and lattice modes are very low ($<40 \text{cm}^{-1}$ and $<100 \text{cm}^{-1}$, respectively) because of the weak molecular coupling (interaction) and the large molecular moment of inertia of fullerene molecules.[28,30,37,46,52–60] While some experimental investigations[61] attributed nine new modes, with frequencies ranging between 300 and 1620cm^{-1}, observed during their investigation of pressure-induced phase transition in C_{60} crystals to crystal modes, the frequencies are too high to fit the interpretation. Interaction between fullerene molecules and the solvent used as a pressure-transmission medium in the study is most plausibly the reason for the observed new modes. Solvent interactions and solubility of the fullerenes are discussed in a following section. During the "Fullerene Rush" in the early 1990s, several investigations led to conclusions that were later revised, or are still a controversy. For example, the origin of the non-polarized 1458cm^{-1} shoulder sometimes observed in the C_{60} spectrum under different measurement conditions had several interpretations. Such interpretations included association with the a downshift of the intrinsic pentagonal-pinch mode,[62–65] association with a photo-polymerized state of fullerene,[36–37,43,44,46–48] association with oxygen adsorption,[66,67] and the presence of two different phases of the fullerite crystal.[68,69] This controversy, and many others as we will discuss later, is not completely resolved.[40,70,71]

The power of Raman spectroscopy is best demonstrated in its ability to provide information regarding the presence of carbon isotopes (^{13}C) within the fullerene molecule. The isotope ^{13}C exists with a natural abundance of 1.1%. Hence, for C_{60} molecules, almost half of the molecules would have one or more ^{13}C isotope atoms in their structure.[26,46,72,73] Owing to the presence of the isotope atom(s) within the structure, the symmetry of the fullerene will be lowered, with an expected effect on Raman modes intensity and position. We mentioned earlier [Section 2.8.4, Equation (2.12)] that the frequency of a vibration mode (v) is actually scaled to the reduced mass of the atoms involved in the mode. Hence, the presence of one or two heavier isotopes in the fullerene cage is expected to lower the frequency of the mode involving such isotopes. Experimentally, Raman lines associated with $^{12}C_{60}$, $^{13}C_1{}^{12}C_{59}$, and $^{13}C_2{}^{12}C_{58}$ could be resolved with a separation of $1 \pm 0.02 \text{cm}^{-1}$ between these Raman peaks. Figure 4.4(a) shows high resolution non-polarized Raman spectrum around the 1469cm^{-1} "pentagon pinch" mode of a frozen

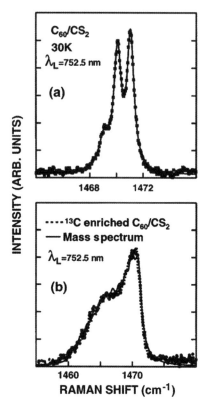

Figure 4.4 (a) High-resolution non-polarized Raman spectrum around the 1469 cm^{-1} "pentagon pinch" mode of a frozen solution of C_{60} in CS_2 at 30 K. The spectrum shows a three-Lorentzian fit to the experimental data. The highest wavenumber peak is assigned to the totally symmetric pentagonal pinch $A_g(2)$ mode in $^{12}C_{60}$. The other two lines are assigned to the pentagonal-pinch mode in molecules containing one and two ^{13}C atoms, respectively. (b) Measured unpolarized Raman spectrum in the pentagonal-pinch region for a frozen solution of ^{13}C-enriched C_{60} in CS_2 at 30 K (•) as well as the theoretical spectrum computed using the sample's mass spectrum (solid line). (Reproduced with kind permission from Guha et al., ref. 73. Copyright American Physical Society 1997.)

solution of C_{60} in CS_2 at 30 K. The spectrum shows a three-Lorentzian fit to the experimental data. The highest wavenumber peak is assigned to the totally symmetric pentagonal-pinch $A_g(2)$ mode in $^{12}C_{60}$. The other two lines are assigned to the pentagonal-pinch mode in molecules containing one and two ^{13}C atoms, respectively. Figure 4.4(b) shows the measured non-polarized Raman spectrum in the pentagonal-pinch region for a frozen solution of ^{13}C-enriched C_{60} in CS_2 at 30 K (points) as well as the theoretical spectrum computed using the sample's mass spectrum (solid line).

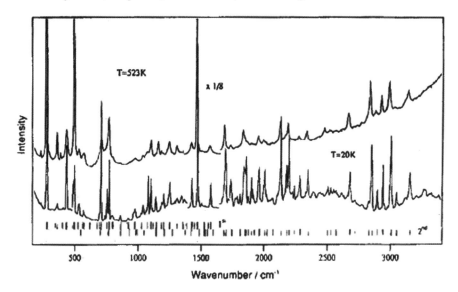

Figure 4.5 Full-range Raman spectra of C_{60} showing all first order in addition to overtones and combination modes of C_{60} at low ($T = 20$ K) and high ($T = 523$ K) temperatures. (Reproduced with kind permission from Dong et al., ref. 43. Copyright American Physical Society 1993.)

Higher-order Raman bands, including overtones and recombination, in C_{60} were also experimentally observed.[43] Figure 4.5 shows the spectrum recorded in the spectral range 100–3500 cm^{-1}.

4.2.2 Raman Scattering of C_{70}

Raman scattering of C_{70} has also been investigated extensively both theoretically[14,21,24,25,44,47,57,74–85] and experimentally.[47,56,81–83,85–94] The Raman spectrum of C_{70} is much more complicated than that of C_{60} because of the lower symmetry of the C_{70} molecule (D_{5h} point group, Section 2.6.1), and contains 53 Raman active modes. Figure 4.6 shows one of the earliest reported non-polarized Raman spectrum of C_{70} film on a silicon substrate.

Table 4.2 shows the assignment of mode symmetries for C_{70}. The table lists two assignments, one from the very early work on the subject in 1991 and the other is from one of the latest reports in 2008. The slight difference in the results is interesting. In general, it can be noticed that modes below *ca.* 900 cm^{-1} tend to have predominantly radial displacements, while the higher wavenumber modes have predominantly tangential displacements.

Regarding the Raman scattering of higher fullerenes, very little has been reported about the details of their vibrational spectra.[15,80] Several factors have contributed to this state of little knowledge, including lower symmetry, larger number of degrees of freedom, and the many possible isomers of such higher

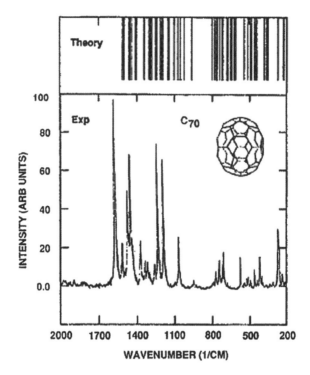

Figure 4.6 Non-polarized Raman spectrum of C_{70} film deposited on a silicon substrate. Superimposed are the calculated Raman active modes. (Reproduced with kind permission from Meilunas *et al.*, ref. 89. Copyright American Institute of Physics 1991.)

fullerenes. More importantly, the lack of adequate quantities of well-separated and characterized samples of higher fullerenes was a major reason for the dearth of experimental investigations.

4.2.3 Raman Scattering of Single-walled Carbon Nanotubes

The Raman spectrum of single-walled carbon nanotubes (SWCNTs) has also been investigated both theoretically and experimentally. Lattice dynamics and phonon symmetry investigations[95–102] for SWCNTs show that there are eight Raman active modes for achiral tubes and 14 Raman active modes for chiral tubes. The symmetry species of the Raman active modes can be classified as follows:

$$\text{Zigzag tubes}: \quad \Gamma = 2A_{1g} + 3E_{1g} + 3E_{2g} \tag{4.2}$$

$$\text{Armchair tubes}: \quad \Gamma = 2A_{1g} + 2E_{1g} + 4E_{2g} \tag{4.3}$$

$$\text{Chiral tubes}: \quad \Gamma = 3A_1 + 5E_1 + 6E_2 \tag{4.4}$$

Table 4.2 Symmetry assignment and frequency (cm^{-1}) of Raman active modes in C$_{70}$. Early and current state on knowledge.

Meilunas et al. (1991)[89]		Wang and Fang (2008)[408]	
V_{Raman}	Mode	V_{Raman}	Mode
224(S)			
		227(s)	E_2'
229 5			
252(S)		253(S)	A_1'
260	A_1'	259(S)	A_1'
		303(m)	E_2'
		309(S)	E_2'
		327(m)	A_1'
		361(w)	A_1'
		382(w)	A_1'
400		396(m)	A_1'
		410(m)	E_1''
413			
		419(w)	E_1''
		431(m)	E_2''
436			
457		457(S)	A_1'
		481(w)	E_1''
		490(vw)	E_2''
508		509(w)	E_1''
521		520(w)	E_2'
		535(vw)	E_2''
		568(S)	A_1'
572	A_1'		
		578(vw)	A_1'
		640(vw)	E_2'
		675(m)	A_1'
		700(S)	E_1''
704	E_1'		
715			
		721(m)	A_1'
		737(S)	E_1''
740	E_1'' or E_2'		
		796(m)	E_2'
771	E_1'' or E_2'	801(m)	E_2'
		898(m)	E_2'
		947(vw)	E_2'
		1012(w)	E_1''
1053(S)			
		1061(S)	A_1'
1063	A_1'		
		1086(w)	E_1''
1167	A_2''		
		1182(S)	E_1''
1187	A_1'		
		1228(S)	E_2'
1232	A_1'		
		1256(w)	E_1''
1259	E_1'' or E_2'		

(*Continued*)

Table 4.2 Continued.

Meilunas et al. (1991)[89]		Wang and Fang (2008)[408]	
V_{Raman}	Mode	V_{Raman}	Mode
		1294(m)	E_1''
1301			
1316	E_1'' or E_2'		
		1332(m)	E_2'
1335	E_1'' or E_2'		
		1349(m)	A_1'
		1367(m)	A_1'
1371	E_1'		
		1373(w)	
1439(S)	E_1'		
1443(S)			E_1''
		1445(S)	E_2'
1449	A_1'		
1461 (S)			
1463(S)			
		1469(S)	E_1''
1471	A_1'		
		1512(S)	E_1''
1515	E_1'' or E_2'		
		1565(S)	E_2'
1569	A_1'		

Figure 4.7 shows the displacement vector for Raman active modes of (a) armchair and (b) zigzag SWCNTs. As shown in the figure, for armchair tubes the two totally symmetric (A_{1g}) modes originate from the radial breathing mode (RBM) and the displacement of atoms in the circumferential (transverse) directions. For zigzag tubes, the two totally symmetric (A_{1g}) modes originate from the radial breathing mode and the longitudinal (shear) displacement of the carbon atoms. In chiral tubes, however, both longitudinal and transverse phonons are fully symmetric (A_1).[101–106]

In general, the Raman spectrum[103–112] of SWCNTs shows two classes of peaks: a low wavenumber class (usually less than 300 cm^{-1}) resulting from radial displacements of the carbon atoms (radial breathing mode of the tube), and a high wavenumber class resulting from tangential displacements of the carbon atoms (tangential or surface modes) in the range 1350–1580 cm^{-1}. Second-order modes can also be observed in the Raman spectrum around 2700 cm^{-1}. Figure 4.8 shows a typical spectrum of SWCNTs, showing the RBM, the tangential modes, and the second-order modes as measured using a near-infrared 785 nm wavelength laser excitation and recorded at ambient conditions. The higher wavenumber modes include the D-band (around 1360 cm^{-1}), which is related to structural defects in the nanotube, the G-band (three modes belonging to the A, E_1, and E_2 symmetry in the range 1550–1605 cm^{-1}), and the G′ band (a second order mode around 2700 cm^{-1}). Features marked with '*' in the figure at 303, 521, and 963 cm^{-1} are from the Si/SiO substrate.[103]

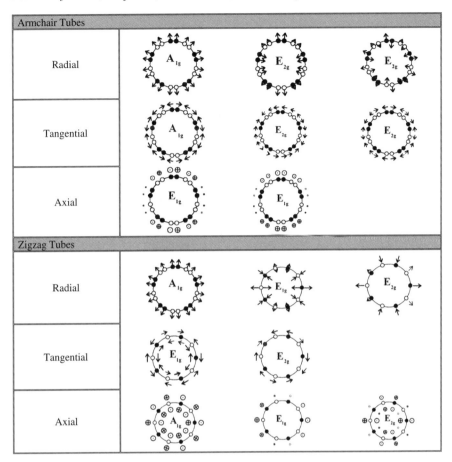

Figure 4.7 Schematic diagrams depicting the atomic displacement vectors corresponding to all Raman active vibration modes in armchair (n,n) and in zigzag (n,0) SWCNTs.

Several studies have investigated the correlation between Raman spectrum characteristics and SWCNTs structure and properties. The position of the RBM (v) in SWCNT bundles was reported to correlate linearly with the tube diameter (d) according to the relationship:[113,114]

$$v(\text{cm}^{-1}) = \frac{223.75}{d(\text{nm})} + 14 \tag{4.5}$$

When isolated individual SWCNTs were investigated, the correlation between the RBM position and tube diameter was reported to be:[98,103,110]

$$v(\text{cm}^{-1}) = \frac{248}{d(\text{nm})} \tag{4.6}$$

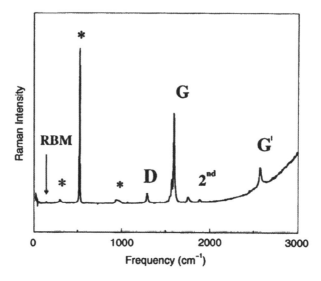

Figure 4.8 Full Raman spectrum of SWCNT taken with excitation laser (785 nm). (Reproduced with kind permission from Jorio et al., ref. 103. Copyright American Physical Society 2001.)

and later reported as:[115]

$$v(\text{cm}^{-1}) = \frac{214.4 \pm 2}{d(\text{nm})} + 18.7 \pm 2 \tag{4.7}$$

The difference in Raman position of the RBM was related to tube–tube interactions within a tube bundle. Tube interaction with their supporting substrate or their environment[116–119,98] would also cause a shift in the tube RBMs due to perturbation effects as we discussed earlier (Section 2.11). Such perturbation effects can cause significant shifts in the RBM of SWCNTs, rendering Equations (4.5)–(4.7) useless.

Intensities of RBMs were also found to depend on the energy of the excitation laser due to Raman resonance effects (Section 2.12). Metallic tubes with $|n-m| = 3q$ and semiconducting tubes with $|n-m| = 3q \pm 1$ (where q is an integer, see Section 3.9.1) would resonate at different excitation laser energies according to their diameter, which controls their electronic structure. It has been established[117,120–122] that once the excitation laser energy is within 0.1 eV of the van Hove singularity[i] inter-band transition (E_{ii}) for a particular tube, the radial breathing mode of that particular tube will exhibit a resonance Raman effect and its RBM band will be prominent in the Raman spectrum. Hence

---

[i] For readers without physics background, an elementary solid state physics reference can be consulted for further information on van Hove singularities. However, the point here is to realize that depending on the excitation laser wavelength certain types (metallic versus semiconducting) with certain diameter nanotubes will be in resonance; hence their RBM will be more prominent in the spectrum.

certain nanotubes can be more observable using the typical argon ion green laser (wavelength 514.5 nm) while other nanotubes can be more observable using a typical solid state laser (wavelength 780 nm).

Raman dispersion effects (Section 2.15) were clearly observed in the D-band and its second-order G'-band features of both metallic and semiconducting SWCNT Raman spectra.[98,101,108,121,103–105,123–137] Owing to dispersion effects, the observed frequency of the D-band ranges between 1250 and 1450 cm^{-1}. The frequency of the second-order feature G', however, was reported to range between 2500 and 2900 cm^{-1}, also due to dispersion effects. The observed dispersion effect in the frequency of both bands showed a linear dependence on the laser excitation energy (E_L) within the range $1.0 < E_L < 4.5$ eV with a plateau or step-like feature around excitation energy of 2 eV.[98,111,138,139] The step-like feature was related to electron–phonon coupling effects. The frequency of the G' ($\omega_{G'}$, in cm^{-1}) linear dependence on the excitation laser energy (E_L in eV) can be fitted to the equation:

$$\omega_{G'} = 2420 + 106 E_L \tag{4.8}$$

Figure 4.9 shows the experimental results measured for the dispersion effect on the peak position of the G' band for laser excitation energies ranging between 1.5 and 3.0 eV.

More recently, dispersion effect in graphitic material in general, and in carbon nanotubes in particular, was related to a double resonance scattering effect.[105,136,137a,140–142] Figure 4.10 shows the experimental results *versus* calculations – based upon double-resonance theory – for the dependence of the D-band on laser excitation energy. Clearly, the calculations are in good agreement with experimental results reported by several groups.

Another important capability of the Raman technique is its ability to discriminate metallic and semiconducting nanotubes based on the shape of the G-band.[140,143–148] The crucial point to note here is that, in a nanosystem such as a SWCNT, slight changes in the system features, chirality for example, lead to measurable changes in the system's properties that can be investigated using the Raman technique. Notably, while Raman investigation of SWCNTs started as early as their discovery in the early 1990s, the richness of the technique capabilities as a diagnostic tool for carbon nanotubes is still being explored.[106,131,148–181]

Polarized Raman also plays a major role in investigating the structure of carbon nanotube based systems.[182,183] The technique provides invaluable information regarding the orientation of nanotubes. Table 4.3 shows the Raman scattering tensors[184–190] of the phonons in carbon nanotubes. The intensity of Raman active mode depends on the polarization scattering settings (Section 2.14) as measured in reference to the tube axial direction. Figure 4.11 shows polarized Raman spectra (collected in a backscattered ||/|| geometry) of an ultrathin film (<60 nm) of oriented SWCNTs. Clearly, from the figure, the intensity of the Raman active modes is highest when the laser polarization direction is parallel to the tube axis direction ($\theta = 0°$). The Raman active modes, however, disappear when the laser polarization direction is normal to the tube

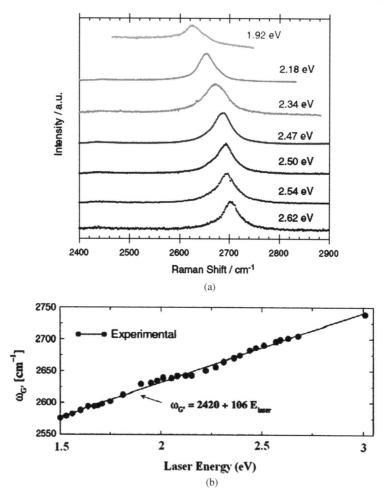

Figure 4.9 Dispersion effect in the G'-band of isolated SWCNTs: (a) Raman spectra; (b) G'-band position plotted as a function of excitation laser energy. [(a) Reproduced with kind permission from Shimada et al., ref. 106. Copyright Elsevier 2005. (b) Reproduced with kind permission from Souza Filho et al., ref. 138. Copyright American Physical Society 2001.]

axis ($\theta = 90°$). Figure 4.12 shows the experimentally measured intensities (collected in a backscattered ||/|| geometry, a VV geometry) of all Raman active bands in an isolated SWCNT as a function of orientation angle (θ) between the excitation laser polarization and the tube axial directions.

The ability of the Raman technique to measure orientation of nanotubes is essential to investigating nanostructured mesosystems. Figure 4.13 shows color-coded polarized Raman maps depicting the orientation of SWCNTs within two different films.[191] The map in Figure 4.13(a) shows a very well-aligned film, while that in Figure 4.13(b) shows a film that contains domains of

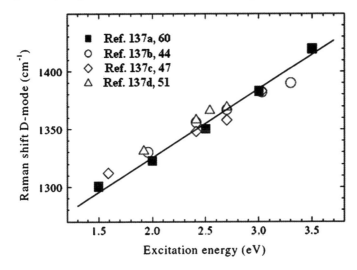

Figure 4.10 Measured and calculated frequencies of the D band as a function of the excitation energy. The open symbols correspond to experimental data and the closed squares to the calculated phonon energies in double resonance. The line is a linear fit to the theoretical values. The numbers following the references give the corresponding slopes (in cm^{-1}/eV) for the data reported by each research group. (Adapted with kind permission from Thomsen and Reich, Ref. 137a. Copyright American Physical Society, 2000.)

Table 4.3 Raman scattering tensors for different modes in carbon nanotubes.

$A_{1(g)}$	$A_{2(g)}$	$E_{1(g)}$	$E_{1(g)}$	$E_{2(g)}$	$E_{2(g)}$
$\begin{pmatrix} a & 0 & 0 \\ 0 & a & 0 \\ 0 & 0 & b \end{pmatrix}$	$\begin{pmatrix} 0 & e & 0 \\ -e & 0 & 0 \\ 0 & 0 & 0 \end{pmatrix}$	$\begin{pmatrix} 0 & 0 & c \\ 0 & 0 & 0 \\ d & 0 & 0 \end{pmatrix}$	$\begin{pmatrix} 0 & 0 & 0 \\ 0 & 0 & -c \\ 0 & -d & 0 \end{pmatrix}$	$\begin{pmatrix} 0 & f & 0 \\ f & 0 & 0 \\ 0 & 0 & 0 \end{pmatrix}$	$\begin{pmatrix} -f & 0 & 0 \\ 0 & f & 0 \\ 0 & 0 & 0 \end{pmatrix}$

well-aligned tubes separated by boundaries of misaligned tubes. Based on our previous discussions, it can be predicted that the electrical, thermal, optical, and most probably chemical properties of these two films will be different. However, this particular area is still a frontier to explore by systematic scientific investigation driven by very promising evidence, as we discussed in Chapter 1.

4.2.4 Raman Scattering of Double- and Multi-walled Carbon Nanotubes

The Raman spectrum of multi-walled carbon nanotubes (MWCNTs) typically shows the D, G, and G' bands. Radial breathing modes typical of SWCNTs are not part of a multi-walled tube. The simplest explanation for this is that the radial breathing mode requires all carbon atoms to translate in-phase in the

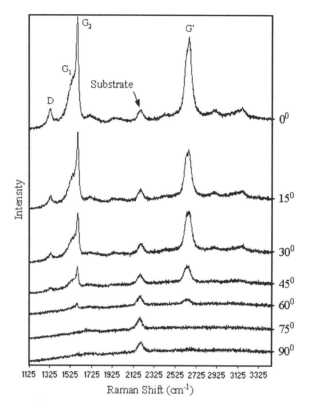

Figure 4.11 Polarized Raman spectra collected in a backscattered ||/|| setup of an ultrathin film ($t < 60$ nm) of oriented SWCNTs deposited on a substrate. Each spectrum is collected at a different angle (θ) between the polarization direction and the tube axis direction. (Reproduced from Reber and Amer, ref. 182. Courtesy of Professor M. Amer.)

radial direction. The possibility for such a mode to occur in multi-concentric tubes is very low. In 2008, however, while investigating the Raman spectra of MWCNTs with argon ion (514.5 nm) laser power ranging between 5 and 20 mW, Rai et al.[192] observed RBMs characteristic of SWCNTs as the laser power is increased to 10, 15, and 20 mW. The RBMs appear in the range 200–610 cm^{-1}. They were attributed to the local synthesis of SWCNTs at the top surface of the samples due to higher laser power. Interestingly, the spectrum recorded at a laser power of 20 mW shows radial breathing modes only! All higher frequency modes are not observed in the spectrum recorded at such high laser power. More interestingly, exposing the sample to 20 mW of laser power and then recording the Raman spectrum at 5 mW resulted in, basically, a similar spectrum to that recorded at 20 mW of laser power. These results require further investigation to better understand such a phenomenon. Figure 4.14 shows the Raman spectra for MWCNTs recorded using different laser powers.

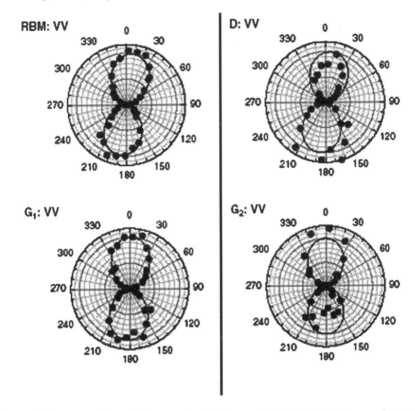

Figure 4.12 Intensities of Raman bands for an isolated SWCNT experimentally measured in a VV configuration as a function of orientation angle (θ) between tube axis and laser polarization direction. (Reproduced with kind permission from Duesberg et al., ref. 183. Copyright American Physical Society 2000.)

Double-walled carbon nanotubes (DWCNTs), in contrast, provide the simplest system to study the perturbation effects on nanotubes, especially the confinement due to interaction between concentric nano-cylinders. Since SWCNTs can be either metallic (M) or semiconducting (S), DWCNTs can assume any of the four possible configurations (i.e., M@M, M@S, S@S, and S@M). Each of these configurations would have different electronic properties and, hence, a different Raman spectrum. The configurations available for MWCNTs are yet more versatile and sophisticated. Several investigations have utilized Raman scattering to understand the dependence of the electronic and optical properties of double- and multi-walled carbon nanotubes on their configuration.[135,192–199] Considering the recently developed techniques to remove end caps and shorten nanotubes,[200] or, in other words, tailoring nanotubes, and to selectively synthesize different types of single and multi-layered nanotubes,[201] such investigations definitely have important implications in the fabrication of electronic devices using different types of

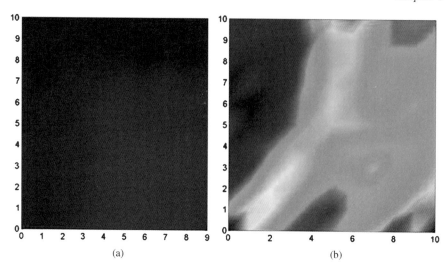

Figure 4.13 Polarized Raman maps of SWCNT films on silicon substrates. (a) A 10 × 10 μm orientation map of a well-aligned homogeneous film and (b) a 10 × 10 μm map showing multi-domains with boundaries of unaligned regions. (Reproduced from Amer, ref. 191. Courtesy of Professor M. Amer.)

semiconducting and metallic nanotubular interconnects. Figure 4.15 shows Raman spectra of the RBM, G band, and G′ band features taken from the same isolated DWCNT consisting of a semiconducting tube in a metallic one. By using two laser energies (shaded regions), the inner semiconducting ($E_L = 1.91$ eV) and the outer metallic ($E_L = 1.61$ eV) constituents of the DWCNT can be detected. The peak frequencies are marked followed by the corresponding full-width at half maximum (FWHM) in parenthesis. A blank frame in the figure means that no G′ data was acquired due to luminescence from the silicon substrate. Notably, once the structure of the DWCNT was reversed, *i.e.*, a metallic tube inside a semiconducting tube, the Raman spectrum of the tube was different. In this case both metallic and semiconducting tubes were excited – meaning that their RBMs were observable – at the same excitation laser energy. Figure 4.16 shows the Raman spectrum of an isolated M@S DWCNT ($E_L = 2.33$ eV) depicting the RBM, G, and G′ bands of both tubes.

Inter-tube interaction and its effect on tube properties (especially electrical, thermal, and optical) as reflected in the tube Raman activity and band shape and position is also illustrated by the observed shifts in the G′-band positions of single-walled and multi-walled carbon nanotubes once they are mixed together. Amer[202] found that once SWCNTs are mixed with multi-walled nanotubes, the position of the G′-band of both types of tubes exhibits a blue-shift. The shift is linear as a function of SWCNT weight (%) in the mixture. The mixtures were prepared by dissolving both types of tubes in dimethylformamide (DMF), mixing the solutions and sonicating the mixed solution for 30 min. Raman

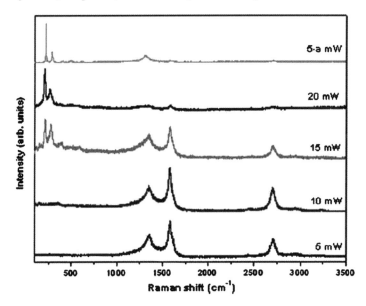

Figure 4.14 Micro-Raman spectra recorded using a 514.5 nm laser at laser power ranging between 5 and 20 mW on the top surface of the MWCNT samples. Spectrum 5-a was recorded at 5 mW power again after exposing the sample to 20 mW laser power. (Reproduced with kind permission from Rai et al., ref. 192. Copyright Elsevier 2008.)

spectra were recorded at ambient conditions using a 514.5 nm argon ion laser after the mixtures were allowed to completely dry on glass slides. Figure 4.17 shows the blue-shift exhibited by both single- and multi-walled carbon nanotubes once allowed to mix together reflecting their interaction effects on their electronic and optical behaviour.

4.2.5 Raman Scattering of Graphene

Owing to its potential numerous applications and predicted and demonstrated superior electrical and thermal properties, graphene has been very extensively investigated both theoretically and experimentally in the past five years.[203–241] Raman scattering, with its powerful ability to investigate vibration modes and perturbation effects, is the natural technique to investigate such a building block and has been heavily utilized in investigating fullerene structural features and superior electrical and thermal performance.[134,221,233,237,242–255] The Raman spectrum of graphene mainly exhibits two major Raman active modes: a G-band around 1580 cm^{-1} and a G'-band (also referred to as a 2D-band in the literature) around 2700 cm^{-1}. Figure 4.18 shows Raman spectra of graphene and graphite for comparison.

Raman scattering features of graphene were found to depend sharply on the number of graphene layers in the sample, thermal effects,[244,247] and, more

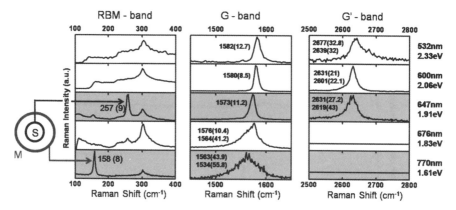

Figure 4.15 Raman spectra of the RBM, G band, and G' band features taken from an isolated S@M DCWNT. By using two laser energies (shaded regions), we can detect the inner semiconducting ($E_L = 1.91$ eV) and the outer metallic ($E_L = 1.61$ eV) constituents of the same DWCNT. Peak frequencies are marked followed by the corresponding FWHM in parenthesis. A blank frame means that no G' data was acquired for 1.61 and 1.83 eV due to luminescence from the Si substrate. (Reproduced with kind permission from Villalpando-Paez et al., ref. 135. Copyright American Chemical Society 2008.)

interestingly, on the interaction between the graphene sample and its supporting substrate[243,249] and interaction with any dopants.[250] Figure 4.19 shows the evolution of the graphene Raman spectrum as the number of graphene layers increases. High-frequency first- and second-order Raman spectra of graphene films supported on a SiO_2:Si substrate and that of HOPG (highly-ordered pyrolytic graphite). The data were collected using 514.5 nm radiation under ambient conditions. Interestingly, the shape and position of the Raman bands are sensitive to the number of layers (n) in the graphene film. Figure 4.20 shows the evolution of the G-band peak intensity and position as a function of the film's number of graphene layers (n). It was reported that the G-band position shifts *apparently linearly* to lower wavenumbers as the number of graphene layers increases (inset in Figure 4.20). The observed dependence of G-band intensity on the number of graphene layers (Figure 4.20) was utilized recently in generating Raman images of graphene sheets, distinguishing the number of layers of the graphene sheet based on the integrated intensity of its G-band. Figure 4.21 shows optical image of graphene with 1, 2, and 3 layers (upper), and Raman image plotted by intensity of the G band (lower). It is important to emphasize that while Raman intensity of the G-band (with its reported linear dependence on the number of graphene layers up to ten layers) was suggested as an *excellent, precise and quick* method to determine the number of layers in graphene, extreme caution should be taken while applying such a method. Based on our previous and following discussions, it should be clear that unaccounted for perturbation effects could easily render such a method misleading. In addition, it is important to note that the exact

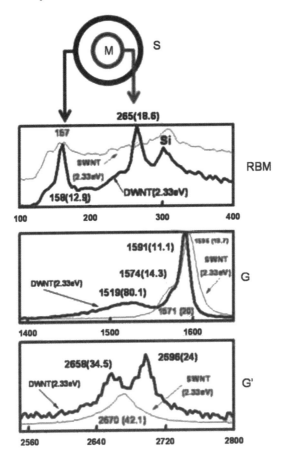

Figure 4.16 Raman spectra of the RBM, G band, and G′ band features taken (using $E_L = 2.33$ eV) from an isolated M@S DWCNT. The RBM region shows that the inner and the outer tubes are simultaneously in resonance with the same laser line. (Reproduced with kind permission from Villalpando-Paez et al., ref. 135. Copyright American Chemical Society 2008.)

interaction between successive graphene layers (certainly including the exact orientation of the graphene layers) will definitely affect the collective Raman spectrum of the graphene sheet. Investigation[256] of single, double, and two (folded) graphene layers showed that the Raman spectrum of the folded two-layer graphene is different from a double layer graphene (Figure 4.22).

A disorder (or double resonance) induced D-band around 1360 cm^{-1} was also observed in graphene Raman characteristic spectra. As shown in Figure 4.23, the D-band intensity changes with the number of graphene layers in the film. Strains induced in the graphene film due to substrate roughness (1–2 nm) are thought to be the reason for the observed D-band.

The perturbation effect of the supporting substrate on Raman active modes of graphene was also investigated and reported.[233,241,243,249,257] Room-temperature

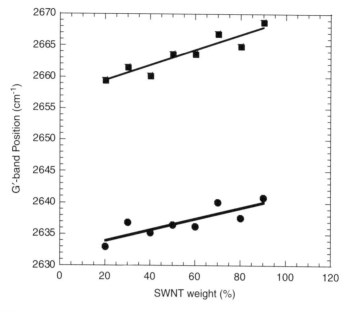

Figure 4.17 Measured G′-band peak position for SWNT (●) and MWNT (■) mixtures as a function of SWNT wt% in the mixture. (Reproduced from Amer, ref. 202. Courtesy of Professor M. Amer 2008.)

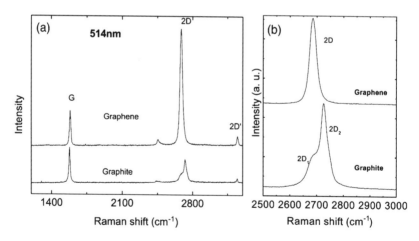

Figure 4.18 (a) Comparison of the Raman spectra of graphene and graphite measured using a 514.5 nm excitation laser. (b) Comparison of the 2D peaks in graphene and graphite. (Reproduced with kind permission from A. Ferrari, ref. 245. Copyright Elsevier 2007.)

peak positions of Raman active modes (G-band and G′-band) from single layer graphene interacting with different substrates showed significant shifts (up to $11\,\text{cm}^{-1}$) depending on the substrate. Figure 4.24 shows Raman spectra of single layer graphene interacting with different substrates. Table 4.4 shows

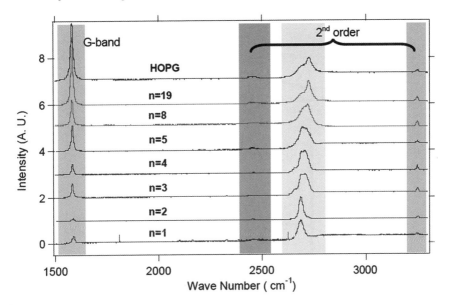

Figure 4.19 High-frequency first- and second-order Raman spectra of (n) graphene layer films supported on a SiO_2:Si substrate and that of HOPG. Data were collected using 514.5 nm radiation under ambient conditions. The spectra are scaled to produce an approximate match in intensity for the $\sim 2700\,cm^{-1}$ band. (Reproduced with kind permission from Gupta et al., ref. 242. Copyright American Chemical Society 2006.)

the exact positions and FWHM of the G-, and the G'-bands. The precise nature of the substrate perturbation effect is still to be explored and better understood.

Another important perturbation effect to consider is the effect of an applied electric field (applied voltage) across graphene films. Graphene is very promising for electronic device applications and has been shown to demonstrate unique electronic performance best demonstrated by unique properties of charge carriers in graphene such as particle–hole symmetry of Dirac fermions. The coupling of long wavelength optical phonons (the G-band) with Dirac fermions was found to display remarkable changes in frequency and line-width that are tunable by the electron field effects.[246] Applying a voltage across a single-layer graphene sheet significantly shifts the G-band position of the graphene. Shift *versus* applied voltage enable the determination of the Dirac point of the graphene sheet. Figure 4.25 shows Raman spectra of single-layer graphene sheets depicting the G-band at different applied gate voltages at temperatures of 10 K (a) and 300 K (b).

Once again, Raman spectroscopy proved to be *the technique* to investigate fullerene building blocks. As we discussed before (Section 3.11) structural defects (or imperfections) in graphene sheets would definitely affect the graphene's electronic and optical properties and, hence, are expected to impact the

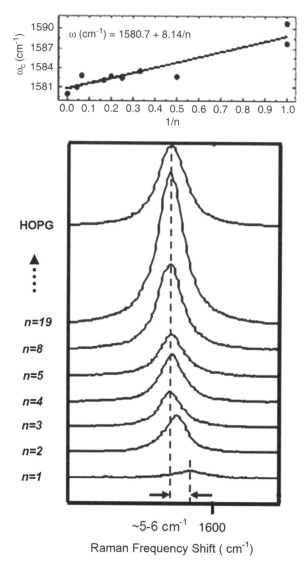

Figure 4.20 Evolution of the G-line in graphene films with different number of layers. Data were collected using 514.5 nm radiation under ambient conditions. Inset: apparent linear shift in G-band position with increasing number of graphene layers.(Reproduced with kind permission from Gupta *et al.*, ref. 242. Copyright American Chemical Society 2006.)

characteristics of its Raman spectrum. Recent density functional theory (DFT) based calculations[254] investigating the impact of different types of point and topological defects, namely mono-vacancy (MV), di-vacancy (DV), and Stone–Wales (SW) defect (see Figure 3.46), on Raman spectra of graphene single

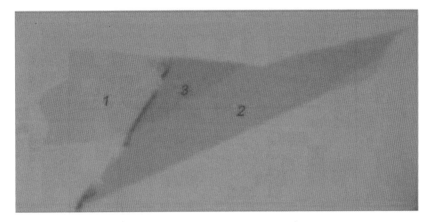

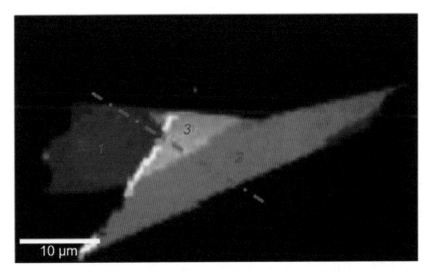

Figure 4.21   Optical image of graphene with 1, 2, and 3 layers (a) and Raman image plotted by intensity of the G band (b). (Reproduced with kind permission from Ni *et al.*, ref. 251. Copyright Springer 2008.)

sheets showed that such defects would induce new phonons, resulting in a new Raman active mode in the graphene Raman spectrum. Based on the calculated Raman spectra, Raman lines that can serve as signatures of the specific type of investigated defects can be determined. In addition, it was predicted (based on the calculations) that the presence of defects would enhance the intensity of the G-band of graphene up to one order of magnitude compared to defect-free graphene. Figure 4.26 shows the calculated Raman spectra of graphene with different point defects for five laser photon energies (E_L): (a) MV, (b) DV, and (c) SW. Only the frequency region with more prominent Raman features is shown, and some of the spectra are scaled for better presentation. Note that the

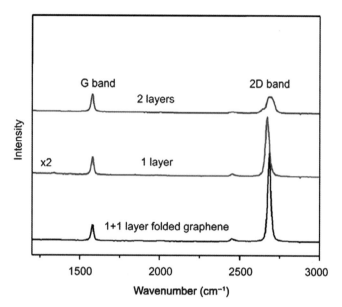

Figure 4.22 Raman spectra of double-layer graphene (DLG), single-layer graphene (SLG), and 1 + 1 layer folded graphene. Spectra are normalized to have a similar G band. (Reproduced with kind permission from Ni et al., ref. 256. Copyright American Physical Society 2008.)

calculation results presented in Figure 4.26 are for a free-standing single graphene sheet in pure vacuum. The positions, intensities, and resonance ranges for the predicted Raman modes would definitely be affected by substrate interaction, environment, and other experimentally unavoidable conditions.

4.2.6 Thermal Effects on Raman Scattering

Thermal properties of carbon nanotubes and the temperature dependence of their Raman active modes have frequently been investigated experimentally[111,258–267] and theoretically.[268,269] In general, all investigations have consistently reported two observations: first, that the intensities of all Raman active modes decrease with increasing the measurement temperature (Figure 4.27) and, second, that the frequencies of all Raman active modes exhibit a linear redshift (softening) as the temperature is increased (Figure 4.28). It was also reported[264,265] that while the slope of the linear frequency dependence on temperature ($d\omega/dT$) is independent of the tube diameter for the G-band, it increases nonlinearly as the tube diameter increases for the radial breathing mode (Figure 4.29). Similar linear dependence of the G-band position on temperature was reported for single- and multi-layer graphene.[244,247] The thermal coefficient ($d\omega/dT$) reported for bi-layer graphene ($0.015 \, cm^{-1} \, K^{-1}$) is slightly higher than that reported for HOPG ($0.011 \, cm^{-1} \, K^{-1}$).

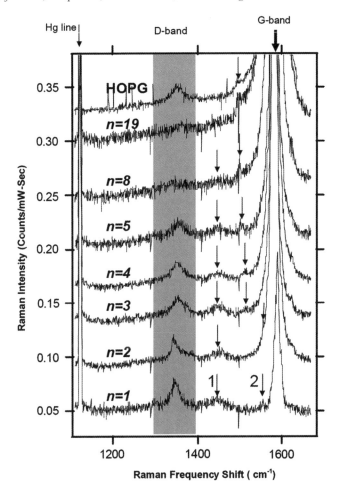

Figure 4.23 Raman spectra (expanded intensity scale) of supported graphene films with (n) layers and HOPG, showing the D-band. Data were collected using 514.5 nm radiation under ambient conditions. Arrows in the figure identify weak scattering features. The mercury (Hg) line was used for calibration purposes. (Reproduced with kind permission from Gupta et al., ref. 242. Copyright American Chemical Society 2006.)

Recalling from Section 2.10 that the Stokes/anti-Stokes intensity ratio (I_S/I_{AS}) is related to the local temperature of a materials system according to Equation (2.16), it could be tempting to assume that determining the local temperature in nanotubes is straightforward. The fact that radial breathing modes are typically located at wavenumbers lower than 300 cm^{-1} making their experimental intensity ratio measurements possible and accurate increases such temptation. However, the resonance dependence of the radial breathing mode of SWCNTs on excitation laser energy, discussed earlier, puts serious

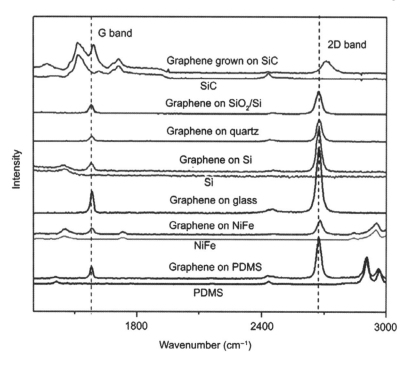

Figure 4.24 Raman spectra of single-layer graphene deposited on different substrates. (Reproduced with kind permission from Wang et al., ref. 257. Copyright American Chemical Society 2008.)

Table 4.4 Position and full width at half maximum (FWHM) of Raman active modes in graphene monolayer on various substrates.

Substrate	G-band Position (cm^{-1})	G-band FWHM (cm^{-1})	G'-band position (cm^{-1})	G'-band FWHM (cm^{-1})
SiO_2/Si	158,[a] 1580.5	15,[a] 288.2[b]	2710.5[b]	59.0[b]
GaAs	1580[a,b]	15[a,b]		
Sapphire	1575[a]	20[a]		
Glass	1580,[a,c] 1582.5[b]	35,[a] 16.8[b]	2672.8[b]	30.8[b]
Si	1580[b]	16[b]	2672[b]	28.3[b]
Quartz	1581.9[b]	15.6[b]	2674.6[b]	29.0[b]
NiFe	1582.5[b]	288.9[b]	2678.6[b]	30.8[b]
PDMS	1581.6[b]	15.6[b]	2673.6[b]	27[b]

[a]Calizo et al.[243]
[b]Wang et al.[257]
[c]G-band split on this substrate, the value represent the middle frequency.

restrictions on such scientific temptation. Fantini et al.[111] reported on the Stokes and anti-Stokes intensities of the radial breathing mode of in-solution individual as well as in-air bundles of SWCNTs. The intensities were measured using different excitation laser energies, covering the range 1.9–2.4 eV. They

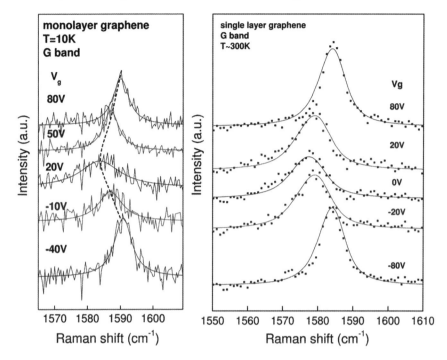

Figure 4.25 Raman spectra of single-layer graphene sheets, depicting the G-band at different applied gate voltages (V_g) at temperatures of 10 K (a) and 300 K (b). (Reproduced with kind permission from Yan et al., ref. 246. Copyright Elsevier 2007.)

show an interesting dependence of both intensities on laser energy. Figure 4.30 shows Stokes (solid symbols) and anti-Stokes (open symbols) experimental resonance windows obtained for the same (n,m) SWCNT (RBM 244.4 cm^{-1}) dispersed in aqueous solution and wrapped with SDS (a) and in a bundle (b). Clearly, from the figure, the I_S/I_{AS} ratio does, indeed, depend on the excitation laser wavelength. Comparison of the two graphs also indicates that perturbation effects from the tube environment also have an effect on the measured ratio. This clearly indicates that due to resonance and perturbation effects the I_S/I_{AS} ratio may not be characteristic of the tube temperature. While suggestions to utilize the calibrated linear dependence of phonons' frequencies on temperature as a method to obtain sample temperature[101] have been given, it is highly recommended to avoid the radial breathing modes due to their thermal coefficient dependence on exact tube diameter (Figure 4.29).

To this end we can summarize the benefits of Raman scattering investigations of carbon nanospecies as follows: Raman scattering provides a great investigating tool capable of identifying different types of fullerene molecules. It can discriminate the exact type of fullerene with different dimensionalities. To clarify, it can distinguish between C_{60} and C_{70} and also distinguish between

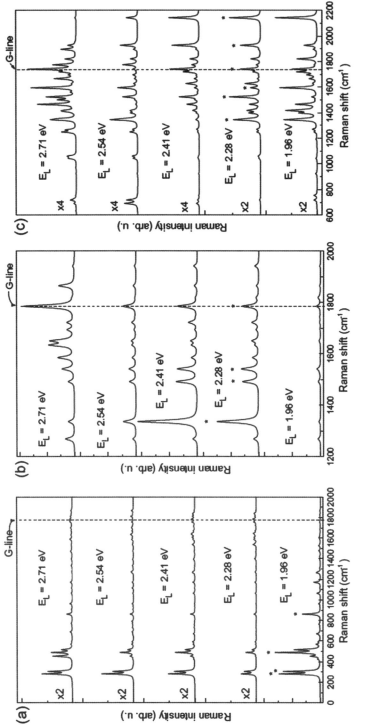

Figure 4.26 Calculated Raman spectra of graphene with different point defects for five laser photon energies EL: (a) MV, (b) DV, and (c) SW. Only the frequency region with more prominent Raman features is shown. Some of the spectra are scaled for better presentation. The most intense Raman lines are denoted by asterisks. (Reproduced with kind permission from Popov *et al.*, ref. 254. Copyright, Elsevier 2009.)

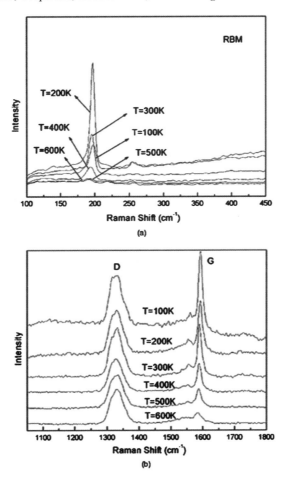

Figure 4.27 Variation of relative intensity of (a) radial breathing mode (RBM) and (b) D- and G-bands with changing temperatures for individual SWCNTs. (Reproduced with kind permission from Zhou *et al.*, ref. 264. Copyright American Chemical Society 2005.)

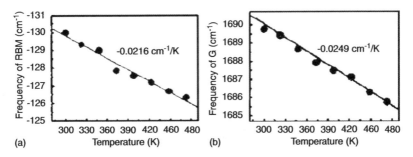

Figure 4.28 Temperature dependence of the RBM (a) and G-band (b) of a suspended individual SWCNT. (Reproduced with kind permission from Zhang *et al.*, ref. 265. Copyright American Chemical Society 2007.)

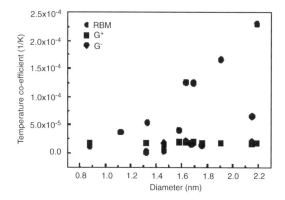

Figure 4.29 Slope of frequency dependence on temperature (temperature coefficient $d\omega/dT$) as a function of tube diameter for RBM (•) and the high and low frequency G-bands (■ and ▼, respectively) in SWCNTs. (Reproduced with kind permission from Zhang et al., ref. 265. Copyright American Chemical Society 2007.)

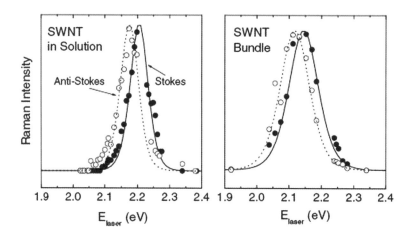

Figure 4.30 Stokes (filled symbols) and anti-Stokes (open symbols) experimental resonance windows obtained for the same (n,m) SWCNT (RBM 244.4 cm^{-1}) dispersed in aqueous solution and wrapped with SDS (sodium dodecyl sulfate) (a) and in a bundle (b). (Reproduced with kind permission from Fantini et al., ref. 111. Copyright American Physical Society 2004.)

single-, double-, and multi-walled nanotubes. In addition, polarized Raman spectroscopy has proven to be a powerful tool capable of measuring the orientation of carbon nanotubes. More importantly, from the viewpoint of nanostructured systems investigation, Raman scattering can be considered as the best technique to investigate fullerene interactions with their environment.

Finally, we emphasize an important point regarding tube diameter determination based on its radial breathing mode position. As discussed earlier, while many papers are still utilizing any of Equations (4.5)–(4.7) in determining the tube diameter based only on its radial breathing mode position, it should be emphasized that without cautious consideration of perturbation effects such equations can be very misleading.

4.3 Fullerene Solubility and Solvent Interactions

Solubility is perhaps the most important property of a substance to investigate if chemical performance of the substance to be understood.[270] In 1912, D. Tyrer reviewed the solubility theory as understood then and stated that:

> "The maximum amount of a substance that a liquid will hold in homogeneous solution at a given temperature, or as it is termed, the solubility of that substance in the given liquid, represents a physical constant which, *as yet*, it has been found impossible to connect in any consistent manner with any other physical constants or properties of either the solvent or the solute."[271]

Despite the fact that during the 90 years followed our understanding of "solubility" has advanced dramatically, it can be said that we are still on a steep learning curve once fullerenes, and nano-moieties in general, are considered. In the early 1990s fullerene solubility in various solvents was investigated extensively.[272–276] Of all fullerene family members, C_{60} and C_{70} solubility were most investigated and reported in the literature.[277–296] The results, in general, support the age-old principle "*similia similibus solvuntur*", which is Latin for "like dissolves like".[274] Table 4.5 shows the solubility of C_{60} and C_{70} in various solvents. It is interesting to note the wide range of solubility limits in certain solvents. Perhaps, the most important point to emphasize regarding the data presented in Table 4.5 is the huge reported difference in solubility limit between solvents that are not essentially different in their chemical structure or properties. For example, the solubility limit of C_{60} was reported to be 4.702 and 17.928 mg mL^{-1} in 1,2,3-trimethylbenzene, and 1,2,4-trimethylbenzene, respectively.[279] In addition, attempts to find a quantitative relation between the solubility limit (in molar fractions or in mg mL^{-1}) and traditional characteristics of solvents such as polarizability, polarity, molar volume, or Hildebrand's solubility parameter (δ) have not been very successful.[283,284] In addition, notably, while SWCNT is fairly soluble in dimethylformamide (DMF),[297] C_{60} is barely soluble in this solvent, clearly emphasizing the molecular shape and entropic effects on fullerene–solvent interactions. These and many other interesting phenomena observed for fullerene behavior in solutions added richness to the subject and sparked an interest that is still active and is expected to remain so for some time.[298–305]

Table 4.5 Solubility limits of C_{60}, and C_{70}, in different solvents.

Solvent	Solubility of C_{60} (mg mL^{-1})	Solubility of C_{70} (mg mL^{-1})	Reference
Benzene	1.699	1.116[a]	280
	1.7		290
	1.498		281
	1.440		283
	0.878		285
	1.397		285[b]
	1.858		286
Toluene	2.801		280
	2.8		290
	2.902		281
	2.290	0.92884	291
	2.153	1.2024[a]	283
	2.268		287
	2.902	1.224	278
	2.398		285[b]
	3.197		286
1,2-Dimethylbenzene	8.712		281
	7.344		288
	9.288	13.392	278
1,3-Dimethylebenzene	1.397		281
	2.8290		288
1,4-Dimethylbenzene	5.897	3.4128[a]	281
	3.139		288
1,2,3-Trimethylbenzene	4.702		281
1,2,4-Trimethylbenzene	17.928		281
1,3,5-Trimethylbenzene	1.498	1.26[a]	280
	0.994		283
	1.699		281
1,2,3,4-Tetramethylbenzene	5.803		281
1,2,3,5-Tetramethylbenzene	20.880		281
Tetralin	15.984		280
	288.616		289
	15.696		285[b]
Ethylbenzene	2.599		281
	2.160		288
n-Propylbenzene	1.498		281
Isopropylbenzene	1.202		281
n-Butylbenzene	1.901		281
sec-Butylbenzene	1.102		281
Fluorobenzene	0.590		280
	1.202		281
Chlorobenzene	6.998		280
	5.702		281
Bromobenzene	3.298		280
	2.801		281
Iodobenzene	2.102		281
1,2-Dichlorobenzene	27.000		280
	24.624	31.032[a]	281
	22.896	22.32	289
	23.400	25.704	292

Table 4.5 Continued.

Solvent	Solubility of C_{60} (mg mL^{-1})	Solubility of C_{70} (mg mL^{-1})	Reference
1,2-Dibromobenzene	13.824		281
1,3-Dichlorobenzene	2.398		281
	5.033	16.056	292
1,3-Dichlorobenzene	13.824		281
1,2,4-Trichlorobenzene	8.496		280
	10.368		281
	4.846		293
	21.312		286
Styrene	3.751		293
o-Cresol	0.0288		280
Benzonitrile	0.410		280
Nitrobenzene	0.799		280
Anisole	5.602		280
	6.696		285[b]
p-Bromoanisole	16.776		285[b]
m-Bromoanisole	16.200		285[b]
Benzaldehyde	0.389		288
Phenyl isocyanate	2.441		288
Thiophenol	6.912		293
1-Methyl 2-nitrobenzene	2.434		288
1-Methyl 3-nitrobenzene	2.362		288
Benzyl chloride	2.398		288
Benzyl bromide	4.939		288
1,1,1-Trichloromethylbenzene	4.802		288
1-Methylaphthalane	32.976		280
	33.192		281
1-Phenylnaphthalene	49.968		280
1-Chloronaphthalene	50.976		280
1-Bromo,2-methylnaphthalene	34.776		281
Xylenes	5.2		280,290
Mesitylene	1.5		280,290
Tetralin	16.00		280,290
Alkanes			
n-Pentane	0.005	0.001728[a]	280
	0.005		290
	0.004		283
	0.003		294
	0.007		286
n-Hexane	0.043	0.0108[a]	280
	0.040		283
	0.037		294
	0.052		287
	0.046		286
2-Methylpentane	0.019		294
3-Methylpentane	0.025		294
n-Heptane	0.048	0.04032[a]	294
	0.2902		285[b]
n-Octane	0.025	0.036[a]	283
	0.025[b]		280,290
	0.020		294

(*Continued*)

Table 4.5 Continued.

Solvent	Solubility of C_{60} (mg mL^{-1})	Solubility of C_{70} (mg mL^{-1})	Reference
	0.2902		285[b]
	0.025		286
Isooctane	0.026		283
	0.026[b]		280,290
	0.028		286
n-Nonane	0.062		294
	0.034		286
n-Decane	0.071	0.04536[a]	280
	0.070[b]		280,290
	0.070		283
	0.072		286
Dodecane	0.091	0.0864[a]	283
	0.091[b]		280,290
	0.103		286
Tetradecane	0.126		283
	0126[b]		280,290
	0.168		286
Cyclic alkanes			
Cyclopentane	0.002		280
Cyclohexane	0.036	0.0684[a]	280
	0.036[b]		290
	0.051		283
	0.036		288
	0.036		287
	0.035		295
	0.054		286
Cyclohexene	1.210		293
1-Methyl,1-cyclohexene	1.0290		293
Methylcyclohexane	17.280		293
1,2-Dimethylcyclohexane, mixture of cis and trans	0.1290		293
Ethylcyclohexane	0.252		293
3.7 Mixture of cis and trans decalins	4.601		280
	1.872		293
cis-Decalin	2.232		280
trans-Decalin	1.296		280
Haloalkanes			
Dichloromethane	0.259	0.0684[a]	280
	0.252		283
	0.2290		288
Trichloromethane	0.158		280
	0.173		293
	0.511		286
Tertrachloromethane	0.317		280
	0.446		283
	0.101		296[c]
Dibromomethane	0.360		288
Tribromomethane	5.638		288
Iodomethane	0.770		288
Di-iodomethane	0.122		293

Table 4.5 Continued.

Solvent	Solubility of C_{60} (mg mL^{-1})	Solubility of C_{70} (mg mL^{-1})	Reference
Bromochloromethane	0.749		293
Bromoethane	0.072		293
Iodoethane	0.281		280
Trichloroethylene	1.397		280
	1.4		290
Tetrachloroethylene	1.202		280
	1.2		290
Dichlorodifloroethane	0.022		280
1,1,2-Trichlorotrifloroethane	0.0288		280
1,1,2,2-Tetrachloethane	5.299		288
	5.3		280,290
1,2-Dibromoethylene	1.879		288
1,2-Dichloroethane	0.086		280
1,2-Dibromoethane	0.497		288
	0.540		288
1,1,1-Trichloroethane	0.151		293
1-Chloropropane	0.022		288
1-Bromopropane	0.058		288
1-Iodopropane	0.2940		288
2-Iodopropane	0.122		288
1,2-Dichloropropane	0.101		288
1,3-Dichloropropane	0.360		288
1,2-Dibromopropane	0.403		293
1,3-Diiodopropane	2.765		288
1,2,3-Trichloropropane	0.778		288
1,2,3-Tribromopropane	7.013		288
1-Chloro-2-methylpropane	0.029		288
1-Bromo-2-methylpropane	0.086		293
1-Iodo-2-methylpropane	0.338		288
2-Chloro-2-methylpropane	0.010		293
2-Bromo-2-methylpropane	0.060		293
2-Iodo-2-methylepropane	0.2290		293
Bromobutane	1.202		285[b]
Cyclopentyl bromide	0.410		288
Cyclohexyl chloride	0.533		288
Cyclohexyl bromide	2.203		288
Cyclohexyl iodide	8.064		288
Bromoheptane	2.297		285[b]
Bromooctane	3.398		285[b]
1-Bromotetradecane	6.192		285[b]
1-Bromooctadecane	6.192		285[b]
Methylene chloride[b]	0.254		280,290
Alcohols			
Methanol	0.000		294
Ethanol	0.001		280
	0.001		294
1-Propanol	0.004		294
1-Butanol	0.009		294
1-Pentanol	0.0290		294
1-Hexanol	0.042		294
1-Octanol	0.047		294

(*Continued*)

Table 4.5 Continued.

Solvent	Solubility of C_{60} (mg mL^{-1})	Solubility of C_{70} (mg mL^{-1})	Reference
2-Propanol	0.002		294
2-Butanol	0.004		294
2-Pentanol	0.0294		294
3-Pentanol	0.029		294
1,3-Propandiol	0.001		294
1,4-Butandiol	0.002		294
1,5-Pentadiol	0.004		294
N-Methyl-2-pyrrolidone	0.89		
Naphthalenes			
1-Methylnaphthalene	33.0		280,290
Dimethylnaphthalene	36.0		280,290
1-Phenylnaphthalene	50.0		280,290
1-Chloronaphthalene	51.0		280,290
Other polar solvents			
Nitromethane	0.000		280
	0.216		285[b]
Nitroethane	0.002		280
Acetone	0.001	0.001656[a]	280
Acetonitrile	0.000		280
Acrylonitrile	0.004		293
n-Butylamine	3.686		293
2-Methyloxyethyl ether	0.032		288
N,N-Dimethylflormamide	0.027		288
Dioxane	0.041		283
Water	0.000	1.15E-10	294
Miscellaneous			
Carbon disulfide	7.920	8.496[a]	280
	5.162	13.104	283
	7.704		278
	7.488		293
	11.808		286
Thiophene	0.403		281
	0.238		293
Tetrahydrofuran	0.058		280
	0.576		285[b]
	0.360		286
2-Methylthiophene	6.797		280
Pyridine	0.893		280
	0.2902		281
	0.2902		285[b]
Piperidine	53.280		285[b]
2,4,6-Trimethylpyridine	8.712		285[b]
Pyrrolidine	47.520		285[b]
N-Methyl-2-pyrrolidone	0.893		280
Tetrahydrothiophene	0.0290		280
	0.1288		293
Quinoline	7.200		281
Isopropanol		0.00294[a]	
Carbon tetrachloride	0.32	0.1008[a]	280,290

[a]Measurements taken at 298 K.
[b]Measurements taken at 291 K.
[c]Measurements taken at 303 K.

4.3.1 Solvent Effects on Fullerenes

Many investigations have observed the solvent effect on the fullerene properties as detected by optical absorption[306] and nuclear magnetic resonance (NMR);[307] however, very few investigations have reported on solvent effects on Raman frequencies of fullerenes.[81,308,309] Even fewer investigations have been devoted to calculating the change in Raman frequencies as the fullerene molecule interacts with different solvating environments.[310] Amer *et al.* have utilized semi-empirical molecular simulation methods to predict the effect of water interaction on the Raman frequencies of C_{60}. Figure 4.31 shows the calculated Raman frequency shifts for the three Raman active surface modes [$A_g(2)$, $H_g(7)$, and $H_g(8)$] of a C_{60} molecule as a result of interaction with increasing number (from 1 to 64) of water molecules.

As shown in Figure 4.31, at a low number of interacting water molecules no significant shifts in the Raman frequencies are predicted. The frequencies start to shift at increasing rate as the number of interacting molecules increases until it reaches about 40 water molecules. The Raman shift forms a plateau beyond this level before increasing again as the number of interacting molecules reaches 64. The calculated changes in the Raman frequencies were explained as follows: at low number of interacting water molecules, and due to the hydrophobic nature of the fullerene molecule, water molecules will interact weakly with the fullerene and, hence, little or no change in the Raman frequencies of the

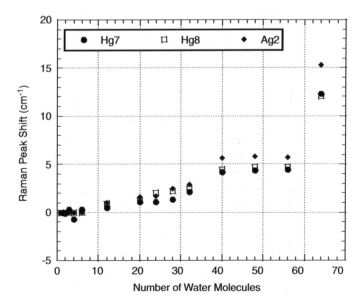

Figure 4.31 Raman Peak shifts of fullerene surface modes for different numbers of interacting water molecules. (Reproduced with kind permission from Amer *et al.*, ref. 310. Copyright Elsevier 2005.)

fullerene molecule is predicted. As the number of water molecules increases to a certain limit, the water molecules around the fullerene form a hydrogen-bonded network that forces the interaction with the fullerene, causing the sharp increase in the Raman peak position predicted at 40 interacting water molecules. Addition of more water molecules beyond this limit will not lead to increasing interaction. The plateau in the peak position can be rationalized in terms of water molecules that are not directly interacting with the fullerene, which is shielded by the water cage already formed around it. As the number of water molecules is further increased to the limit that an additional tight-bonded cage can be formed, a further increase in the water/fullerene interaction occurs causing the calculated blue-shift in Raman peak position.

This study draws attention to a very important point: the mutual water–water and water–fullerene interactions not only affect the physicochemical properties of the fullerene molecules (as reflected in their NMR, optical absorption properties, and Raman characteristics) but also affect the solvent structure. Figure 4.32 shows the calculated equilibrium structure of a fullerene molecule interacting with 12 water molecules. Clearly, from Figure 4.32, the mutual and simultaneous interaction between the water and fullerene molecules resulted in the shown thread-like structure not common in pure water.

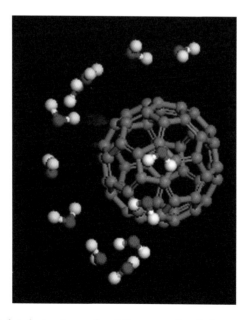

Figure 4.32 Simulated structure of a fullerene molecule interacting with 12 water molecules. (Reproduced with kind permission from Amer *et al.*, ref. 310. Copyright Elsevier 2005.)

More recent investigations further elucidate the important and unique role played by the solvent in the formation of versatile fullerene-based meso-systems and devices with unique structures and, hence, properties and performance.[311–317] For example, it was shown that the performance of fullerene-based solar cells is largely affected by the fullerene structure which, in turn, is controlled by the solvent used in their processing.[318] In addition, it has been reported that interaction between 3-aminopropyl-triethoxysilane (APTS) and carbon nanotubes modifies the physicochemical properties of the tubes, making them a better CO_2 absorbent than many previously reported types of modified carbon or silica.[319]

In 2009, in a breakthrough, Park et al.[314] reported the critical effect of solvent geometry on the determination of [60] fullerene self-assembly into dot, wire, and disk structures *via* a droplet drying process. Figure 4.33 schematically shows the solvent structure and the self-assembled fullerene structures resulting upon evaporation.

The study shows that the morphologies of C_{60} crystalline structures formed *via* evaporation are precisely determined by the molecular geometries of solvents. It experimentally establishes the specific correlation of solvent molecular geometry to directed assembly of C_{60} structure. The study demonstrates that while solvents with *pseudo* three-dimensional (*p*3D) molecules direct C_{60} molecules to assemble into *pseudo*-two-dimensional hexagonal disk structures, solvents with *p*2D and *p*1D molecular structures result in *p*1D C_{60} (wires) and

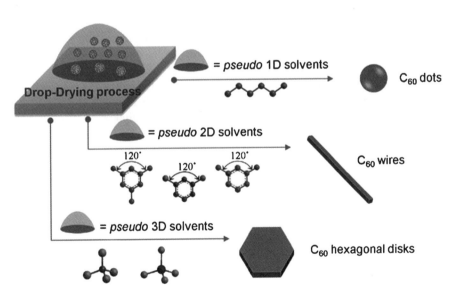

Figure 4.33 Self-assembly of dimensionally-confined C_{60} structures *via* a solution drop-drying process. Systematic correlations in the formation of *pseudo*-2D (hexagonal disk), *pseudo*-1D (wire), and *pseudo*-0D (dot) C_{60} structures from C_{60} solutions of *p*3D, *p*2D and *p*1D solvents, respectively. (Reproduced with kind permission from Park *et al.*, ref. 314. Copyright Royal Society of Chemistry 2009.)

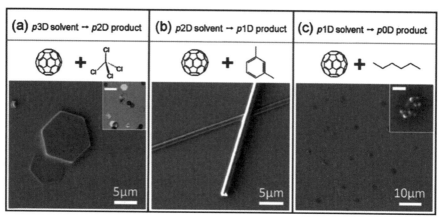

Figure 4.34 SEM and optical microscope images of C_{60} crystalline structures obtained from C_{60} solutions in (a) p3D CCl$_4$, (b) p2D m-xylene, and (c) p1D hexane. Scale bars of the insets in (a) and (c) are 30 mm and 500 nm, respectively. (Reproduced with kind permission from Park *et al.*, ref. 314. Copyright Royal Society of Chemistry 2009.)

p0D C_{60} spheroidal structures, respectively. Figure 4.34 shows scanning electron microscope (SEM) and optical microscope images of C_{60} crystalline structures obtained from C_{60} solutions in (a) p3D CCl$_4$, (b) p2D m-xylene and (c) p1D hexane.

Despite such a major leap in correlation finding, details of the mechanism responsible for such a correlation are still to be understood. Understanding the physics of such correlations is essential for our ability to further develop nanotechnology. In addition to its importance in enabling the development of technologically advanced nanostructured devices with unique properties, the ability to understand the correlation between the environment and self-assembly or directed assembly of structures would, definitely, have a major impact on our ability to understand the mechanisms of many biological processes involving growth and coagulation. A crucial point to realize in the aforementioned study is that the geometrical shape of the solvent molecules was the major factor in determining the structure of the directed assembly. This emphasizes the critical role played by molecular geometry that is related to entropic effects in the assembly of nanosystems. Molecular geometry as related to entropic effects on the directed assembly of nanostructures is a major frontier that is currently being explored.[320–326] Knowledge generated in colloidal science investigations should be a solid base to build upon.[327–330]

Another interesting issue to point out is the finding that while free evaporation of n-dimensional solvents containing C_{60} molecules led to the assembly of an $(n–1)$-dimensional C_{60} structures, coagulation of C_{60}-saturated toluene (two-dimensional solvent) solution into alcohols (one-dimensional liquid) (*via* liquid–liquid interface precipitation, LLIP, technique) led to the assembly of C_{60} molecules into one-dimensional (wire, whiskers, or rod)

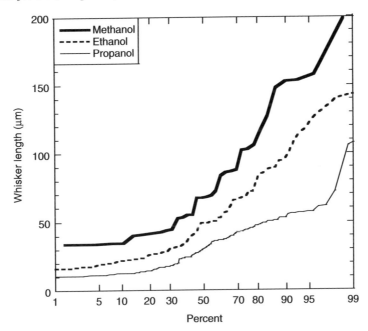

Figure 4.35 Length distribution for C_{60} wires directly assembled by LLIP (liquid–liquid interface precipitation) in three different alcohols. (Reproduced with kind permission from Amer, 2009, ref. 334. Courtesy of Professor M. Amer.)

structures.[317,331–333] The length of the assembled one-dimensional structures was recently found to depend on the molecular *length* of the alcohol. Coagulation in methanol, ethanol, and propanol led to the formation of one-dimensional assemblies with average length shortest for propanol and longest for methanol.[334] Figure 4.35 shows the length distribution for C_{60} wires directly assembled by LLIP technique in three different alcohols. Clearly, from the figure, while 95% of the wires directly assembled by methanol are less than 150 μm long, 95% of the wires assembled in ethanol and propanol are less than 100 and 50 μm long, respectively.

4.3.2 Fullerene Effects on Solvents

When a fullerene interacts with a solvent, not only does the solvent affect the fullerene properties and behavior, as we discussed in the previous section, but it was reported that the fullerene molecule also affects the solvent. Mainly, it was reported that a [60] fullerene molecule affects the structure of aromatic solvents it interacts with.[335–343] In fact, the high solubility of C_{60} in benzene, toluene, *para*-xylene, and other aromatic solvents, itself, indicates strong interaction and, hence, suggests that the structure of the aromatic solvent can change in response to the presence of the dissolved fullerene. Structuring of aromatic solvents interacting with C_{60} was assumed based on observed anomalies in many of the physical properties and behavior of the solvent. Ginzburg et al.[335–343] have reported non-monotonic changes in the solution density with an increasing concentration of fullerene molecules. Their small- and wide-angle X-ray diffraction, and permittivity, experimental results pointed out the possibility of solvent structuring around the spheroidal fullerene. Figure 4.36 shows wide-angle X-ray diffraction scattering curves ($Cu_{K\alpha}$ radiation) of fullerene C_{60} solutions in (a) p-xylene, (b) toluene, and (c) benzene with different fullerene concentrations.

Amer et al.[344,345] have shown that Brillouin scattering measurements of adiabatic compressibility and evaporation kinetic measurements on C_{60}/toluene solutions of different concentrations can also be interpreted in terms of solvent structuring around the fullerene molecule. Figure 4.37 shows the adiabatic compressibility of C_{60}/toluene solutions as measured using Brillouin

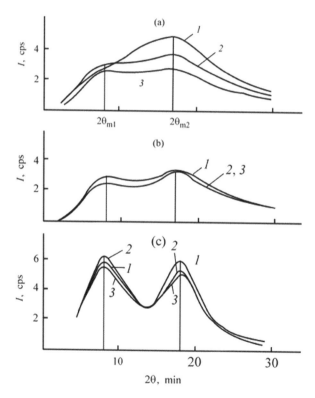

Figure 4.36 Wide-angle X-ray diffraction scattering curves ($Cu_{K\alpha}$ radiation) of fullerene C60 solutions in (a) p-xylene, (b) toluene, and (c) benzene. Fullerene concentration (%): (a) (1) 0 (pure p-xylene), (2) 0.005, and (3) 0.05; (b) (1) 0 (pure toluene), (2) 0.005, and (3) 0.05; and (c) (1) 0 (pure benzene), (2) 0.001, and (3) 0.075. (Reproduced with kind permission from Ginzburg and Tuichiev, ref. 340. Copyright Pleiades Publishing 2008.)

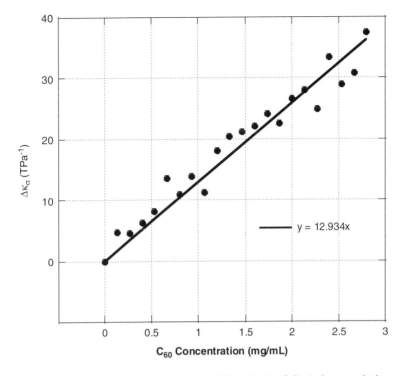

Figure 4.37 Change in adiabatic compressibility ($\Delta\kappa_\sigma$) of C_{60}/toluene solution as a function of C_{60} concentration. (Reproduced with kind permission from Amer et al., ref. 345. Copyright Elsevier 2008.)

spectroscopy. The observed linear increase in the solution compressibility is indicative of solvent structuring rendering an open structure with higher compressibility.[345]

While the exact nature of the produced solvent structures is not yet clear, it is most plausible that once a [60] fullerene molecule is introduced in the toluene (or similar aromatic solvents) the presence of C_{60} and its strong interaction with solvent molecules causes a sort of molecular structuring in the toluene, leading to the formation of a tightly bound toluene shell around the nanospheres (about 1 nm thick). Such a C_{60}/toluene clathrate would be covered by a solvophobic (open structure) shell. The toluene shell around the fullerene molecule was reported to be strongly bound to the extent that it survives thermal evaporation up to the point where C_{60} starts to sublime.[346] Amer et al.[344] have recently estimated the value of the outer radius of the solvophobic shell to be 6.6 nm. This means that the *long-range interaction distance* around the fullerene molecule in toluene is around 6.6 nm. Notably, the estimated value of 6.6 nm represents the minimum interaction distance because the authors used the density of undisturbed bulk toluene in their calculations. Disturbed (solvophobic) toluene is expected to have a lower density; hence the interaction range is expected to be longer than the estimated 6.6 nm. Such a structuring interpretation of the observed experimental results also agrees with recent thermodynamic investigation of enthalpic interaction between C_{60} and aromatic solvents, which concluded that interaction between C_{60} and aromatic solvents is mainly controlled by entropic effects.[303,304]

The interaction of fullerene molecules with different liquids (or solvents) leading to structuring of the solvent has also been confirmed by molecular dynamic simulation results.[347] To investigate the effect of fullerene on the molecular structure of their environments, molecular dynamic simulations were utilized to calculate the equilibrium molecular structure of three different types of environments: a non-interacting environment represented by an inert gas, a non-associated, non-polar fluid represented by carbon tetrachloride (CCl_4), and an associated, polar fluid represented by water (H_2O). Figure 4.38 shows the equilibrated structures of molecular dynamic simulation results for the three aforementioned cases. Clearly, from the figure, the presence of the tubular fullerene (nanotube) does, indeed, have an effect on the molecular structure of the surrounding fluid. Layered molecular rings of the surrounding fluid can be clearly detected around the nanotube. The effect of the nature (or in other words the intramolecular interactions within the fluid) is clearly demonstrated in the fact that the *long-range interaction distance* is different from one environment to the other. Equilibrated structures shown in Figure 4.38(a)–(c)

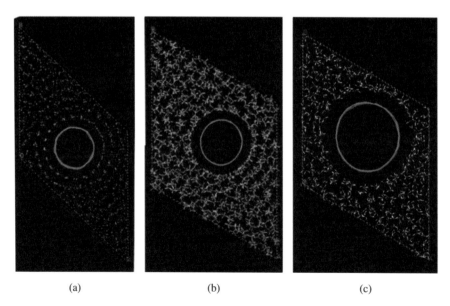

(a) (b) (c)

Figure 4.38 Equilibrated structures of molecular dynamic simulation results for a SWCNT inserted in three different environments: (a) a Lennard-Jones fluid (Ar, Kr), (b) a non-polar fluid (CCl_4), and (c) a polar fluid (H_2O). (Reproduced from ref. 347, Courtesy of Dr J. Elliott, University of Cambridge, England.)

indicate that the long-range interaction effect of the nanotube on its environment decreases as the association among the environment molecules increases. Such results are perfectly in agreement with expectation since associative interactions among the fluid molecules would definitely resist any external influence arising from the presence of the nanotube.

4.4 Fullerenes under Pressure

Numerous investigations have been, and still are, devoted to investigating mechanical properties of different types of fullerenes[348–360] – including C_{60}, C_{70}, and single- and multi-walled carbon nanotubes – as well as other nanostructured materials systems.[361–365] In fact, the mechanical properties of SWCNTs, as predicted by molecular mechanics methods, have been thoroughly reviewed recently.[366] In this section we will not attempt to review or cover the mechanical properties of spheroids, tubular, or sheet fullerenes, instead we will focus on the nanophenomena observed in fullerene-based systems once exposed to mechanical perturbation fields such as hydrostatic pressures.

Strikingly, while investigations of fullerenes under pressure started as early as the beginning of the fullerene rush in the early 1990s, the subject is far from being well understood. Surprising results, regarding what the scientific community considers physical constants, are still reported, necessitating further investigations of what seemed to be resolved scientific issues. Based on theoretical calculations (including continuum theory, *ab initio*, tight binding, force constant, and molecular dynamic methods) and indirect measurements involving micromechanical manipulation using atomic force microscope (AFM) tips, the elastic modulus of SWCNTs is believed to be around 1–1.4 TPa. There is also a general agreement that size and chirality would affect the elastic constant. However, no one, thus far, can precisely tell what the elastic modulus of a (10,10) or a (5,5) SWCNT would be. In addition, the bulk modulus of a C_{60} molecule is not yet precisely known. Values ranging between 200 and 1400 GPa have been reported in the literature.[64,349,367–375] Hence, one could say that we do not yet have an answer. So as not to lead the reader into despair, we should emphasize that we do not have an answer because the question itself is not correct.

Based on all our previous discussions so far, we understand that fullerenes are nanosystems, the properties and performance of which are controlled by quantum phenomena, entropic effects, fluctuations, and perturbation effects. Hence, one should not suffer the illusion that mechanical constants traditionally observed and determinable in bulk systems would be similarly determinable for a nanosystem. Their values will definitely depend on the exact fluctuations, perturbation effects, *etc.* affecting the nanosystem. For example, Brillouin spectroscopy measurements of the adiabatic speed of sound performed on a fly wing showed that the elastic constant of the wing (a natural nanostructured system) can change by a factor of 4 depending on the moisture content of the wing.[376] Clearly, nature has figured out such a nanophenomenon and has utilized it in the superior flying capabilities of many insects. For almost two centuries, the scientific community kept correlating mechanical hardness of materials to the materials yield and tensile strength. It was only 20 years ago that Pearson[377] and Gillman[378,379] pointed out that mechanical hardness is actually correlated to the electronic structure of the material, namely its energy gap. Such a finding should also point out the possibility that mechanical constants of nanosystems could be correlated to their electronic structures, which have been shown to be largely affected by external effects.

Returning to the main point of this section – fullerenes under pressure – one can say that since their discovery, the mechanical properties and performance of fullerenes, including carbon nanotubes, have been a subject of intense research and investigation because of their numerous potential applications. The behavior of SWCNTs under high hydrostatic pressure has been the focus of several theoretical[380–386] as well as experimental[61,112,165,351,387–393] investigations. Experimental investigations have employed several powerful techniques to elucidate the behavior of such carbon nanospecies under high hydrostatic pressure. These include X-ray diffraction, neutron diffraction, optical absorption spectra, and Raman spectroscopy. Diamond anvil cells (DACs) have been routinely used to conduct high-pressure investigations of different materials for the past few decades.[394] In such experiments, a sample is placed inside a micro-hole in a metal gasket that becomes a pressure microchamber pressurized between two diamond anvils. The micro-chamber is filled with a liquid that is used as a pressure transmission fluid (PTF) to impart a hydrostatic pressure on the solid sample as the diamond anvils are pressed together. In the usual mode of operation, the primary requirement of the PTF is to maintain its liquid status within the pressure and temperature range of the experiment to assure a hydrostatic pressure condition during the investigation.[395]

Raman results reported by different research groups that have investigated SWNT under high hydrostatic pressure show two distinctive characteristics: first, the peak positions of the radial breathing modes shift linearly to higher wavenumbers (blue-shift) and their intensities decrease as the applied hydrostatic pressure increases, eventually disappearing (or becoming too weak to be detected) around an applied pressure of 3 GPa; second, the peak position of the tangential mode (especially the second-order mode around 2660 cm^{-1}, the G′-band) is also blue-shifted as the applied hydrostatic pressure increases, reaching a plateau at the same applied pressures at which the radial modes starts to disappear. Figure 4.39 shows typical SWCNT Raman spectral behavior for the peak position of both the radial breathing mode and the second-order tangential mode (G′-band). Figure 4.40 shows the RBM and G′-band peak position as a function of applied hydrostatic pressure.

The disappearance of the radial breathing mode and the independence of the second-order peak position on applied pressure have both been interpreted previously as attributable to lateral deformation of the nanotube as its cross-section changes from a circular into elliptical, hexagonal, and eventually flattens in what has been termed as ovalization, polygonization, or collapse of the

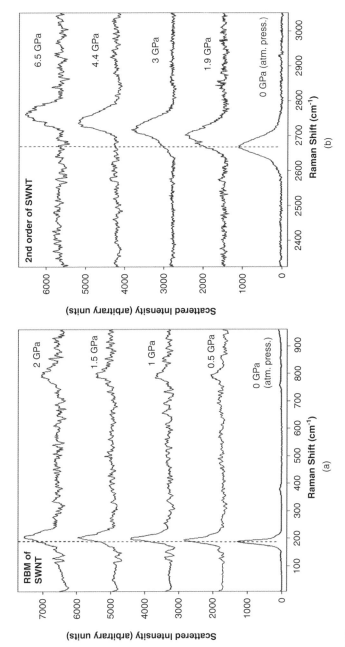

Figure 4.39 Raman spectra for SWCNTs under successively increasing hydrostatic pressure: (a) radial breathing mode (RBM) and (b) G′-band. (Reproduced with kind permission from Amer et al., ref. 112. Copyright American Chemical Society 2004.)

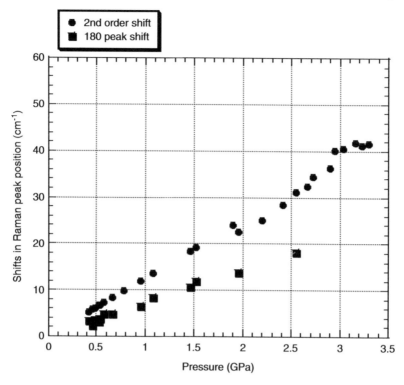

Figure 4.40 Radial breathing mode and second-order G'-band Raman peak positions as a function of applied hydrostatic pressure in pure methanol. (Reproduced with kind permission from Amer *et al.*, ref. 112. Copyright American Chemical Society 2004.)

nanotube. The same interpretation in terms of cross-section deformation has also been invoked to interpret X-ray and neutron diffraction results as well as several theoretical and simulation investigations.

A different interpretation for the observed high-pressure Raman behavior of SWCNT was introduced in 2004, relating the Raman behavior to molecular perturbation effects resulting from pressure transmission fluid molecular adsorption and interaction on the nanotube surface.[112,165,396,397] The observed disappearance of the radial breathing mode according to this interpretation would be due to the possibility that adsorbed molecules on the nanotube surface take the RBM out of resonance, causing its disappearance. This would definitely depend on the exact nature of the adsorbing molecule and the exact nature of its interaction with the nanotube. Figure 4.41 shows the peak position of the radial breathing mode for three different SWCNTs [(10,10), (9,9), and (8,8)] dispersed in either methanol or water (using the known 1% SDS surfactant method) as a function of applied hydrostatic pressure. Notably, while the RBM mode for the (10,10) dispersed in methanol disappears around 3 GPa of applied pressure, it does not disappear up to 14 GPa of applied

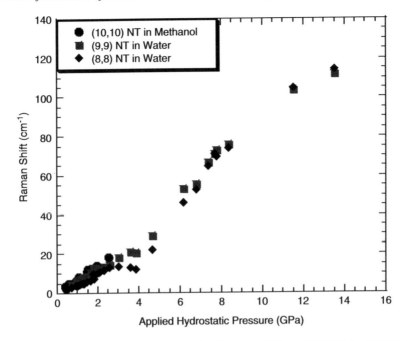

Figure 4.41 RBM (radial breathing mode) for three different SWCNTs: (10,10) dispersed in methanol, (9,9) dispersed in water, and (8,8) dispersed in water, as a function of applied hydrostatic pressure. Note that the RBM for CNTs dispersed in water does not disappear up to 14 GPa of applied pressure. (Reproduced from ref. 397. Courtesy of Professor M. Amer 2004.)

pressure for the nanotubes dispersed in water. Interestingly, the reversible disappearance of radial breathing mode under temperature effects was reported in 2005 (Figure 4.27a).

Importantly, on the molecular level, the molecules in the pressure transmission fluid intimately interact with the surface of the pressurized sample, thereby transmitting the pressure to the bulk of the solid. For a typical high-pressure Raman investigation on macroscopic samples of characteristic length scale (L) in the range of millimeters or even micrometers (a bulk system), the interactions between the liquid phase (PTF) and the sample surface are not observable since the spectrum is completely dominated by the bulk phonons of the sample. In such a case, due to the relatively large interaction volume between the excitation laser and the sample (typically a few cubic microns), none of the perturbation effects reflected in the surface phonons can be detected. However, as the sample characteristic length scale (L) is greatly reduced into the nanometer range, surface effects become increasingly more important. Changes in surface phonons due to solvent interactions start to have a measurable effect on the observed spectral response of the sample. In particular, mesoscopic molecules like SWCNTs and spheroidal fullerenes, when dispersed as individual nano-clusters or molecules surrounded by solvent molecules (the PTF), have essentially no bulk phase as such but have only surface. In this case, the solvent interactions on the surface of the nanoparticles are expected to play a major role in Raman spectral changes. In fact, the Raman spectra in such a case should reflect, to a large extent, the perturbation effects due to solvent interactions on the surface of fullerenes (Section 4.2.1).

Amer et al.,[112] investigating the pressure behavior of C_{60} molecules in pure methanol and methanol–water mixtures of different compositions using Raman spectroscopy, have reported a behavior that further supports the solvent effect interpretation mentioned above. The dependence of the Raman peak position of the pentagon shear mode around $1420\,cm^{-1}$ on applied hydrostatic pressure very closely resembles a type VI or a "step like" adsorption isotherm (Figure 4.42). Type VI adsorption isotherms are predicted for adsorption on very homogenous surfaces where the adsorbed molecules laterally interact. Such very rare adsorption isotherms have been observed before for argon and krypton adsorbed on the surface of highly-oriented graphite crystals[398] and

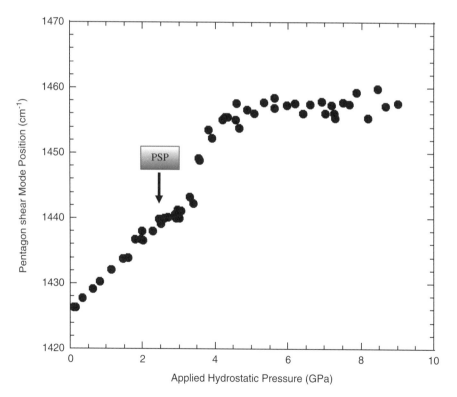

Figure 4.42 Pentagon shear mode position as a function of applied hydrostatic pressure for C_{60} in pure methanol (PSP = plateau starting pressure). (Reproduced with kind permission from Amer et al., ref. 112. Copyright American Chemical Society 2004.)

have been observed for the adsorption of methane and krypton on SWCNTs.[399] In this type of adsorption isotherm, a blue-shift in the Raman peak position indicates an increasing number of solvent molecules adsorbed on the fullerene surface. The onset of a plateau indicated the formation of a complete molecular layer of adsorbed molecules. The plateau indicates a phase transition into the adsorbed layer, and the start of a second blue-shift indicates a second layer of molecules being adsorbed, and so on.

Figure 4.42 shows the Raman peak position of the pentagon shear mode of C_{60} in pure methanol as a function of applied hydrostatic pressure. The pressure at which the first observed plateau starts, *i.e.*, plateau starting pressure (PSP, see Figure 4.42) for both C_{60} and SWCNTs pressurized in water–methanol mixtures of different compositions is shown in Figure 4.43 as a function of methanol mole fraction in the solvent (pressure transmission fluid). It is interesting to realize that the pressure starting plateau (which represents the pressure at which a complete molecular layer has been adsorbed on the fullerene surface) does, indeed, depend on the solvent composition. Figure 4.43 clearly shows that the plateau starting pressure (PSP) depends almost linearly on the methanol mole fraction in the methanol–water pressurizing solvent up to a certain mole fraction and then becomes independent of the composition. According to the results, the mole fraction at which PSP becomes independent

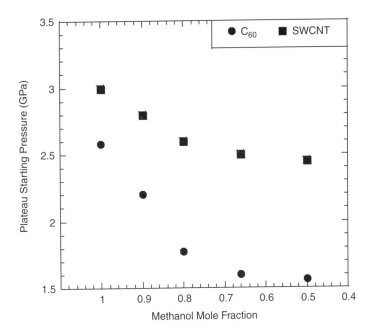

Figure 4.43 Plateau starting pressures (PSP) for C_{60} and SWCNT in water–methanol mixtures as a function of methanol mole fraction in the pressure transmission fluid. (Reproduced with kind permission from Amer *et al.*, ref. 112. Copyright American Chemical Society 2004.)

of the solvent composition is around 0.8 for SWCNTs (this fraction corresponds to a 4:1 methanol : water molecular ratio), and 0.66 for C_{60} fullerene, corresponding to a 2:1 methanol : water molecular ratio. Taking into consideration that methanol and water molecules can have up to two and four hydrogen bonds, respectively, these results point out the possibly important role played by solvent effects and structuring on observed physical behavior of such carbon nanospecies. The different behaviors observed for the same solvent interacting with spheroidal and tubular fullerenes also draws attention to the importance of nanosystem shape on its performance as we discussed earlier. High-pressure Raman investigations of different types of fullerenes, especially single- and double-walled carbon nanotubes, are still in the focus of scientific community.[150,157,167,170,194,195,400–406]

In 2008, Gao *et al.*[401] investigating the effect of pressure-transmitting fluid on the Raman features of SWCNTs using five different organic solvents, namely methanol, propanol, 1-butanol, hexane, and octane, as PTFs, reported a linear dependence of the pressure coefficient of the G-band on the molecular weight of the organic solvent used for transmitting the pressure. The pressure coefficient of the G-band (the slope of the linear dependence of the G-band position on applied hydrostatic pressure) was found to increase as the molecular weight of the organic solvent increased (Figure 4.44).

Recently, molecular dynamic simulations investigating solvent effects on compressibility of individual [60] fullerene molecules reported that interaction with water and methanol significantly affects the fullerene pressure–volume thermodynamic equation of state (EOS), leading to a significantly less compressible fullerene molecule.[367] Figure 4.45(a) shows the simulated P–V (EOS) for an isolated C_{60} molecule not interacting with a solvent and those interacting with water and methanol, separately, and Figure 4.45(b) shows the fullerene bulk modulus calculated based on the simulated EOS. The study draws attention to the important point that mechanical *"constants"* of fullerene, as a nanosystem, cannot be considered out of their chemical environment context.

We conclude this section by drawing attention to the point that while perturbation effects originating from the chemical potentials of adsorbed molecules on the fullerene surface (solvent effect) on the mechanical performance of fullerenes still represent a point of debate in the scientific community, their effects on fullerene electrical and optical properties are more widely accepted.[175,407–416]

4.5 Overview, Potentials, Challenges, and Concluding Remarks

This book started with the *supposedly* simple question *what is nanotechnology?* The answer we strived to provide, with more than seventy thousand words, implies that the question was, indeed, not so simple. In fact, here, we tried to capture the essence of what most plausibly would shape the future of humankind. We defined nanotechnology, in non-scientific terms, as the

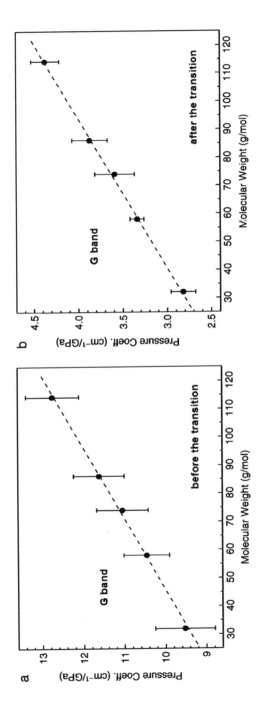

Figure 4.44 Pressure coefficient of the G-band of a SWCNT under hydrostatic pressure as a function of the molecular weight of different organic solvents used as pressure transmitting fluids, (a) before and (b) after the plateau appears. (Reproduced with kind permission from Gao et al., ref. 401. Copyright Elsevier 2008.)

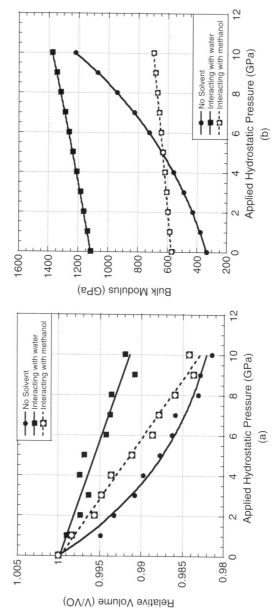

Figure 4.45 Molecular dynamic simulation results for (a) $P-V$ equation of state calculated for single fullerene molecule interacting with different chemical moieties and (b) calculated values for the fullerene bulk modulus based on the simulated EOS. (Reproduced with kind permission from Amer et al., ref. 367. Copyright Elsevier 2009.)

technology that would enable us to bridge what is natural and what is human-made. We have pointed out that nanostructured systems perform in a manner that is neither customary to us nor can be explained within the frame of knowledge we developed for *bulk* systems.

We showed that within the frame of our knowledge, the best tools – thermodynamic and statistical mechanics – we have developed to describe the behavior of materials are incapable once the system approaches certain length-scales. We showed that once the system approaches such length-scales, fundamental thermodynamic quantities such as temperature, pressure, free energy, chemical potential, and surface tension become indefinable, and therefore new physics, chemistry, and materials science are needed to enable describing the new system behavior. These combined new fields, in fact, represent a new interdisciplinary field – nanoscience and technology – that ought to be developed and understood. We also showed that while critical or characteristic length-scales can be on the order of few millions of light years, the most interesting length-scale for scientist and engineers on planet earth is the nanometer. The nanometer gains its magic power from the fact that it is the distance over which atoms and molecules on our planet can correlate together *via* several energetic interactions, forcing the system to enter the nano-domain. Once in the nano-domain, a material system's behavior becomes dominated by quantum effects, thermal fluctuations, and entropic effects. Thus the system starts to behave unconventionally and exhibits size and shape dependent chemical and physical properties. We showed how a slight (ten extra atoms) change in molecular size between C_{60} and C_{70} would alter its interaction with the same solvent. We also showed how changing the shape of a molecule from spherical (C_{60}) into cylindrical (single-walled nanotube) would turn it insoluble in the same solvent. In addition, we showed how optical and electrical properties of a nanosystem would significantly depend on the system's size and shape. We also showed how a nanosystem would, amazingly, interact with its environment. It was shown that as much as the fullerene properties are altered by perturbation effects imposed by a liquid environment around it, it also affects the structure, and hence the properties of such a liquid. Future investigations will definitely reveal more amazing phenomena of the nano-domain, or more amazing nanophenomena.

Nanophenomena represent a new domain of human knowledge. However, it is nothing that we can claim to have invented. Nature has, for over 3 billion years, been utilizing nanophenomena and nano-manufacturing in creating the most versatile and efficient system that is capable of sustaining itself – life – as we know it and as we are still trying to reveal it secrets. In his remarkable book *What is Life?* Erwin Schrödinger[417] pointed out that life[ii] is based upon "aperiodic crystals" while non-living material systems, as we know them, are based on "periodic crystals". Within the frame of bulk system laws, it is not simple to comprehend the formation mechanism or the stability of aperiodic

---

[ii] Here, life is used to denote living systems and not the ideologies and principles shaping human behavior.

crystals. This is why we refer to it as "the miracle of life". We currently refer to aperiodicity in a material system as "faults" or "defects". However, with our current crumb of knowledge in nanoscience, we are in a position allowing us to appreciate nature's marvelous aperiodic crystal-based or nanodomain-based designs. In fact, we are currently in a position enabling us to add to Schrödinger's viewpoint by stating that "life is based on nanosystems".

Once we moot many of the miracles of life, we cannot avoid realizing directed-assembly[iii] of nanosystems reacting significantly to minute changes in their environment. More importantly and amazingly, we realize that such systems possess a unique ability of self-healing within a certain limit. Examples are numerous. The ability of plant roots to grow in the correct direction under the effect of gravitational field, and the ability of sun flowers to control their stem stiffness to enable them to follow the sun under photo-effects are among the well known ones.

Nowadays, we have started to appreciate the potential of directed assembly on nanosystems. We have shown how the shape of solvent molecules would direct fullerene molecules to assemble themselves into zero-, one-, or two-dimensional shapes. We also started to investigate self-healing systems. While we are still far from being capable of assembling systems resembling those carried out by nature, it will not be exaggerating to say that future generations of scientists and engineers will be well positioned for that task.

Such foreseen ability to generate self-assembled, nanostructured systems capable of self-healing, indeed, provides humankind with unprecedented power. This necessitates overcoming numerous scientific and non-scientific challenges. Perhaps the most important challenge to overcome is learning how to harness the power of nanotechnology.[iv] After all, regardless of any other considerations, it is important to emphasize that the wealth of information that has been generated has significantly enhanced our general understanding of how to develop nanostructured systems that, if utilized wisely, would definitely shape our future.

References

1. D. Vlassopoulos, G. Fytas, J. Roovers, T. Pakula and G. Fleischer, *Faraday Discuss.*, 1999, **112**, 225.
2. I. Soten and G. Ozin, *Curr. Opin. Colloid Interface Sci.*, 1999, **4**, 325.
3. J. Tan and W. Saltzman, *Biomaterials*, 2004, **25**, 3593.
4. N. Dan, *Trends Biotechnol.*, 2000, **18**, 370.
5. C. Kirschhock, S. Kremer, J. Vermant, G. Van Tendeloo, P. Jacobs and J. Martens, *Chem.–Eur. J.*, 2005, **11**.

[iii] Self-assembly is the more frequently used term in the literature. Directed-assembly is used here to emphasize that the atoms and molecules would assemble themselves under the effect of external fields in a way that is directed by the exact nature of the interaction with the field. Both terms are accepted in the scientific community.

[iv] While the subject of nano-ethics, discussing ethical issues and societal impact of nanotechnology, is beyond the scope of this book, relevant texts should be regarded as "a must read".

6. R. Tu and M. Tirrell, *Adv. Drug Deliv. Rev.*, 2004, **56**, 1537.
7. W. Ni, J. Chen and Q. Xu, *BioResources*, 2008, **3**, 461.
8. A. A. Zakhidov, R. H. Baughman, Z. Iqbal, C. Cui, I. Khayrullin, S. O. Dantas, J. Marti and V. G. Ralchenko, *Science*, 1998, **282**, 897.
9. J. Fabian, *Phys. Rev. B*, 1996, **53**, 13864.
10. M. Schlüter, M. Lannoo, M. Needels, G. Baraff and D. Tomanek, *Phys. Rev. Lett.*, 1992, **68**, 526.
11. M. Schlüter, M. Lannoo, M. Needels, G. Baraff and D. Tomanek, *J. Phys. Chem. Solids*, 1992, **53**, 1473.
12. M. S. Dresselhaus, G. Dresselhaus and P. C. Eklund, *Science of Fullerenes and Carbon Nanotubes*, Academic Press, San Diego, 1996, p. 329.
13. K. Gregory, L. Korst, P. Cane, L. Platt, K. Kahn, P. Rafailov, V. Hadjiev, H. Jantoljak and C. Thomsen, *Solid State Commun.*, 1999, **112**, 517.
14. D. Bethune, G. Meijer, W. Tang and H. Rosen, *Phys. Chem. Fullerenes: a reprint collection*, 1993, 31.
15. S. Lebedkin, A. Gromov, S. Giesa, R. Gleiter, B. Renker, H. Rietschel and W. Krätschmer, *Chem. Phys. Lett.*, 1998, **285**, 210.
16. S. Guha, J. Menendez, J. Page, G. Adams, G. Spencer, J. Lehman, P. Giannozzi and S. Baroni, *Phys. Rev. Lett.*, 1994, **72**, 3359.
17. V. Antropov, O. Gunnarsson and A. Liechtenstein, *Phys. Rev. B*, 1993, **48**, 7651.
18. Z. Wu, D. Jelski and T. George, *Chem. Phys. Lett.*, 1987, **137**, 291.
19. W. Krätschmer, L. D. Lamb, K. Fostiropoulos and D. R. Huffman, *Nature*, 1990, **347**, 354.
20. R. E. Stanton and M. D. Newton, *J. Phys. Chem.*, 1988, **92**, 2141.
21. Z. Slanina, J. M. Rudzinski, M. Togasi and E. Osawa, *J. Mol. Struct.*, 1989, **202**, 169.
22. F. Negri, G. Orlandi and F. Zerbetto, *Chem. Phys. Lett.*, 1988, **144**, 31.
23. R. A. Jishi, R. M. Mirie and M. S. Dresselhaus, *Phys. Rev. B*, 1992, **45**, 13685.
24. A. Quong, M. Pederson and J. Feldman, *Solid State Commun.*, 1993, **87**, 535.
25. J. L. Feldman, J. Q. Broughton, L. L. Boyer, D. E. Reich and M. D. Kluge, *Phys. Rev. B*, 1992, **46**, 12731.
26. S. Guha, J. Menéndez, J. B. Page, G. B. Adams, G. S. Spencer, J. P. Lehman, P. Giannozzi and S. Baroni, *Phys. Rev. Lett.*, 1994, **72**, 3359.
27. G. B. Adams, J. B. Page, M. O'Keeffer and O. F. Sankey, *Chem. Phys. Lett.*, 1994, **228**, 485.
28. V. Schettino, M. Pagliai, L. Ciabini and G. Cardini, *J. Phys. Chem. A*, 2001, **105**, 11192.
29. C. H. Choi, M. Kertesz and L. Mihaly, *J. Phys. Chem. A*, 2000, **104**, 102.

30. V. Schettino, P. R. Salvi, R. Bini and G. Cardini, *J. Chem. Phys.*, 1994, **101**, 11079.
31. G. B. Adams, J. B. Page, O. F. Sankey and M. O'Keeffe, *Phys. Rev. B*, 1994, **50**, 17471.
32. R. Garrell, T. Herne, C. Szafranski, F. Diederich, F. Ettl and R. Whetten, *J. Am. Chem. Soc.*, 1991, **113**, 6302.
33. Y. Zhang, G. Edens and M. Weaver, *J. Am. Chem. Soc.*, 1991, **113**, 9395.
34. T. Wågberg, P. Persson, B. Sundqvist and P. Jacobsson, *Appl. Phys. A*, 1997, **64**, 223.
35. J. Arvanitidis, K. Meletov, K. Papagelis, A. Soldatov, K. Prassides, G. Kourouklis and S. Ves, *Phys. Status Solidi B*, 1999, **215**.
36. G. Chambers and H. Byrne, *Chem. Phys. Lett.*, 1999, **302**, 307.
37. F. Cataldo, *Eur. Polym. J.*, 2000, **36**, 653.
38. H. Grebel, Z. Iqbal and A. Lan, *Appl. Phys. Lett.*, 2001, **79**, 3194.
39. V. Schettino, M. Pagliai and G. Cardini, *J. Phys. Chem. A*, 2002, **106**, 1815.
40. M. Baibarac, L. Mihut, N. Preda, I. Baltog, J. Mevellec and S. Lefrant, *Carbon*, 2005, **43**, 1.
41. Z. Niu and Y. Fang, *Vibrational Spectrosc.*, 2007, **43**, 415.
42. Y. Hamanaka, M. Norimoto, S. Nakashima and M. Hangyo, *Chem. Phys. Lett.*, 1995, **233**, 590.
43. Z. Dong, P. Zhou, J. Holden, P. Eklund, M. Dresselhaus and G. Dresselhaus, *Phys. Rev. B*, 1993, **48**, 2862.
44. K. A. Wang, A. M. Rao, P. C. Eklund, M. S. Dresselhaus and G. Dresselhaus, *Phys. Rev. B*, 1993, **48**, 11375.
45. P. Giannozzi and S. Baroni, *J. Chem. Phys.*, 1994, **100**, 8537.
46. M. S. Dresselhaus, G. Dresselhaus and P. C. Eklund, *J. Raman Spectrosc.*, 1996, **27**, 351.
47. P. Eklund, A. Rao, Y. Wang, P. Zhou, K. Wang, J. Holden, M. Dresselhaus and G. Dresselhaus, *Thin Solid Films*, 1995, **257**, 211.
48. P. Eklund, Z. Ping, W. Kai-An, G. Dresselhaus and M. Dresselhaus, *J. Phys. Chem. Solids*, 1992, **53**, 1391.
49. S. Huant, J. B. Robert, G. Chouteau, P. Bernier, C. Fabre and A. Rassat, *Phys. Rev. Lett.*, 1992, **69**, 2666.
50. D. W. Snoke, Y. S. Raptis and K. Syassen, *Phys. Rev. B*, 1992, **45**, 14419.
51. J. Copley, W. David and D. Neumann, *Neutron News*, 1993, **4**, 20.
52. L. Kavan, L. Dunsch, H. Kataura, A. Oshiyama, M. Otani and S. Okada, *J. Phys. Chem. B*, 2003, **107**, 7666.
53. H. Kuzmany, R. Pfeiffer, M. Hulman and C. Kramberger, *Philos. Trans. R. Soc. London, Ser. A*, 2004, **362**, 2375.
54. M. L. McGlashen, M. E. Blackwood and T. G. Spiro, *J. Am. Chem. Soc.*, 1993, **115**, 2074.
55. Y. Zhang, G. Edens and M. J. Weaver, *J. Am. Chem. Soc.*, 1991, **113**, 9395.

56. V. Schettino, M. Pagliai and G. Cardini, *J. Phys. Chem. A*, 2002, **106**, 1815.
57. B. Chase, N. Herron and E. Holler, *J. Phys. Chem.*, 1992, **96**, 4262.
58. S. Chase, W. Bacsa, M. Mitch, L. Pilione and J. Lannin, *Phys. Rev. B*, 1992, **46**, 7873.
59. D. Palles, A. Marucci, A. Penicaud and G. Ruani, *J. Phys. Chem. B*, 2003, **107**, 4904.
60. W. Que and M. B. Walker, *Phys. Rev. B*, 1993, **48**, 13104.
61. G. A. Kourouklis, S. Ves and K. P. Meletov, *Physica B (Amsterdam)*, 1999, **265**, 214.
62. S. J. Duclos, R. C. Haddon, S. Glarum and A. F. Hebard, *Science*, 1991, **254**, 1625.
63. H. Kuzmany, M. Matus, T. Pichler and J. Winter, in *Physics and Chemistry of Fullerenes*, ed. K. Prassides, Kluwer, Dordrecht, 1993, p. 287.
64. S. J. Duclos, K. Brister, R. C. Haddon, A. R. Kortan and F. A. Thiel, *Nature*, 1991, **351**, 380.
65. S. H. Tolbert, A. P. Alivisatos, H. E. Lorenzana, M. B. Kruger and R. Jeanloz, *Chem. Phys. Lett.*, 1992, **188**, 163.
66. R. M. Lynden-Bell and K. H. Michel, *Rev. Mod. Phys.*, 1994, **66**, 721.
67. A. Rosenberg and D. P. DiLella, *Chem. Phys. Lett.*, 1994, **223**, 76.
68. K. L. Akers, C. Douketis, T. L. Haslett and M. Moskovits, *J. Phys. Chem.*, 1994, **98**, 10824.
69. C. A. Mirkin and W. B. Caldwell, *Tetrahedron*, 1996, **52**, 5113.
70. K. Meletov, V. Davydov, A. Rakhmanina, V. Agafonov and G. Kourouklis, *Chem. Phys. Lett.*, 2005, **416**, 220.
71. P. Verma, K. Yamada, H. Watanabe, Y. Inouye and S. Kawata, *Phys. Rev. B*, 2006, **73**, 45416.
72. M. S. Dresselhaus, G. Dresselhaus and P. C. Eklund, *Science of Fullerenes and Carbon Nanotubes*, Academic Press, San Diego, 1996.
73. S. Guha, J. Menéndez, J. B. Page and G. B. Adams, *Phys. Rev. B*, 1997, **56**, 15431.
74. J. Baker, P. W. Fowler, P. Lazzeretti, M. Malagoli and R. Zanasi, *Chem. Phys. Lett.*, 1991, **184**, 182.
75. S. Saito and A. Oshiyama, *Phys. Rev. B*, 1991, **44**, 11532.
76. G. E. Scuseria, *Chem. Phys. Lett.*, 1991, **180**, 451.
77. D. R. McKenzie, C. A. Davis, D. J. H. Cockayne, D. A. Muller and A. M. Vassallo, *Nature*, 1992, **355**, 622.
78. M. Sprik, A. Cheng and M. L. Klein, *Phys. Rev. Lett.*, 1992, **69**, 1660.
79. C. Z. Wang, C. T. Chan and K. M. Ho, *Phys. Rev. B*, 1992, **46**, 9761.
80. K. Nakao, N. Kurita and M. Fujita, *Phys. Rev. B*, 1994, **49**, 11415.
81. K. Lynch, C. Tanke, F. Menzel, W. Brockner, P. Scharff and E. Stumpp, *J. Phys. Chem.*, 1995, **99**, 7985.
82. G. Sun and M. Kertesz, *J. Phys. Chem. A*, 2002, **106**, 6381.

83. V. Blank, G. Dubitsky, N. Serebryanaya, B. Mavrin, V. Denisov, S. Buga and L. Chernozatonskii, *Physica B (Amsterdam)*, 2003, **339**, 39.
84. M. Kalbac, L. Kavan, M. Zukalova and L. Dunsch, *J. Phys. Chem. C*, 2006, **111**, 1079.
85. J. Du and P. Zeng, *Eur. J. Mech./A Solids*, 2009.
86. D. W. Snoke, Y. S. Raptis and K. Syassen, *Phys. Rev. B*, 1992, **45**, 14419.
87. P. H. M. van Loosdrecht, P. J. M. van Bentum and G. Meijer, *Phys. Rev. Lett.*, 1992, **68**, 1176.
88. J. B. Howard, J. T. McKinnon, Y. Makarovsky, A. L. Lafleur and M. E. Johnson, *Nature*, 1991, **352**, 139.
89. R. Meilunas, R. P. H. Chang, S. Liu, M. Jensen and M. M. Kappes, *J. Appl. Phys.*, 1991, **70**, 5128.
90. Y. Quo, N. Karasawa and W. A. Goddard, *Nature*, 1991, **351**, 464.
91. N. Chandrabhas, K. Jayaram, D. V. S. Muthu, A. K. Sood, R. Seshadri and C. N. R. Rao, *Phys. Rev. B*, 1993, **47**, 10963.
92. N. Chandrabhas, A. K. Sood, D. V. S. Muthu, C. S. Sundar, A. Bharathi, Y. Hariharan and C. N. R. Rao, *Phys. Rev. Lett.*, 1994, **73**, 3411.
93. H. Liu, B. Taheri and W. Jia, *Phys. Rev. B*, 1994, **49**, 10166.
94. A. Avent, P. Birkett, A. Darwish, H. Kroto, R. Taylor and D. Walton, *Tetrahedron*, 1996, **52**, 5235.
95. M. Damnjanovic, I. Bozovic and N. Bozovic, *J. Phys. A: Math. Gen.*, 1984, **17**, 747.
96. M. Damnjanovic, I. Miloševic, T. Vukovic and R. Sredanovic, *Phys. Rev. B*, 1999, **60**, 2728.
97. M. Damnjanovi, *Phys. Lett. A*, 1983, **94**, 337.
98. M. S. Dresselhaus, G. Dresselhaus, A. Jorio, A. G. Souza Filho and R. Saito, *Carbon*, 2002, **40**, 2043.
99. R. A. Jishi, D. Inomata, K. Nakao, M. Dresselhaus and G. Dresselhaus, *J. Phys. Soc. Jpn.*, 1994, **63**, 2252.
100. E. Di Donato, M. Tommasini, C. Castiglioni and G. Zerbi, "The electronic structure of achiral nanotubes: a symmetry based treatment," in *Electronic Properties of Synthetic Nanostructures: XVIII International Winterschool/Euroconference on Electronic Properties of Novel Materials*, ed. H. Kuzmany, J. Fink and M. Mehring, American Institute of Physics, 2004, AIP Conference Proceedings 723, pp. 359–363.
101. S. Reich, C. Thomsen and J. Maultzsch, *Carbon Nanotubes, Basic Concepts and Physical Properties*, Wiley-VCH, Weinheim, 2004.
102. V. N. Popov, V. E. Van Doren and M. Balkanski, *Phys. Rev. B*, 1999, **59**, 8355.
103. A. Jorio, R. Saito, J. H. Hafner, C. M. Lieber, M. Hunter, T. McClure, G. Dresselhaus and M. S. Dresselhaus, *Phys. Rev. Lett.*, 2001, **86**, 1118.
104. A. Jorio, G. Dresselhaus and M. Dresselhaus (eds.), *Carbon Nanotubes: Advanced Topics in the Synthesis, Structure, Properties And Applications*, Springer, 2008.

105. M. S. Dresselhaus, A. Jorio and M. A. Pimenta, *An. Acad. Bras. Cienc.*, 2006, **78**, 423.
106. T. Shimada, T. Sugai, C. Fantini, M. Souza, L. G. Cançado, A. Jorio, M. A. Pimenta, R. Saito, A. Grüneis, G. Dresselhaus, M. S. Dresselhaus, Y. Ohno, T. Mizutani and H. Shinohara, *Carbon*, 2005, **43**, 1049.
107. P. Corio, A. Jorio, N. Demir and M. S. Dresselhaus, *Chem. Phys. Lett.*, 2004, **392**, 396.
108. K. Sato, R. Saito, Y. Oyama, J. Jiang, L. G. Cançado, M. A. Pimenta, A. Jorio, G. G. Samsonidze, G. Dresselhaus and M. S. Dresselhaus, *Chem. Phys. Lett.*, 2006, **427**, 117.
109. C. Fantini, A. Jorio, M. Souza, M. S. Strano, M. S. Dresselhaus and M. A. Pimenta, *Phys. Rev. Lett.*, 2004, **93**, 147406.
110. A. Jorio, C. Fantini and M. A. Pimenta, *Phys. Rev. B*, 2005, **71**, 075401.
111. C. Fantini, A. Jorio, M. Souza, M. Strano, M. Dresselhaus and M. Pimenta, *Phys. Rev. Lett.*, 2004, **93**, 147406.
112. M. S. Amer, M. M. El-Ashry and J. F. Maguire, *J. Chem. Phys.*, 2004, **121**, 2752.
113. A. M. Rao, E. Richter, S. Bandow, B. Chase, P. C. Eklund, K. A. Williams, S. Fang, K. R. Subbaswamy, M. Menon, A. Thess, R. E. Smalley, G. Dresselhaus and M. S. Dresselhaus, *Science*, 1997, **275**, 187.
114. K. Williams, M. Tachibana, J. Allen, L. Grigorian, S. Cheng, S. Fang, G. Sumanasekera, A. Loper, J. Williams and P. Eklund, *Chem. Phys. Lett.*, 1999, **310**, 31.
115. H. Telg, J. Maultzsch, S. Reich, F. Hennrich and C. Thomsen, *Phys. Rev. Lett.*, 2004, **93**, 177401.
116. M. S. Amer, 2004, unpublished, Max Planck Institute for Solid State Physics.
117. R. Saito, G. Dresselhaus and M. S. Dresselhaus, *Phys. Rev. B*, 2000, **61**, 2981.
118. Y. Zhang, J. Zhang, H. Son, J. Kong and Z. Liu, *J. Am. Chem. Soc.*, 2005, **127**, 17156.
119. A. Rao, J. Chen, E. Richter, U. Schlecht, P. Eklund, R. Haddon, U. Venkateswaran, Y. Kwon and D. Tomanek, *Phys. Rev. Lett.*, 2001, **86**, 3895.
120. A. G. Souza Filho, A. Jorio, G. G. Samsonidze, G. Dresselhaus, M. S. Dresselhaus, A. K. Swan, M. S. Ünlü, B. B. Goldberg, R. Saito, J. H. Hafner, C. M. Lieber and M. A. Pimenta, *Chem. Phys. Lett.*, 2002, **354**, 62.
121. M. S. Dresselhaus, G. Dresselhaus and R. Saito, *Carbon*, 1995, **33**, 883.
122. T. W. Odom, J.-L. Huang, P. Kim and C. M. Lieber, *Nature*, 1998, **391**, 62.
123. V. W. Brar, G. G. Samsonidze, A. P. Santos, S. G. Chou, D. Chattopadhyay, S. N. Kim, F. Papadimitrakopoulos, M. Zheng, A. Jagota,

G. B. Onoa, A. K. Swan, M. S. Unlü, B. B. Goldberg, G. Dresselhaus and M. S. Dresselhaus, *J. Nanosci. Nanotechnol.*, 2005, **5**, 209.
124. S. D. M. Brown, P. Corio, A. Marucci, M. A. Pimenta, M. S. Dresselhaus and G. Dresselhaus, *Phys. Rev. B*, 2000, **61**, 7734.
125. P. Corio, A. P. Santos, P. S. Santos, M. L. A. Temperini, V. W. Brar, M. A. Pimenta and M. S. Dresselhaus, *Chem. Phys. Lett.*, 2004, **383**, 475.
126. P. Corio, M. L. A. Temperini, P. S. Santos, J. V. Romero, J. G. Huber, C. A. Luengo, S. D. M. Brown, M. S. Dresselhaus, G. Dresselhaus, M. S. S. Dantas, C. F. Leite, F. Matinaga, J. C. Gonzalez and M. A. Pimenta, *Chem. Phys. Lett.*, 2001, **350**, 373.
127. S. Cronin, A. Swan, M. Ünlü, B. Goldberg, M. Dresselhaus and M. Tinkham, *Compos., Part A Phys. Rev. B*, 2001, **72**, 035425.
128. A. C. Dillon, M. Yudasaka and M. S. Dresselhaus, *J. Nanosci. Nanotechnol.*, 2004, **4**, 691.
129. M. S. Dresselhaus, *Carbon*, 1995, **33**, 871.
130. M. S. Dresselhaus and G. Dresselhaus, *Nanostruct. Mater.*, 1997, **9**, 33.
131. X. Duan, H. Son, B. Gao, J. Zhang, T. Wu, G. G. Samsonidze, M. S. Dresselhaus, Z. Liu and J. Kong, *Nano Lett.*, 2007, **7**, 2116.
132. A. Grüneis, R. Saito, J. Jiang, G. G. Samsonidze, M. A. Pimenta, A. Jorio, A. G. Souza Filho, G. Dresselhaus and M. S. Dresselhaus, *Chem. Phys. Lett.*, 2004, **387**, 301.
133. J. Jiang, R. Saito, A. Grüneis, G. Dresselhaus and M. S. Dresselhaus, *Chem. Phys. Lett.*, 2004, **392**, 383.
134. J. S. Park, A. Reina, R. Saito, J. Kong, G. Dresselhaus and M. S. Dresselhaus, *Carbon*, 2009, **47**, 1303.
135. F. Villalpando-Paez, H. Son, D. Nezich, Y. P. Hsieh, J. Kong, Y. A. Kim, D. Shimamoto, H. Muramatsu, T. Hayashi, M. Endo, M. Terrones and M. S. Dresselhaus, *Nano Lett.*, 2008, **8**, 3879.
136. S. Reich and C. Thomsen, *Philos. Trans. R. Soc. London, Ser. A*, 2004, **362**, 2271.
137. (a) C. Thomsen, S. Reich, *Phys. Rev. Lett.*, 2000, **85**, 5214; (b) I. Póscik, M. Hundhausen, M. Koos, O. Berkese and L. Ley, in *Proceedings of the XVI International Conference on Raman Spectroscopy*, ed. A. M. Heyns, Wiley-VCH, Berlin, 1998, p. 64; (c) Y. Wang, D. C. Aolsmeyer and R. L. McCreery, *Chem. Mater.*, 1990, **2**, 557; (d) M. J. Matthews, M. A. Pimenta, G. Dresselhaus, M. S. Dresselhaus and M. Endo, *Phys. Rev. B.*, 1999, **59**, R6585.
138. A. G. Souza Filho, A. Jorio, J. H. Hafner, C. M. Lieber, R. Saito, M. A. Pimenta, G. Dresselhaus and M. S. Dresselhaus, *Phys. Rev. B*, 2001, **63**, 241404.
139. C. Fantini, A. Jorio, M. Souza, R. Saito, G. G. Samsonidze, M. S. Dresselhaus and M. A. Pimenta, *Phys. Rev. B*, 2005, **72**, 085446.
140. J. Maultzsch, S. Reich, U. Schlecht and C. Thomsen, *Phys. Rev. Lett.*, 2003, **91**, 087402.

141. J. Maultzsch, S. Reich and C. Thomsen, *Phys. Rev. B*, 2001, **64**, 121407.
142. R. Saito Jr, A. Neis, G. G. Samsonidze, V. W. Brar, G. Dresselhaus, M. S. Dresselhaus, A. Jorio Can, L. G. Ado, C. Fantini, M. A. Pimenta and A. G. S. Filho, *New J. Phys.*, 2003, **5**, 157.
143. S. M. Bose, S. Gayen and S. N. Behera, *Phys. Rev. B*, 2005, **72**, 153402.
144. Y. Wu, J. Maultzsch, E. Knoesel, B. Chandra, M. Huang, M. Y. Sfeir, L. E. Brus, J. Hone and T. F. Heinz, *Phys. Rev. Lett.*, 2007, **99**, 027402.
145. M. Lazzeri, S. Piscanec, F. Mauri, A. C. Ferrari and J. Robertson, *Phys. Rev. B*, 2006, **73**, 155426.
146. M. Paillet, P. Poncharal, A. Zahab, J. L. Sauvajol, J. C. Meyer and S. Roth, *Phys. Rev. Lett.*, 2005, **94**, 237401.
147. M. Oron-Carl, F. Hennrich, M. M. Kappes, H. v. Lohneysen and R. Krupke, *Nano Lett.*, 2005, **5**, 1761.
148. V. N. Popov, L. Henrard and P. Lambin, *Phys. Rev. B*, 2005, **72**, 035436.
149. G. Li, Z. Han, G. Piao, J. Zhao, S. Li and G. Liu, *Mater. Sci. Eng., B*, 2009, **163**, 161.
150. M. Yao, Z. Wang, B. Liu, Y. Zou, S. Yu, W. Lin, Y. Hou, S. Pan, M. Jin, B. Zou, T. Cui, G. Zou and B. Sundqvist, *Phys. Rev. B*, 2008, **78**, 205411.
151. S. G. Stepanian, M. V. Karachevtsev, A. Glamazda, V. A. Karachevtsev and L. Adamowicz, *Chem. Phys. Lett.*, 2008, **459**, 153.
152. M. Oron-Carl and R. Krupke, *Phys. Rev. Lett.*, 2008, **100**, 127401.
153. A. W. Musumeci, E. R. Waclawik and R. L. Frost, *Spectrochim. Acta, Part A*, 2008, **71**, 140.
154. S. Huang, Y. Qian, J. Chen, Q. Cai, L. Wan, S. Wang and W. Hu, *J. Am. Chem. Soc.*, 2008, **130**, 11860.
155. J.-B. C. Dokyung Yoon, C.-S. Han, Y.-J. Kim and S. Baik, *Carbon*, 2008, **46**, 1530.
156. S. Dittmer, N. Olofsson, J. E. Weis, O. A. Nerushev, A. V. Gromov and E. E. B. Campbell, *Chem. Phys. Lett.*, 2008, **457**, 206.
157. Y. Zou, B. Liu, M. Yao, Y. Hou, L. Wang, S. Yu, P. Wang, B. Li, B. Zou, T. Cui, G. Zou, T. Wagberg and B. Sundqvist, *Phys. Rev. B*, 2007, **76**, 195417.
158. N. Takeda and K. Murakoshi, *Anal. Bioanal. Chem.*, 2007, **388**, 103.
159. A. P. Shreve, E. H. Haroz, S. M. Bachilo, R. B. Weisman, S. Tretiak, S. Kilina and S. K. Doorn, *Phys. Rev. Lett.*, 2007, **98**, 037405.
160. M. Seifi, D. K. Ross and A. Giannasi, *Carbon*, 2007, **45**, 1871.
161. U. Ritter, P. Scharff, O. P. Dmytrenko, N. P. Kulish, Y. I. Prylutskyy, N. M. Belyi, V. A. Gubanov, L. A. Komarova, S. V. Lizunova, V. V. Shlapatskaya and H. Bernas, *Chem. Phys. Lett.*, 2007, **447**, 252.
162. S. W. Lee, G.-H. Jeong and E. E. B. Campbell, *Nano Lett.*, 2007, **7**, 2590.

163. G. Fanchini, H. E. Unalan and M. Chhowalla, *Nano Lett.*, 2007, **7**, 1129.
164. N. Anderson, A. Hartschuh and L. Novotny, *Nano Lett.*, 2007, **7**, 577.
165. S. Amer, *J. Raman Spectrosc.*, 2007, **38**, 721.
166. P. Leyton, J. S. Gómez-Jeria, S. Sanchez-Cortes, C. Domingo and M. Campos-Vallette, *J. Phys. Chem. B*, 2006, **110**, 6470.
167. S. Lebedkin, K. Arnold, O. Kiowski, F. Hennrich and M. M. Kappes, *Phys. Rev. B*, 2006, **73**, 094109.
168. L. Kavan, M. Kalbáč, M. Zukalová and L. Dunsch, *Phys. Status Solidi B*, 2006, **243**, 3130.
169. A. Ilie, J. S. Bendall, D. Roy, E. Philp and M. L. H. Green, *J. Phys. Chem. B*, 2006, **110**, 13848.
170. D. Christofilos, J. Arvanitidis, C. Tzampazis, K. Papagelis, T. Takenobu, Y. Iwasa, H. Kataura, C. Lioutas, S. Ves and G. A. Kourouklis, *Diamond Relat. Mater.*, 2006, **15**, 1075.
171. J. L. Blackburn, C. Engtrakul, T. J. McDonald, A. C. Dillon and M. J. Heben, *J. Phys. Chem. B*, 2006, **110**, 25551.
172. E. Anglaret, F. Dragin, A. Pénicaud and R. Martel, *J. Phys. Chem. B.*, 2006, **110**, 3949.
173. V. Skákalová, A. B. Kaiser, U. Dettlaff-Weglikowska, K. Hrncariková and S. Roth, *J. Phys. Chem. B.*, 2005, **109**, 7174.
174. K. Shen, S. Curran, H. Xu, S. Rogelj, Y. Jiang, J. Dewald and T. Pietrass, *J. Phys. Chem. B*, 2005, **109**, 4455.
175. L. Lafi, D. Cossement and R. Chahine, *Carbon*, 2005, **43**, 1347.
176. A. Kukovecz, M. Smolik, S. Bokova, H. Kataura, Y. Achiba and H. Kuzmany, *J. Nanosci. Nanotechnol.*, 2005, **5**, 204.
177. S. K. Doorn, L. Zheng, M. J. O'Connell, Y. Zhu, S. Huang and J. Liu, *J. Phys. Chem. B*, 2005, **109**, 3751.
178. S. K. Doorn, M. J. O'Connell, L. Zheng, Y. T. Zhu, S. Huang and J. Liu, *Phys. Rev. Lett.*, 2005, **94**, 016802.
179. S. K. Doorn, *J. Nanosci. Nanotechnol.*, 2005, **5**, 1023.
180. M. Baibarac, I. Baltog, L. Mihut, N. Preda, T. Velula, C. Godon, J. Y. Mevellec, J. Wery and S. Lefrant, *J. Optoelectron. Adv. Mater.*, 2005, **7**, 2173.
181. A. Verma, M. Kauser and P. Ruden, *Appl. Phys. Lett.*, 2005, **87**, 123101.
182. J. C. Reber and M. S. Amer, *Master Thesis*, Wright State University, 2001.
183. G. S. Duesberg, I. Loa, M. Burghard, K. Syassen and S. Roth, *Phys. Rev. Lett.*, 2000, **85**, 5436.
184. H. Suzuura and T. Ando, *Phys. Rev. B*, 2002, **65**, 235412.
185. T. O. Shegai and G. Haran, *J. Phys. Chem. B*, 2006, **110**, 2459.
186. L. Jensen, O. H. Schmidt, K. V. Mikkelsen and P.-O. Astrand, *J. Phys. Chem. B.*, 2000, **104**, 10462.

187. S. Reich, C. Thomsen, G. Duesberg and S. Roth, *Phys. Rev. B*, 2001, **63**, 41401.
188. R. Saito, T. Takeya, T. Kimura, G. Dresselhaus and M. Dresselhaus, *Phys. Rev. B*, 1998, **57**, 4145.
189. G. Y. Slepyan, M. V. Shuba, S. A. Maksimenko and A. Lakhtakia, *Phys. Rev. B*, 2006, **73**, 195416.
190. T. Vukovi, I. Miloševi and M. Damnjanovi, *Phys. Rev. B*, 2002, **65**, 45418.
191. M. S. Amer, 2002, unpublished work, Wright State University.
192. P. Rai, D. R. Mohapatra, K. S. Hazra, D. S. Misra, J. Ghatak and P. V. Satyam, *Chem. Phys. Lett.*, 2008, **455**, 83.
193. G. M. do Nascimento, T. Hou, Y. A. Kim, H. Muramatsu, T. Hayashi, M. Endo, N. Akuzawa and M. S. Dresselhaus, *Nano Lett.*, 2008.
194. K. Papagelis, J. Arvanitidis, D. Christofilos, K. S. Andrikopoulos, T. Takenobu, Y. Iwasa, H. Kataura, S. Ves and G. A. Kourouklis, *Phys. Status Solidi B*, 2007, **244**, 116.
195. V. Gadagkar, S. Saha, D. V. S. Muthu, P. K. Maiti, Y. Lansac, A. Jagota, A. Moravsky, R. O. Loutfy and A. K. Sood, *J. Nanosci. Nanotechnol.*, 2007, **7**, 1753.
196. Y. A. Kim, M. Kojima, H. Muramatsu, S. Umemoto, T. Watanabe, K. Yoshida, K. Sato, T. Ikeda, T. Hayashi, M. Endo, M. Terrones and M. S. Dresselhaus, *Small*, 2006, **2**, 667.
197. Q.-H. Yang, P.-X. Hou, M. Unno, S. Yamauchi, R. Saito and T. Kyotani, *Nano Lett.*, 2005, **5**, 2465.
198. S. Bandow, M. Takizawa, K. Hirahara, M. Yudasaka and S. Iijima, *Chem. Phys. Lett.*, 2001, **337**, 48.
199. G. Picardi, M. Chaigneau and R. Ossikovski, *Chem. Phys. Lett.*, 2009, **469**, 161.
200. R. Marega, G. Accorsi, M. Meneghetti, A. Parisini, M. Prato and D. Bonifazi, *Carbon*, 2009, **47**, 675.
201. Q. Zhang, M. Zhao, J. Huang, W. Qian and F. Wei, *Chin. J. Catal.*, 2008, **29**, 1138.
202. M. S. Amer, 2008, unpublished work, Wright State University.
203. K. S. Novoselov, A. K. Geim, S. V. Morozov, D. Jiang, M. I. Katsnelson, I. V. Grigorieva, S. V. Dubonos and A. A. Firsov, *Nature*, 2005, **438**, 197.
204. H. Yanagisawa, T. Tanaka, Y. Ishida, M. Matsue, E. Rokuta, S. Otani and C. Oshima, *Surf. Interface Anal.*, 2005, **37**, 133.
205. H. C. Schniepp, J.-L. Li, M. J. McAllister, H. Sai, M. Herrera-Alonso, D. H. Adamson, R. K. Prud'homme, R. Car, D. A. Saville and I. A. Aksay, *J. Phys. Chem. B.*, 2006, **110**, 8535.
206. Z. Chen, Y. Lin, M. Rooks and P. Avouris, *Physica E (Amsterdam)*, 2007, **40**, 228.
207. A. Fasolino, J. Los and M. Katsnelson, *Nat. Mater.*, 2007, **6**, 858.
208. A. Geim and K. Novoselov, *Nat. Mater.*, 2007, **6**, 183.
209. M. Y. Han, B. Ozyilmaz, Y. Zhang and P. Kim, *Phys. Rev. Lett.*, 2007, **98**, 206805.

210. M. Ishigami, J. Chen, W. Cullen, M. Fuhrer and E. Williams, *Nano Lett.*, 2007, **7**, 1643.
211. M. Katsnelson, *Mater. Today*, 2007, **10**, 20.
212. M. Lemme, T. Echtermeyer, M. Baus, H. Kurz and A. Aachen, *IEEE Electron Device Lett.*, 2007, **28**, 282.
213. J. Meyer, A. Geim, M. Katsnelson, K. Novoselov, T. Booth and S. Roth, *Nature*, 2007, **446**, 60.
214. M. Trushin and J. Schliemann, *Phys. Rev. Lett.*, 2007, **99**, 216602.
215. A. A. Balandin, S. Ghosh, W. Bao, I. Calizo, D. Teweldebrhan, F. Miao and C. N. Lau, *Nano Lett.*, 2008, **8**, 902.
216. X. Du, I. Skachko, A. Barker and E. Andrei, *Nat. Nanotechnol.*, 2008, **3**, 491.
217. S. Ghosh, I. Calizo, D. Teweldebrhan, E. P. Pokatilov, D. L. Nika, A. A. Balandin, W. Bao, F. Miao and C. N. Lau, *Appl. Phys. Lett.*, 2008, **92**, 151911.
218. F. Guinea, *J. Low Temp. Phys.*, 2008, **153**, 359.
219. X. Li, X. Wang, L. Zhang, S. Lee and H. Dai, *Science*, 2008, **319**, 1229.
220. X. Li, G. Zhang, X. Bai, X. Sun, X. Wang, E. Wang and H. Dai, *Nat. Nanotechnol.*, 2008, **3**, 538.
221. Z. H. Ni, T. Yu, Y. H. Lu, Y. Y. Wang, Y. P. Feng and Z. X. Shen, *ACS Nano*, 2008, **2**, 2301.
222. Z. Wang, H. Lim, S. Ng, B. Özyilmaz and M. Kuok, *Carbon*, 2008, **46**, 2133.
223. Z. Wei, D. E. Barlow and P. E. Sheehan, *Nano Lett.*, 2008, **8**, 3141.
224. X. Yang, X. Dou, A. Rouhanipour, L. Zhi, H. J. Rader and K. Mullen, *J. Am. Chem. Soc.*, 2008, **130**, 4216.
225. J. Zhu, *Nat. Nanotechnol.*, 2008, **3**, 528.
226. S. Alwarappan, A. Erdem, C. Liu and C.-Z. Li, *J. Phys. Chem. C.*, 2009, **113**, 8853.
227. J. Campos-Delgado, Y. A. Kim, T. Hayashi, A. Morelos-Gómez, M. Hofmann, H. Muramatsu, M. Endo, H. Terrones, R. D. Shull, M. S. Dresselhaus and M. Terrones, *Chem. Phys. Lett.*, 2009, **469**, 177.
228. J. Hu, X. Ruan and Y. P. Chen, *Nano Lett.*, 2009, **9**, 2730.
229. L. Jiao, L. Zhang, X. Wang, G. Diankov and H. Dai, *Nature*, 2009, **458**, 877.
230. C. Jin, H. Lan, L. Peng, K. Suenaga and S. Iijima, *Phys. Rev. Lett.*, 2009, **102**, 205501.
231. D. V. Kosynkin, A. L. Higginbotham, A. Sinitskii, J. R. Lomeda, A. Dimiev, B. K. Price and J. M. Tour, *Nature*, 2009, **458**, 872.
232. J. Ma, D. Alfe, A. Michaelides and E. Wang, *Phys. Rev. B*, 2009, **80**, 033407.
233. Z. H. Ni, T. Yu, Z. Q. Luo, Y. Y. Wang, L. Liu, C. P. Wong, J. Miao, W. Huang and Z. X. Shen, *ACS Nano*, 2009, **3**, 569.
234. D. L. Nika, S. Ghosh, E. P. Pokatilov and A. A. Balandin, *Appl. Phys. Lett.*, 2009, **94**, 203103.

235. J. Park, Y. H. Ahn and C. Ruiz-Vargas, *Nano Lett.*, 2009, **9**, 1742.
236. F. Scarpa, S. Adhikari and A. Srikantha Phani, *Nanotechnology*, 2009, **20**, 065709.
237. S. Shivaraman, M. V. S. Chandrashekhar, J. J. Boeckl and M. G. Spencer, *J. Electron. Mater.*, 2009, **38**, 725.
238. M. L. Teague, A. P. Lai, J. Velasco, C. R. Hughes, A. D. Beyer, M. W. Bockrath, C. N. Lau and N. C. Yeh, *Nano Lett.*, 2009, **9**, 2542.
239. Z. -S. Wu, S. Pei, W. Ren, D. Tang, L. Gao, B. Liu, F. Li, C. Liu and H.-M. Cheng, *Adv. Mater.*, 2009, **21**, 1756.
240. G. D. Yuan, W. J. Zhang, Y. Yang, Y. B. Tang, Y. Q. Li, J. X. Wang, X. M. Meng, Z. B. He, C. M. L. Wu, I. Bello, C. S. Lee and S. T. Lee, *Chem. Phys. Lett.*, 2009, **467**, 361.
241. C. Casiraghi, S. Pisana, K. S. Novoselov, A. K. Geim and A. C. Ferrari, *Appl. Phys. Lett.*, 2007, **91**, 233108.
242. A. Gupta, G. Chen and P. Joshi, *Nano Lett.*, 2006, **6**, 2667.
243. I. Calizo, W. Bao, F. Miao, C. N. Lau and A. A. Balandin, *Appl. Phys. Lett.*, 2007, **91**, 201904.
244. I. Calizo, F. Miao, W. Bao, C. N. Lau and A. A. Balandin, *Appl. Phys. Lett.*, 2007, **91**, 071913.
245. A. C. Ferrari, *Solid State Commun.*, 2007, **143**, 47.
246. J. Yan, Y. Zhang, S. Goler, P. Kim and A. Pinczuk, *Solid State Commun.*, 2007, **143**, 39.
247. D. Abdula, T. Ozel, K. Kang, D. G. Cahill and M. Shim, *J. Phys. Chem. C*, 2008, **112**, 20131.
248. S. P. Berciaud, S. Ryu, L. E. Brus and T. F. Heinz, *Nano Lett.*, 2008, **9**, 346.
249. A. Das, B. Chakraborty and A. K. Sood, *Bull. Mater. Sci.*, 2008, **31**, 579.
250. A. Das, S. Pisana, B. Chakraborty, S. Piscanec, S. Saha, U. Waghmare, K. Novoselov, H. Krishnamurthy, A. Geim and A. Ferrari, *Nat. Nanotechnol.*, 2008, **3**, 210.
251. Z. Ni, Y. Wang, T. Yu and Z. Shen, *Nano Res.*, 2008, **1**, 273.
252. Y. Y. Wang, Z. H. Ni, Z. X. Shen, H. M. Wang and Y. H. Wu, *Appl. Phys. Lett.*, 2008, **92**, 043121.
253. Y. You, Z. Ni, T. Yu and Z. Shen, *Appl. Phys. Lett.*, 2008, **93**, 163112.
254. V. N. Popov, L. Henrard and P. Lambin, *Carbon*, 2009, **47**, 2448.
255. K. Nakada, M. Fujita, G. Dresselhaus and M. Dresselhaus, *Phys. Rev. B*, 1996, **54**, 17954.
256. Z. Ni, Y. Wang, T. Yu, Y. You and Z. Shen, *Phys. Rev. B*, 2008, **77**, 235403.
257. Y. y. Wang, Z. h. Ni, T. Yu, Z. X. Shen, H. m. Wang, Y. h. Wu, W. Chen and A. T. Shen Wee, *J. Phys. Chem. C*, 2008, **112**, 10637.
258. F. Huang, K. T. Yue, P. Tan, S.-L. Zhang, Z. Shi, X. Zhou and Z. Gu, *J. Appl. Phys.*, 1998, **84**, 4022.
259. S. Link and M. A. El-Sayed, *J. Phys. Chem. B*, 1999, **103**, 4212.

260. H. D. Li, K. T. Yue, Z. L. Lian, Y. Zhan, L. X. Zhou, S. L. Zhang, Z. J. Shi, Z. N. Gu, B. B. Liu, R. S. Yang, H. B. Yang, G. T. Zou, Y. Zhang and S. Iijima, *Appl. Phys. Lett.*, 2000, **76**, 2053.
261. N. Raravikar, P. Keblinski, A. Rao, M. Dresselhaus, L. Schadler and P. Ajayan, *Phys. Rev. B*, 2002, **66**, 235424.
262. L. Ci, Z. Zhou, L. Song, X. Yan, D. Liu, H. Yuan, Y. Gao, J. Wang, L. Liu, W. Zhou, G. Wang and S. Xie, *Appl. Phys. Lett.*, 2003, **82**, 3098.
263. T. Uchida, M. Tachibana, S. Kurita and K. Kojima, *Chem. Phys. Lett.*, 2004, **400**, 341.
264. Z. Zhou, X. Dou, L. Ci, L. Song, D. Liu, Y. Gao, J. Wang, L. Liu, W. Zhou, S. Xie and D. Wan, *J. Phys. Chem. B*, 2005, **110**, 1206.
265. Y. Zhang, L. Xie, J. Zhang, Z. Wu and Z. Liu, *J. Phys. Chem. C*, 2007, **111**, 14031.
266. L. Song, W. Ma, Y. Ren, W. Zhou, S. Xie, P. Tan and L. Sun, *Appl. Phys. Lett.*, 2008, **92**, 121905.
267. P. W. Chiu, S. F. Yang, S. H. Yang, G. Gu and S. Roth, *Appl. Phys. A*, 2003, **76**, 463.
268. A. Jorio, C. Fantini, M. S. S. Dantas, M. A. Pimenta, A. G. Souza Filho, G. G. Samsonidze, V. W. Brar, G. Dresselhaus, M. S. Dresselhaus, A. K. Swan, M. S. Ünlü, B. B. Goldberg and R. Saito, *Phys. Rev. B*, 2002, **66**, 115411.
269. M. Osman and D. Srivastava, *Nanotechnology*, 2001, **12**, 21.
270. J. H. Hildebrand, *J. Am. Chem. Soc.*, 1916, **38**, 1452.
271. D. Tyrer, *J. Phys. Chem.*, 1912, **16**, 69.
272. M. Fedurco, M. M. Olmstead and W. R. Fawcett, *Inorg. Chem.*, 1995, **34**, 390.
273. S. H. Gallagher, R. S. Armstrong, P. A. Lay and C. A. Reed, *J. Am. Chem. Soc.*, 1994, **116**, 12091.
274. S. H. Gallagher, R. S. Armstrong, P. A. Lay and C. A. Reed, *J. Phys. Chem.*, 1995, **99**, 5817.
275. R. S. Armstrong, S. H. Gallagher, P. A. Lay, J. R. Reimers and C. A. Reed, *J. Phys. Chem.*, 1996, **100**, 5604.
276. S. Gallagher, R. Armstrong, P. Lay and C. Reed, *Chem. Phys. Lett.*, 1996, **248**, 353.
277. A. L. Smith, E. Walter, M. V. Korobov and O. L. Gurvich, *J. Phys. Chem.*, 1996, **100**, 6775.
278. X. Zhou, J. Liu, Z. Jin, Z. Gu, Y. Wu and Y. Sun, *Fullerene Sci. Technol.*, 1997, **5**, 285.
279. W. A. Scrivens and J. M. Tour, *J. Chem. Soc., Chem. Commun.*, 1993, **15**, 1207.
280. R. S. Ruoff, D. S. Tse, R. Malhotra and D. C. Lorents, *J. Phys. Chem.*, 2002, **97**, 3379.
281. W. A. Scrivens, A. Cassell, B. North and J. Tour, *J. Am. Chem. Soc.*, 1994, **116**, 6939.
282. M. T. Beck and G. Mandi, *Fullerene Sci. Technol.*, 1997, **5**, 291.

283. N. Sivaraman, R. Dhamodaran, I. Kaliappan, T. G. Srinivasan, P. R. V. Rao and C. K. Mathews, *J. Org. Chem.*, 2002, **57**, 6077.
284. R. G. Makitra, R. E. Pristanskii and R. I. Flyunt, *Russ. J. Gen. Chem.*, 2003, **73**, 1227.
285. S. Talukdar, P. Prandhan and A. Banerji, *Fullerene Sci. Technol.*, 1997, **5**, 547.
286. K. Kimata, T. Hirose, K. Moriuchi, K. Hosoya, T. Araki and N. Tanaka, *Anal. Chem.*, 2002, **67**, 2556.
287. W. Chen and Z. Xu, *Fullerene Sci. Technol.*, 1998, **6**, 695.
288. M. T. Beck, G. Mandi and S. Keki, in *Recent Advances in the Chemistry and Physics of Fullerenes and Related Materials*, ed. R. S. Ruoff and K. M. Kadish, The Electrochemical Society, Pennington, NJ, 1995, **vol. 2**, p. 1510.
289. R. J. Doome, S. Dermaunt, A. Fonseca, M. Hammida and B. Nagy, *Fullerene Sci. Technol.*, 1997, **5**, 1593.
290. T. Anderson, K. Nillson, M. Sundahl, G. Westman and O. Wennerstrom, *J. Chem. Soc., Chem. Commun.*, 1992, 604.
291. T. Tomiyama, S. Uchiyama and H. Shinohara, *Chem. Phys. Lett.*, 1997, **264**, 143.
292. M. V. Korobov and A. L. Smith, in *Fullerenes; Chemistry, Physics, and Technology*, ed. K. M. Kadish and R. S. Ruoff, Wiley-Interscience, New York, 2000, p. 53.
293. M. T. Beck, G. Mandi and S. Keki, in *Recent Advances in the Chemistry and Physics of Fullerenes and Related Materials*, ed. R. S. Ruoff and K. M. Kadish, The Electrochemical Society, Pennington, NJ, 1996, **vol. 3**, p. 32.
294. D. Heymann, *Carbon*, 1996, **34**, 627.
295. Y. Nagano, T. Tamura and T. Kiyobayashi, *Chem. Phys. Lett.*, 1994, **228**, 125.
296. Y. Nagano and T. Nakamura, *Chem. Phys. Lett.*, 1997, **265**, 358.
297. Y.-L. Zeng, Y.-F. Huang, J.-H. Jiang, X.-B. Zhang, C.-R. Tang, G.-L. Shen and R.-Q. Yu, *Electrochem. Commun.*, 2007, **9**, 185.
298. S. Sawamura, *Pure Appl. Chem.*, 2007, **79**, 861.
299. S. Sawamura and N. Fujita, *Carbon*, 2007, **45**, 965.
300. K. N. Semenov, N. A. Charykov and O. V. Arapov, *Russ. J. Phys. Chem. A*, 2008, **82**, 1318.
301. K. Semenov, N. Charykov and O. Arapov, *Fullerenes, Nanotubes Carbon Nanostruct.*, 2009, **17**, 230.
302. K. Semenov, N. Charykov, O. Arapov, N. Alekseev and M. Trofimova, *Russ. J. Phys. Chem. A*, 2009, **83**, 59.
303. M. H. Herbst, G. H. M. Dias, J. G. Magalhães, R. B. Tôrres and P. L. O. Volpe, *J. Mol. Liq.*, 2005, **118**, 9.
304. A. Kolker, N. Islamova, N. Avramenko and A. Kozlov, *J. Mol. Liq.*, 2007, **131**, 95.
305. S. J. Yeo, J. O. Cha, R. Pode, J. S. Ahn, H. M. Kim and B. R. Rhee, *J. Korean Phys. Soc.*, 2008, **53**, 2677.

306. I. F. Gun'kin and N. Y. Loginova, *Russ. J. General Chem.*, 2006, **76**, 1911.
307. F. Langa, P. de la Cruz, J. L. Delgado, E. Espíldora, M. J. Gómez-Escalonilla and A. de la Hoz, *J. Mater. Chem.*, 2002, **12**, 2130.
308. J. F. Maguire, M. S. Amer and J. Busbee, *Appl. Phys. Lett.*, 2003, **82**, 2592.
309. P. Scharff, K. Risch, L. Carta-Abelmann, I. M. Dmytruk, M. M. Bilyi, O. A. Golub, A. V. Khavryuchenko, E. V. Buzaneva, V. L. Aksenov, M. V. Avdeev, Y. I. Prylutskyy and S. S. Durov, *Carbon*, 2004, **42**, 1203.
310. M. S. Amer, J. A. Elliott, J. F. Maguire and A. H. Windle, *Chem. Phys. Lett.*, 2005, **411**, 395.
311. R. F. Enes, A. C. Tome, J. A. S. Cavaleiro, A. El-Agamey and D. J. McGarvey, *ChemInform*, 2006, 37.
312. K. Kawano, J. Sakai, M. Yahiro and C. Adachi, *Solar Energy Mater. Solar Cells*, 2009, **93**, 514.
313. W. Liu, R. Liu, W. Wang, W. Li, W. Liu, K. Zheng, L. Ma, Y. Tian, Z. Bo and Y. Huang, *J. Phys. Chem. C*, 2009, **113**, 11385.
314. C. Park, H. J. Song and H. C. Choi, *Chem. Commun.*, 2009, 4803.
315. J. Geng, I. A. Solovyov, W. Zhou, A. V. Solovyov and B. F. G. Johnson, *J. Phys. Chem. C*, 2009, **113**, 6390.
316. J. Wang, Y. Shen, S. Kessel, P. Fernandes, K. Yoshida, S. Yagai, D. G. Kurth, H. Möhwald and T. Nakanishi, *Angew. Chem. Int. Ed.*, 2009, **48**, 2166.
317. S. I. Cha, K. i. Miyazawa and J.-D. Kim, *Chem. Mater.*, 2008, **20**, 1667.
318. F. Zhang, K. Jespersen, C. Björström,, M. Svensson, M. Andersson, V. Sundström, K. Magnusson, E. Moons, A. Yartsev and O. Inganäs, *Adv. Funct. Mater.*, 2006, **16**, 667.
319. F. Su, C. Lu, W. Cnen, H. Bai and J. F. Hwang, *Sci. Total Environ.*, 2009, **407**, 3017.
320. D. di Caprio and J. P. Badiali, *Entropy*, 2009, **11**, 238.
321. A. Zdetsis, *Phys. Rev. B*, 2008, **77**, 115402.
322. D. Marenduzzo, C. Micheletti and P. Cook, *Biophys. J.*, 2006, **90**, 3712.
323. H. Reiss, *J. Non-Cryst. Solids*, 2009, **355**, 617.
324. W. Nowicki, G. Nowicka and J. Narkiewicz-Michalek, *Polymer*, 2009, **50**, 2161.
325. J. C. Dyre, T. Hechsher and K. Niss, *J. Non-Cryst. Solids*, 2009, **355**, 624.
326. R. Pinal, *Entropy*, 2008, **10**, 207.
327. H. Lekkerkerker and A. Stroobants, *Genes Dev*, 1993, **7**, 1949.
328. H. Lekkerkerker and A. Stroobants, *Nature*, 1998, **393**, 305.
329. L. V. Woodcock, *Nature*, 1997, **385**, 141.
330. G. Makhatadze and P. Privalov, *Protein Sci.*, 1996, **5**, 507.
331. J. Minato and K. miyazawa, *J. Mater. Res.*, 2006, **21**, 539.
332. J. Minato and K. Miyazawa, *Carbon*, 2005, **43**, 2837.
333. J. Minato and K. Miyazawa, *Diamond Relat. Mater.*, 2006, **15**, 1151.

334. M. S. Amer, 2009, unpublished work, Wright State University.
335. B. Ginzburg, S. Tuichiev, S. Tabarov and A. Shepelevskii, *Crystallogr. Rep.*, 2005, **50**, 735.
336. B. M. Ginzburg and S. Tuichiev, *J. Macromol. Sci.-Phys.*, 2005, **B44**, 517.
337. B. M. Ginzburg, V. L. Ugolkov, G. N. Fedorova, E. Y. Melenevskaya, L. A. Shibaev and S. Tuichiev, *Russ. J. Appl. Chem.*, 2006, **79**, 1202.
338. B. M. Ginzburg and S. Tuichiev, *Crystallogr. Rep.*, 2007, **52**, 108.
339. B. M. Ginzburg, S. Tuichiev and S. K. Tabarov, *Tech. Phys. Lett.*, 2007, **33**, 639.
340. B. M. Ginzburg and S. Tuichiev, *Crystallogr. Rep.*, 2008, **53**, 645.
341. B. M. Ginzburg and S. Tuichiev, *Russ. J. Appl. Chem.*, 2008, **81**, 618.
342. B. M. Ginzburg, S. Tuichiev and S. Shukhiev, *Tech. Phys. Lett.*, 2009, **35**, 491.
343. B. M. Ginzburg, S. Tuichiev and S. K. Tabarov, *Russ. J. Appl. Chem.*, 2009, **82**, 387.
344. M. S. Amer and M. T. Abdu, *Philos. Mag. Lett.*, 2009, **89**, 615.
345. M. S. Amer, M. Bennett and J. F. Maguire, *Chem. Phys. Lett.*, 2008, **457**, 329.
346. O. Pozdnyakov, B. Redkov, B. Ginzburg and A. Pozdnyakov, *Tech. Phys. Lett.*, 1998, **24**, 916.
347. K. D. Rule, MPhil, University of Cambridge, 2003.
348. D. Brenner, J. Harrison, C. White and R. Colton, *Thin Solid Films*, 1991, **206**, 220.
349. R. Ruoff and D. Lorents, *Carbon*, 1995, **33**, 925.
350. J. Salvetat, J. Bonard, N. Thomson, A. Kulik, L. Forro, W. Benoit and L. Zuppiroli, *Appl. Phys. A*, 1999, **69**, 255.
351. J. R. Wood, M. D. Frogley, E. R. Meurs, A. D. Prins, T. Peijs, D. J. Dunstan and H. D. Wagner, *J. Phys. Chem. B*, 1999, **103**, 10388.
352. M. Yu, O. Lourie, M. Dyer, K. Moloni, T. Kelly and R. Ruoff, *Science*, 2000, **287**, 637.
353. D. Qian, G. Wagner, W. Liu, M. Yu and R. Ruoff, *Appl. Mech. Rev.*, 2002, **55**, 495.
354. T. Chang and H. Gao, *J. Mech. Phys. Solids*, 2003, **51**, 1059.
355. M. -F. Yu, *J. Eng. Mater. Technol.*, 2004, **126**, 271.
356. Q. Lu and B. Bhattacharya, *Nanotechnology*, 2005, **16**, 555.
357. S. B. Sinnott, N. R. Aluru and D. Liming, in *Carbon Nanotechnology*, Elsevier, Amsterdam, 2006, p. 361.
358. Q. Wang, C. B. Wang, Z. Wang, J. Y. Zhang and D. Y. He, *Appl. Phys. Lett.*, 2007, **91**.
359. M. Ashino, D. Obergfell, M. Haluška, S. Yang, A. Khlobystov, S. Roth and R. Wiesendanger, *Nat. Nanotechnol.*, 2008, **3**, 337.
360. J. Geng, W. Zhou, P. Skelton, W. Yue, I. Kinloch, A. Windle and B. Johnson, *J. Am. Chem. Soc*, 2008, **130**, 2527.
361. S. Habelitz, S. J. Marshall, G. W. Marshall and M. Balooch, *Arch. Oral Biol.*, 2001, **46**, 173.

362. K. Kumar, H. Van Swygenhoven and S. Suresh, *Acta Mater.*, 2003, **51**, 5743.
363. M. Meyers, A. Mishra and D. Benson, *Prog. Mater. Sci.*, 2006, **51**, 427.
364. A. Desai and M. Haque, *Sens. Actuators, A*, 2007, **134**, 169.
365. Y. Zhang, N. Tao and K. Lu, *Acta Mater.*, 2008.
366. A. Ávila and G. Lacerda, *Mater. Res.*, 2008, **11**, 325.
367. M. S. Amer and J. F. Maguire, *Chem. Phys. Lett.*, 2009, **476**, 232.
368. R. Poloni, M. Fernandez-Serra, S. Le Floch, S. De Panfilis, P. Toulemonde, D. Machon, W. Crichton, S. Pascarelli and A. San-Miguel, *Phys. Rev. B*, 2008, **77**, 35429.
369. P. Liu, *Carbon*, 2006, **44**, 1484.
370. L. Pintschovius, O. Blaschko, G. Krexner and N. Pyka, *Phys. Rev. B*, 1999, **59**, 11020.
371. L. Girifalco, *Phys. Rev. B*, 1995, **52**, 9910.
372. G. Gadd, S. Kennedy, S. Moricca, C. Howard, M. Elcombe, P. Evans and M. James, *Phys. Rev. B*, 1997, **55**, 14794.
373. G. Gadd, P. Evans, S. Kennedy, M. James, M. Elcombe, D. Cassidy, S. Moricca, J. Holmes, N. Webb and A. Dixon, *Fullerenes, Nanotubes Carbon Nanostruct.*, 1999, **7**, 1043.
374. R. S. Ruoff and A. L. Ruoff, *Nature*, 1991, **350**, 663.
375. J. E. Fischer, P. A. Heiney, A. R. McGhie, W. J. Romanow, A. M. Denenstein, J. P. J. McCauley and A. B. Smith III, *Science*, 1991, **252**, 1288.
376. M. S. Amer, 2007, unpublished work, Wright State University.
377. R. Pearson, *Proc. Natl. Acad. Sci. U.S.A.*, 1986, **83**, 8440.
378. J. Gilman, *Mater. Sci. Eng. A*, 1996, **209**, 74.
379. J. J. Gilman, *Science*, 1993, **261**, 1436.
380. S. Karmakar, S. Sharma, P. Teredesai, D. Muthu, A. Govindaraj, S. Sikka and A. Sood, *New J. Phys.*, 2003, **5**, 143.
381. J. Wu, W. Walukiewicz, W. Shan, E. Bourret-Courchesne, J. Ager Iii, K. Yu, E. Haller, K. Kissell, S. Bachilo and R. Weisman, *Phys. Rev. Lett.*, 2004, **93**, 17404.
382. S. Reich, C. Thomsen and P. Ordejon, *Phys. Status Solidi B*, 2003, **235**, 354.
383. M. H. F. Sluiter, V. Kumar and Y. Kawazoe, *Phys. Rev. B*, 2002, **65**, 161402.
384. S. Okada, A. Oshiyama and S. Saito, *J. Phys. Soc. Jpn.*, 2001, **70**, 2345.
385. Y. Prylutski, S. Durov, A. Oglolya, P. Eklund and L. Grigorian, *Mol. Mater.*, 2000, **13**, 71.
386. Y. I. Prilustski, S. Durov and A. V. Nazarenko, *Phys. Status Solidi B*, 1999, **211**, 213.
387. J. Elliott, J. Sandler, A. Windle, R. Young and M. Shaffer, *Phys. Rev. Lett.*, 2004, **92**, 95501.
388. J. Sandler, M. S. P. Shaffer, A. H. Windle, M. P. Halsall, M. A. Montes-Morán, C. A. Cooper and R. J. Young, *Phys. Rev. B*, 2003, **67**, 035417.

389. S. Rols, I. N. Goncharenko, R. Almairac, J. L. Sauvajol and I. Mirebeau, *Phys. Rev. B*, 2001, **64**, 153401.
390. L. C. Chen, L. J. Wang, D. S. Tang, S. S. Xie and C. Q. Jin, *Chin. Phys. Lett.*, 2001, **18**, 577.
391. N. Minami, S. Kazaoui, R. Jacquemin, H. Yamawaki, K. Aoki, H. Kataura and Y. Achiba, *Synth. Metals*, 2001, **116**, 405.
392. D. S. Tang, L. C. Chen, L. J. Wang, L. Sun, Z. Liu, G. Wang, W. Zhou and S. S. Xie, *J. Mater. Res.*, 2000, **15**, 560.
393. U. D. Venkateswaran, A. M. Rao, E. Richter, M. Menon, A. Rinzler, R. E. Smalley and P. C. Eklund, *Phys. Rev. B*, 1999, **59**, 10928.
394. J. Ferraro, K. Nakamoto and C. Brown, *Introductory Raman Spectroscopy*, Academic Press, London, 2003.
395. A. Jayaraman, *Rev. Sci. Instrum.*, 1986, **57**, 1013.
396. M. Amer, in *Fullerene Research Advances*, ed. C. Kramer, Nova Publishers, Philadelphia, 2007, p. 223.
397. M. S. Amer, 2004, unpublished work, Max Planck Institute for Solid State Physics.
398. S. J. Gregg and K. S. W. Sing, *Adsorption, Surface Area, and Porosity*, Academic Press, New York, 1982.
399. M. Muris, N. Dufau, M. Bienfait, N. Dupont-Pavlovsky, Y. Grillet and J. P. Palmari, *Langmuir*, 2000, **16**, 7019.
400. Z. Q. Zhang, H. W. Zhang, Y. G. Zheng, J. B. Wang and L. Wang, *Curr. Appl. Phys.*, 2008, **8**, 217.
401. K. Gao, R. C. Dai, Z. Zhao, Z. M. Zhang and Z. J. Ding, *Solid State Commun.*, 2008, **147**, 65.
402. W. Yang, R.-Z. Wang, Y.-F. Wang, X.-M. Song, B. Wang and H. Yan, *Phys. Rev. B*, 2007, **76**, 033402.
403. K. Papagelis, K. S. Andrikopoulos, J. Arvanitidis, D. Christofilos, C. Galiotis, T. Takenobu, Y. Iwasa, H. Kataura, S. Ves and G. A. Kourouklis, *Phys. Status Solidi B*, 2007, **244**, 4069.
404. R. S. Kumar, M. G. Pravica, A. L. Cornelius, M. F. Nicol, M. Y. Hu and P. C. Chow, *Diamond Relat. Mater.*, 2007, **16**, 1250.
405. T.-a. Yano, Y. Inouye and S. Kawata, *Nano Lett.*, 2006, **6**, 1269.
406. J. Arvanitidis, D. Christofilos, K. Papagelis, T. Takenobu, Y. Iwasa, H. Kataura, S. Ves and G. A. Kourouklis, *Phys. Rev. B*, 2005, **72**.
407. A. P. Henderson, L. N. Seetohul, A. K. Dean, P. Russell, S. Pruneanu and Z. Ali, *Langmuir*, 2009, **25**, 931.
408. P. Wang and Y. Fang, *J. Chem. Phys.*, 2008, **129**, 134702.
409. U. Burghaus, D. Bye, K. Cosert, J. Goering, A. Guerard, E. Kadossov, E. Lee, Y. Nadoyama, N. Richter and E. Schaefer, *Chem. Phys. Lett.*, 2007, **442**, 344.
410. D. Tang, L. Ci, W. Zhou and S. Xie, *Carbon*, 2006, **44**, 2155.
411. Y. Ding, X.-b. Yang and J. Ni, *Frontiers Phys. Chin.*, 2006, **1**, 317.
412. S. Sharma, D. Singh and Y. Li, *J. Raman Spectrosc.*, 2005, **36**.
413. S. Picozzi, S. Santucci, L. Lozzi, L. Valentini and B. Delley, *J. Chem. Phys.*, 2004, **120**, 7147.

414. L. Zhou and S. Shi, *Carbon*, 2003, **41**, 613.
415. S. Letardi, M. Celino, F. Cleri and V. Rosato, *Surf. Sci.*, 2002, **496**, 33.
416. F. Suárez-García, A. Martínez-Alonso and J. M. D. Tascón, in *Adsorption by Carbons*, ed. J. B. Eduardo and J. M. D. Tascón, Elsevier, Amsterdam, 2008, pp. 329.
417. E. Schrodinger, *What is Life?*, Cambridge University Press, Cambridge, 1944.

APPENDIX 1
Character Tables for Various Point Groups

Based on data published in K. Nakamoto, *Infrared and Raman Spectra of Inorganic and Coordination Compounds*, John Wiley and Sons, Inc., New York, 1978.

C_s	E	$\sigma(xy)$	Raman activity
A'	+1	+1	$\alpha_{xx}, \alpha_{yy}, \alpha_{zz}, \alpha_{xy}$
A''	+1	−1	α_{yz}, α_{xz}

C_2	E	$C_2(z)$	Raman activity
A	+1	+1	$\alpha_{xx}, \alpha_{yy}, \alpha_{zz}, \alpha_{xy}$
B	+1	−1	α_{yz}, α_{xz}

C_i	E	I	Raman activity
A_g	+1	+1	All components of α
A_u	+1	−1	

C_{2v}	E	$C_2(z)$	$\sigma_v(xz)$	$\sigma_v(yz)$	Raman activity
A_1	+1	+1	+1	+1	$\alpha_{xx}, \alpha_{yy}, \alpha_{zz},$
A_2	+1	+1	−1	−1	α_{xy}
B_1	+1	−1	+1	−1	α_{xz}
B_2	+1	−1	−1	+1	α_{yz}

C_{3v}	E	$2C_3(z)$	$3\sigma_v$	Raman activity
A_1	+1	+1	+1	$\alpha_{xx}+\alpha_{yy}, \alpha_{zz}$
A_2	+1	+1	−1	
E	+2	−1	0	$(\alpha_{xx}-\alpha_{yy}, \alpha_{xy}), (\alpha_{yz}, \alpha_{xz})$

C_{4v}	E	$2C_4(z)$	$C_4^2 \equiv C_2''$	$2\sigma_v$	$2\sigma_d$	Raman activity
A_1	+1	+1	+1	+1	+1	$\alpha_{xx}+\alpha_{yy}, \alpha_{zz}$
A_2	+1	+1	+1	−1	−1	
B_1	+1	−1	+1	+1	−1	$\alpha_{xx}-\alpha_{yy}$
B_2	+1	−1	+1	−1	+1	α_{xy}
E	+2	0	−2	0	0	$(\alpha_{yz}, \alpha_{xz})$

$C_{\infty v}$	E	$2C_\infty^\phi$	$2C_\infty^{2\phi}$	$2C_\infty^{2\phi}$	...	$\infty\sigma_v$	Raman activity
Σ^+	+1	+1	+1	+1	...	+1	$\alpha_{xx}+\alpha_{yy}, \alpha_{zz}$
Σ^-	+1	+1	+1	+1	...	−1	
Π	+2	$2\cos\phi$	$2\cos 2\phi$	$2\cos 3\phi$	...	0	$(\alpha_{yz}, \alpha_{xz})$
Δ	+2	$2\cos 2\phi$	$2\cos 2.2\phi$	$3\cos 3.2\phi$	...	0	$(\alpha_{xx}-\alpha_{yy}, \alpha_{xy})$
ϕ	+2	$2\cos 3\phi$	$2\cos 2.3\phi$	$2\cos 3.3\phi$	...	0	
...	...	...	...	...	...	...	

C_{2h}	E	$C_2(z)$	$\sigma_h(xy)$	i	Raman activity
A_g	+1	+1	+1	+1	$\alpha_{xx}, \alpha_{yy}, \alpha_{zz}, \alpha_{xy},$
A_u	+1	+1	−1	−1	
B_g	+1	−1	−1	+1	α_{yz}, α_{xz}
B_u	+1	−1	+1	−1	

D_3	E	$2C_3(z)$	$3C_2$	Raman activity
A_1	+1	+1	+1	$\alpha_{xx}+\alpha_{yy}, \alpha_{zz}$
A_2	+1	+1	−1	
E	+2	−1	0	$(\alpha_{xx}-\alpha_{yy}, \alpha_{xy}), (\alpha_{yz}, \alpha_{xz})$

Character Tables for Various Point Groups

$D_{2d} \equiv V_d$	E	$2S_4(z)$	$S_4^2 \equiv C_2''$	$2C_2$	$2\sigma_d$	Raman activity
A_1	+1	+1	+1	+1	+1	$\alpha_{xx}+\alpha_{yy}, \alpha_{zz}$
A_2	+1	+1	+1	−1	−1	
B_1	+1	−1	+1	+1	−1	$\alpha_{xx}-\alpha_{yy}$
B_2	+1	−1	+1	−1	+1	α_{xy}
E	+2	0	−2	0	0	$(\alpha_{yz}, \alpha_{xz})$

D_{3d}	E	$2S_6(z)$	$2S_6^2 \equiv 2C_3$	$S_6^3 \equiv S_3 \equiv i$	$3C_2$	$3\sigma_d$	Raman activity
A_{1g}	+1	+1	+1	+1	+1	+1	$\alpha_{xx}+\alpha_{yy}, \alpha_{zz}$
A_{1u}	+1	−1	+1	−1	+1	−1	
A_{2g}	+1	+1	+1	+1	−1	−1	
A_{2u}	+1	−1	+1	−1	−1	+1	
E_g	+2	−1	−1	+2	0	0	$(\alpha_{xx}-\alpha_{yy}, \alpha_{xy}), (\alpha_{yz}, \alpha_{xz})$
E_u	+2	+1	−1	−2	0	0	

D_{4d}	E	$2S_8(z)$	$2S_8^2 \equiv 2C_4$	$2S_8^3$	$S_8^4 \equiv C_2''$	$4C_2$	$4\sigma_d$	Raman activity
A_1	+1	+1	+1	+1	+1	+1	+1	$\alpha_{xx}+\alpha_{yy}, \alpha_{zz}$
A_2	+1	+1	+1	+1	+1	−1	−1	
B_1	+1	−1	+1	−1	+1	+1	−1	
B_2	+1	−1	+1	−1	+1	−1	+1	
E_1	+2	$+\sqrt{2}$	0	$-\sqrt{2}$	−2	0	0	
E_2	+2	0	−2	0	+2	0	0	$(\alpha_{xx}-\alpha_{yy}, \alpha_{xy})$
E_3	+2	$-\sqrt{2}$	0	$+\sqrt{2}$	−2	0	0	$(\alpha_{yz}, \alpha_{xz})$

$D_{2h}=V_h$	E	$\sigma(xy)$	$\sigma(xz)$	$\sigma(yz)$	i	$C_2(z)$	$C_2(y)$	$C_2(x)$	Raman activity
A_{1g}	+1	+1	+1	+1	+1	+1	+1	+1	$\alpha_{xx}+\alpha_{yy}, \alpha_{zz}$
A_{1u}	+1	−1	−1	−1	−1	+1	+1	+1	
B_{1g}	+1	+1	−1	−1	+1	+1	−1	−1	α_{xy}
B_{1u}	+1	−1	+1	+1	−1	+1	−1	−1	
B_{2g}	+1	−1	+1	−1	+1	−1	+1	−1	α_{xz}
B_{2u}	+1	+1	−1	+1	−1	−1	+1	−1	
B_{3g}	+1	−1	−1	+1	+1	−1	−1	+1	α_{yz}
B_{3u}	+1	+1	+1	−1	−1	−1	−1	+1	

D_{3h}	E	$2C_3(z)$	$3C_2$	σ_h	$2S_3$	$3\sigma_v$	Raman activity
A'_1	+1	+1	+1	+1	+1	+1	$\alpha_{xx}+\alpha_{yy}, \alpha_{zz}$
A''_1	+1	+1	+1	−1	−1	−1	
A'_2	+1	+1	−1	+1	+1	−1	
A''_2	+1	+1	−1	−1	−1	+1	
E'	+2	−1	0	+2	−1	0	$(\alpha_{xx}-\alpha_{yy}, \alpha_{xy})$
E''	+2	−1	0	−2	−1	0	$(\alpha_{yz}, \alpha_{xz})$

D_{4h}	E	$2C_4(z)$	$C_4^2 \equiv C''_2$	$2C_2$	$2C'_2$	σ_h	$2\sigma_v$	$2\sigma_d$	$2S_4$	$S_2 \equiv i$	Raman activity
A_{1g}	+1	+1	+1	+1	+1	+1	+1	+1	+1	+1	$\alpha_{xx}+\alpha_{yy}, \alpha_{zz}$
A_{1u}	+1	+1	+1	+1	+1	−1	−1	−1	−1	−1	
A_{2g}	+1	+1	+1	−1	−1	+1	−1	−1	+1	+1	
A_{2u}	+1	+1	+1	−1	−1	−1	+1	+1	−1	−1	
B_{1g}	+1	−1	+1	+1	−1	+1	+1	−1	−1	+1	$\alpha_{xx}-\alpha_{yy}$
B_{1u}	+1	−1	+1	+1	−1	−1	−1	+1	+1	−1	
B_{2g}	+1	−1	+1	−1	+1	+1	−1	+1	−1	+1	α_{xy}
B_{2u}	+1	−1	+1	−1	+1	−1	+1	−1	+1	−1	
E_g	+2	0	−2	0	0	−2	0	0	0	+2	$(\alpha_{yz}, \alpha_{xz})$
E_u	+2	0	−2	0	0	+2	0	0	0	−2	

D_{5h}	E	$2C_5(z)$	$2C_5^2$	σ_h	$5C_2$	$5\sigma_v$	$2S_5$	$2S_5^3$	Raman activity
A'_1	+1	+1	+1	+1	+1	+1	+1	+1	$\alpha_{xx}+\alpha_{yy}, \alpha_{zz}$
A''_1	+1	+1	+1	−1	+1	−1	−1	−1	
A'_2	+1	+1	+1	+1	−1	−1	+1	+1	
A''_2	+1	+1	+1	−1	−1	+1	−1	−1	
E'_1	+2	$2\cos72°$	$2\cos144°$	+2	0	0	$+2\cos72°$	$+2\cos144°$	
E''_1	+2	$2\cos72°$	$2\cos144°$	−2	0	0	$-2\cos72°$	$-2\cos144°$	$(\alpha_{yz}, \alpha_{xz})$
E'_2	+2	$2\cos144°$	$2\cos72°$	+2	0	0	$+2\cos144°$	$+2\cos72°$	$(\alpha_{xx}-\alpha_{yy}, \alpha_{xy})$
E''_2	+2	$2\cos144°$	$2\cos72°$	−2	0	0	$-2\cos144°$	$-2\cos72°$	

Character Tables for Various Point Groups

D_{6h}	E	$2C_6(z)$	$C_6^2 \equiv 2C_3$	$C_6^3 \equiv C_2''$	$3C_2$	$3C_2'$	σ_h	$3\sigma_v$	$3\sigma_d$	$2S_6$	$2S_3$	$S_6^3 \equiv S_2 \equiv i$	Raman activity
A_{1g}	+1	+1	+1	+1	+1	+1	+1	+1	+1	+1	+1	+1	$\alpha_{xx}+\alpha_{yy}, \alpha_{zz}$
A_{1u}	+1	+1	+1	+1	+1	+1	−1	−1	−1	−1	−1	−1	
A_{2g}	+1	+1	+1	+1	−1	−1	+1	−1	−1	+1	+1	+1	
A_{2u}	+1	+1	+1	+1	−1	−1	−1	+1	+1	−1	−1	−1	
B_{1g}	+1	−1	+1	−1	+1	−1	−1	+1	−1	+1	−1	+1	
B_{1u}	+1	−1	+1	−1	+1	−1	+1	−1	+1	−1	+1	−1	
B_{2g}	+1	−1	+1	−1	−1	+1	−1	−1	+1	+1	−1	+1	
B_{2u}	+1	−1	+1	−1	−1	+1	+1	+1	−1	−1	+1	−1	
E_{1g}	+2	+1	−1	−2	0	0	−2	0	0	−1	+1	+2	$(\alpha_{yz}, \alpha_{xz})$
E_{1u}	+2	+1	−1	−2	0	0	+2	0	0	+1	−1	−2	
E_{2g}	+2	−1	−1	+2	0	0	+2	0	0	−1	−1	+2	$(\alpha_{xx}-\alpha_{yy}, \alpha_{xy})$
E_{2u}	+2	−1	−1	+2	0	0	−2	0	0	+1	+1	−2	

$D\infty_h$	E	$2C_\infty^\phi$	$2C_\infty^{2\phi}$	$2C_\infty^{3\phi}$	...	σ_h	∞C_2	$\infty \sigma_v$	$2S_\infty^{2\phi}$	$2S_\infty^{2\phi}$	...	$S_2 \equiv i$	Raman activity
Σ_g^+	+1	+1	+1	+1	...	+1	+1	+1	+1	+1	...	+1	$\alpha_{xx}+\alpha_{yy}, \alpha_{zz}$
Σ_u^+	+1	+1	+1	+1	...	−1	−1	+1	−1	−1	...	−1	
Σ_g^-	+1	+1	+1	+1	...	+1	−1	−1	+1	+1	...	+1	
Σ_u^-	+1	+1	+1	+1	...	−1	+1	−1	−1	−1	...	−1	
Π_g	+2	$2\cos\phi$	$2\cos 2\phi$	$2\cos 3\phi$	...	−2	0	0	$-2\cos\phi$	$-2\cos 2\phi$	...	+2	$(\alpha_{yz}, \alpha_{xz})$
Π_u	+2	$2\cos\phi$	$2\cos 2\phi$	$2\cos 3\phi$	...	+2	0	0	$+2\cos\phi$	$+2\cos 2\phi$	...	−2	
Δ_g	+2	$2\cos 2\phi$	$2\cos 4\phi$	$2\cos 6\phi$	...	+2	0	0	$+2\cos 2\phi$	$+2\cos 4\phi$	...	+2	$(\alpha_{xx}-\alpha_{yy}, \alpha_{xy})$
Δ_u	+2	$2\cos 2\phi$	$2\cos 4\phi$	$2\cos 6\phi$	...	−2	0	0	$-2\cos 2\phi$	$-2\cos 4\phi$	...	−2	
Φ_g	+2	$2\cos 3\phi$	$2\cos 6\phi$	$2\cos 9\phi$	...	−2	0	0	$-2\cos 3\phi$	$-2\cos 4\phi$	...	+2	
Φ_u	+2	$2\cos 3\phi$	$2\cos 6\phi$	$2\cos 9\phi$	...	+2	0	0	$+2\cos 3\phi$	$+2\cos 4\phi$	...	−2	
...													

Character Tables for Various Point Groups

T_d	E	$8C_3$	$6\sigma_d$	$6S_4$	$3S_4^2 \equiv 3C_2$	Raman activity
A_1	+1	+1	+1	+1	+1	$\alpha_{xx}+\alpha_{yy}+\alpha_{zz}$
A_2	+1	+1	−1	−1	+1	
E	+2	−1	0	0	+2	$(\alpha_{xx}+\alpha_{yy}-2\alpha_{zz}, \alpha_{xx}-\alpha_{yy})$
F_1	+3	0	−1	+1	−1	
F_2	+3	0	+1	−1	−1	$(\alpha_{xy}, \alpha_{yz}, \alpha_{xz})$

O_h	E	$8C_3$	$6C_2$	$6C_4$	$3C_4^2 \equiv C_2''$	$S_2 \equiv i$	$6S_4$	$8S_6$	$3\sigma_h$	$6\sigma_d$	Raman activity
A_{1g}	+1	+1	+1	+1	+1	+1	+1	+1	+1	+1	$\alpha_{xx}+\alpha_{yy}+\alpha_{zz}$
A_{1u}	+1	+1	+1	+1	+1	−1	−1	−1	−1	−1	
A_{2g}	+1	+1	−1	−1	+1	+1	−1	+1	+1	−1	
A_{2u}	+1	+1	−1	−1	+1	−1	+1	−1	−1	+1	
E_g	+2	−1	0	0	+2	+2	0	−1	+2	0	$(\alpha_{xx}+\alpha_{yy}-2\alpha_{zz}, \alpha_{xx}-\alpha_{yy})$
E_u	+2	−1	0	0	+2	−2	0	+1	−2	0	
F_{1g}	+3	0	−1	+1	−1	+3	+1	0	−1	−1	
F_{1u}	+3	0	−1	+1	−1	−3	−1	0	+1	+1	
F_{2g}	+3	0	+1	−1	−1	+3	−1	0	−1	+1	$(\alpha_{xy}, \alpha_{yz}, \alpha_{xz})$
F_{2u}	+3	0	+1	−1	−1	−3	+1	0	+1	−1	

I_h	E	$12C_5$	$12C_5^2$	$20C_3$	$15C_2$	i	$12S_{10}$	$12S_{10}^3$	$20S_6$	15σ	Raman activity
A_g	1	1	1	1	1	1	1	1	1	1	$\alpha_{xx}+\alpha_{yy}+\alpha_{zz}$
F_{1g}	3	$\tfrac{1}{2}(1+\sqrt{5})$	$\tfrac{1}{2}(1-\sqrt{5})$	0	-1	3	$\tfrac{1}{2}(1-\sqrt{5})$	$\tfrac{1}{2}(1+\sqrt{5})$	0	-1	
F_{2g}	3	$\tfrac{1}{2}(1-\sqrt{5})$	$\tfrac{1}{2}(1+\sqrt{5})$	0	-1	3	$\tfrac{1}{2}(1+\sqrt{5})$	$\tfrac{1}{2}(1-\sqrt{5})$	0	-1	
G_g	4	-1	-1	1	0	4	-1	-1	1	0	
H_g	5	0	0	-1	1	5	0	0	-1	1	$(2\alpha_{zz}-\alpha_{xx}-\alpha_{yy},\ \alpha_{xx}-\alpha_{yy},\ \alpha_{xy},\ \alpha_{yz},\ \alpha_{xz})$
A_u	1	1	1	1	1	-1	-1	-1	-1	-1	
F_{1u}	3	$\tfrac{1}{2}(1+\sqrt{5})$	$\tfrac{1}{2}(1-\sqrt{5})$	0	-1	-3	$-\tfrac{1}{2}(1-\sqrt{5})$	$-\tfrac{1}{2}(1+\sqrt{5})$	0	1	
F_{2u}	3	$\tfrac{1}{2}(1-\sqrt{5})$	$\tfrac{1}{2}(1+\sqrt{5})$	0	-1	-3	$-\tfrac{1}{2}(1+\sqrt{5})$	$-\tfrac{1}{2}(1-\sqrt{5})$	0	1	
G_u	4	-1	-1	1	0	-4	1	1	-1	0	
H_u	5	0	0	-1	1	-5	0	0	1	-1	

APPENDIX 2
General Formula for Calculating the Number of Normal Vibrations in Each Symmetry Species

Based on data published in G. Herzberg, *Molecular Spectra and Molecular Structure, Vol. II: Infrared and Raman Spectra of Polyatomic Molecules*, Van Nostrand, Princeton, New Jersey, 1945.

Table A2.1 Point group including only non-degenerate vibrations.

Point group	Total number of atoms	Species	Number of vibrations[a]
C_2	$2m + m_0$	A	$3m + m_0 - 2$
		B	$3m + 2m_0 - 4$
C_s	$2m + m_0$	A'	$3m + 2m_0 - 3$
		A''	$3m + m_0 - 3$
$C_i \equiv S_2$	$2m + m_0$	A_g	$3m - 3$
		A_u	$3m + m_0 - 3$
C_{2v}	$4m + 2m_{xz} + 2m_{yz} + m_0$	A_1	$3m + 2m_{xz} + 2m_{yz} + 3m + m_0 - 1$
		A_2	$3m + m_{xz} + m_{yz} - 1$
		B_1	$3m + 2m_{xz} + m_{yz} + m_0 - 2$
		B_2	$3m + m_{xz} + 2m_{yz} + m_0 - 2$
C_{2h}	$4m + 2m_h + m_2 + m_0$	A_g	$3m + 2m_h + m_2 - 1$
		A_u	$3m + m_h + m_2 + m_0 - 1$
		B_g	$3m + m_h + 2m_2 - 1$
		B_u	$3m + 2m_h + 2m_2 + 2m_0 - 1$
$D_{2h} \equiv V_h$	$8m + 4m_{xy} + 4m_{xz} + 4m_{yz} + 2m_{2x} + 2m_{2y} + 2m_{2z} + m_0$	A_g	$3m + 2m_{xy} + 2m_{xz} + 2m_{yz} + m_{2x} + m_{2y} + m_{2z}$
		A_u	$3m + m_{xy} + m_{xz} + m_{yz}$
		B_{1g}	$3m + 2m_{xy} + m_{xz} + m_{yz} + m_{2x} + m_{2y} - 1$
		B_{1u}	$3m + m_{xy} + 2m_{xz} + 2m_{yz} + m_{2x} + m_{2y} + m_{2z} + m_0 - 1$
		B_{2g}	$3m + m_{xy} + 2m_{xz} + m_{yz} + m_{2x} + m_{2z} - 1$
		B_{2u}	$3m + 2m_{xy} + m_{xz} + 2m_{yz} + m_{2x} + m_{2y} + m_{2z} + m_0 - 1$
		B_{2g}	$3m + m_{xy} + m_{xz} + 2m_{yz} + m_{2y} + m_{2z} - 1$
		B_{2u}	$3m + 2m_{xy} + 2m_{xz} + m_{yz} + m_{2x} + m_{2y} + m_{2z} + m_0 - 1$

[a] Note that m is always the number of sets of equivalent nuclei not on any element of symmetry; m_0 is the number of nuclei lying on all symmetry elements present; m_{xy}, m_{xz}, m_{yz} are the numbers of sets of nuclei lying on the xy, xz, yz plane, respectively, but not on any axes going through these planes; m_2 is the number of sets of nuclei on two-fold axis but not at the point of intersection with another element of symmetry; m_{2x}, m_{2y}, m_{2z} are the number of sets of nuclei lying on the x, y, z axis if they are two-fold axes, but not on all of them; m_h is the number of sets on nuclei on a plane σ_h but not on the axis perpendicular to this plane.

Table A2.2 Point group including degenerate vibrations.[a]

Point group	Total number of atoms	Species	Number of vibrations[a]
D_3	$6m + 3m_2 + 2m_3 + m_0$	A_1	$3m + m_2 + m_3$
		A_2	$3m + 2m_2 + m3 + m_0 - 2$
		E	$6m + 3m_2 + 2m_3 + m_0 - 2$
C_{3v}	$6m + 2m_v + m_0$	A_1	$3m + 2m_v + m_0 - 1$
		A_2	$3m + m_v - 1$
		E	$6m + 3m_v + m_0 - 2$
C_{4v}	$8m + 3m_v + 4m_d + m_0$	A_1	$3m + 2m_v + m_d + m_0 - 1$
		A_2	$3m + m_v + m_d - 1$
		B_1	$3m + 2m_v + m_d$
		B_2	$3m + m_v + 2m_d$
		E	$6m + 3m_v + 3m_d + m_0 - 2$
$C_{\infty v}$	m_0	Σ^+	$m_0 - 1$
		Σ^-	0
		Π	$m_0 - 2$
		$\Delta, \Phi, \ldots$	0
$D_{2d} \equiv V_d$	$8m + 4m_d + 4m_2 + 2m_4 + m_0$	A_1	$3m + 2m_d + m_2 + m_4$
		A_2	$3m + m_d + 2m_2 - 1$
		B_1	$3m + m_d + m_2$
		B_2	$3m + 2m_d + 2m_2 + m_4 + m_0 - 1$
		E	$6m + 3m_d + 3m_2 + 2m_4 + m_0 - 1$
D_{2d}	$12m + 6m_4 + 6m_2 + 2m_6 + m_0$	A_{1g}	$3m + 2m_d + m_2 + m_6$
		A_{1u}	$3m + m_4 + m_2$
		A_{2g}	$3m + m_d + 2m_2 - 1$
		A_{2u}	$3m + 2m_d + 2m_2 + m_6 + m_0 - 1$
		E_g	$6m + 3m_d + 3m_2 + m_6 - 1$
		E_u	$6m + 3m_d + 3m_2 + m_6 + m_0 - 1$
D_{4d}	$16m + 8m_4 + 8m_2 + 2m_8 + m_0$	A_1	$3m + 2m_d + m_2 + m_6$
		A_2	$3m + m_4 + 2m_2 - 1$
		B_1	$3m + m_d + m_2$

(continued)

Table A2.2 Continued.

Point group	Total number of atoms	Species	Number of vibrations[a]
D_{2h}	$12m + 6m_v + 6m_h + 3m_2 + 2m_3 + m_0$	B_2	$3m + 2m_d + 2m_2 + m_8 + m_0 - 1$
		E_1	$6m + 3m_d + 3m_2 + m_8 + m_0 - 1$
		E_2	$6m + 3m_d + 3m_2$
		E_3	$6m + 3m_d + 3m_2 + m_6 + m_0 - 1$
		A_1'	$3m + 2m_v + 2m_h + m_2 + m_3$
		A_1''	$3m + m_v + m_h$
		A_2'	$3m + 2m_v + 2m_h + m_2 - 1$
		A_2''	$3m + 2m_v + m_h + m_2 + m_3 + m_0 - 1$
		E'	$6m + 3m_v + 4m_h + 2m_2 + m_3 + m_0 - 1$
		E''	$6m + 3m_v + 2m_h + m_2 + m_2 - 1$
D_{4h}	$16m + 8m_v + m_d + 8m_h + 4m_2 + 4m_2'$ $+ 2m_4 + m_0$	A_{1g}	$3m + 2m_v + 2m_d + 2m_h + m_2 + m_2' + m_4$
		A_{1u}	$3m + m_v + m_d + m_h$
		A_{2g}	$3m + m_v + m_d + 2m_h + m_2 + m_2' - 1$
		A_{2u}	$3m + 2m_v + 2m_d + m_h + m_2 + m_2' + m_4 + m_0 - 1$
		B_{1g}	$3m + 2m_v + m_d + 2m_h + m_2 + m_2'$
		B_{1u}	$3m + m_v + 2m_d + m_h + m_2'$
		B_{2g}	$3m + m_v + 2m_d + 2m_h + m_2 + m_2'$
		B_{2u}	$3m + 2m_v + m_d + m_h + m_2$
		E_g	$6m + 3m_v + 3m_d + 2m_h + m_2 + m_2' + m_4 - 1$
		E_u	$6m + 3m_v + 3m_d + 4m_h + 2m_2 + 2m_2' + m_4 + m_0 - 1$
D_{5h}	$20m + 10m_v + 10m_h + 5m_2 + 2m_5 + m_0$	A_1'	$3m + 2m_v + 2m_h + m_2 + m_5$
		A_1''	$3m + m_v + m_h$
		A_2'	$3m + 2m_v + 2m_h + m_2 - 1$
		A_2''	$3m + 2m_v + m_h + m_2 + m_5 + m_0 - 1$
		E_1'	$6m + 3m_v + 4m_h + 2m_2 + m_5 + m_0 - 1$
		E_1''	$6m + 3m_v + 2m_h + m_2 + m_5 - 1$
		E_2'	$6m + 3m_v + 4m_h + 2m_2$
		E_2''	$6m + 3m_v + 2m_h + m_2$
D_{6h}	$24m + 12m_v, 12m_d + 12m_h + 6m_2$ $+ 6m_2' + 2m_6 + m_0$	A_{1g}	$3m + 2m_v + 2m_d + 2m_h + m_2 + m_2' + m_6$
		A_{1u}	$3m + m_v + m_d + m_h$
		A_{2g}	$3m + m_v + m_d + 2m_h + m_2 + m_2' - 1$

General Formula for Calculating the Number of Normal Vibrations

Point group		Formula
$D_{\infty h}$	A_{2u}	$3m + 2m_v + 2m_d + m_h + m_2 + m'_2 + m_6 + m_0 - 1$
	B_{1g}	$3m + m_v + 2m_d + m_h + m'_2$
	B_{1u}	$3m + 2m_v + m_d + 2m_h + m_2 + m'_2$
	B_{2g}	$3m + 2m_v + m_d + m_h + m_2$
	B_{2u}	$3m + m_v + 2m_d + 2m_h + m_2 + m'_2$
	E_{1g}	$6m + 3m_v + 3m_d + 2m_h + m_2 + m'_2 + m_6 - 1$
	E_{1u}	$6m + 3m_v + 3m_d + 4m_h + 2m_2 + 2m'_2 + m_6 + m_0 - 1$
	Σ_g^+	$2m_\infty + m_0$
	Σ_u^+	$m_\infty + m_0$
	Σ_g^-, Σ_u^-	0
	Π_g	$m_\infty - 1$
	Π_u	$m_\infty + m_0 - 1$
	Δ_g, Δ_u	0
	$\Phi_g, \Phi_u, \ldots$	0
T_d	A_1	$3m + 2m_d + m_2 + m_3$
	A_2	$3m + m_d$
	E	$6m + 3m_d + m_2 + m_3$
	F_1	$9m + 4m_d + 2m_2 + m_3 - 1$
	F_2	$9m + 5m_d + 3m_2 + 2m_3 + m_0 - 1$
		$24m + 12m_d + 6m_2 + 2m_6 + m_0$
O_h	A_{1g}	$3m + 2m_h + 2m_d + m_2 + m_3 + m_4$
	A_{1u}	$3m + m_d + m_h$
	A_{2g}	$3m + 2m_h + m_d + m_2$
	A_{2u}	$3m + m_h + 2m_d + m_2 + m_2$
	E_g	$6m + 4m_h + 3m_d + 2m_2 + m_3 + m_4$
	E_u	$6m + 2m_h + 3m_d + m_2 + m_3$
	F_{1g}	$9m + 4m_h + 4m_d + 2m_2 + m_3 + m_4 - 1$
	F_{1u}	$9m + 5m_h + 5m_d + 3m_2 + 2m_3 + m_4 + m_0 - 1$
	F_{2g}	$9m + 4m_h + 5m_d + 2m_2 + 2m_3 + m_4$
	F_{2u}	$9m + 5m_h + 4m_d + 2m_2 + m_3 + m_4$
		$48m + 24m_h + 24m_d + 12m_2 + 8m_3 + 6m_4 + m_0$

[a] Note that m is the number of sets of nuclei not on any element of symmetry; m_0 is the number of nuclei on all elements of symmetry; $m_2, m_3, m_4, \ldots$ are the numbers of sets of nuclei on a two-, three-, four-fold, ... axis but not on any other element of symmetry that does not wholly coincide with that axis; m'_2 is the number of sets of nuclei on the two-fold axis called C'_2 in the proceeding character tables; m_v, m_d, m_h are the numbers of sets of nuclei on planes $\sigma_v, \sigma_d, \sigma_h$, respectively, but not on any other element of symmetry.

APPENDIX 3
Polarizability Tensors for the 32 Point Groups,[a] Including the Icosahedral Group

Directly above the symbol for each irreducible representation is a matrix containing the non-vanishing components of the Raman-scattering tensor.

---

[a] Based on data published by R. Loudon, *Advant. Phys.*, 1964, **13**, 423.

Polarizability Tensors for the 32 Point Groups, Including the Icosahedral Group

System	Class	Raman tensors			
Monoclinic		$\begin{pmatrix} a & 0 & d \\ 0 & b & 0 \\ d & 0 & c \end{pmatrix}$		$\begin{pmatrix} 0 & e & 0 \\ e & 0 & f \\ 0 & f & 0 \end{pmatrix}$	
	2	$A(Y)$		$B(X,Z)$	
	m	$A'(X,Z)$		$A''(Y)$	
	$2/m$	A_g		B_g	
Orthorhombic		$\begin{pmatrix} a & 0 & 0 \\ 0 & b & 0 \\ 0 & 0 & c \end{pmatrix}$	$\begin{pmatrix} 0 & d & 0 \\ d & 0 & 0 \\ 0 & 0 & 0 \end{pmatrix}$	$\begin{pmatrix} 0 & 0 & e \\ 0 & 0 & 0 \\ e & 0 & 0 \end{pmatrix}$	$\begin{pmatrix} 0 & 0 & 0 \\ 0 & 0 & f \\ 0 & f & 0 \end{pmatrix}$
	222	A	$B_1(Z)$	$B_2(Y)$	$B_3(X)$
	$mm2$	$A_1(Z)$	A_2	$B_1(X)$	$B_2(Y)$
	mmm	A_g	B_{1g}	B_{2g}	B_{3g}
Trigonal		$\begin{pmatrix} a & 0 & 0 \\ 0 & a & 0 \\ 0 & 0 & b \end{pmatrix}$	$\begin{pmatrix} c & d & e \\ d & -c & f \\ e & f & 0 \end{pmatrix}$	$\begin{pmatrix} d & -c & -f \\ -c & -d & e \\ -f & e & 0 \end{pmatrix}$	
	3	$A(Z)$	$E(X)$	$E(Y)$	
	$\bar{3}$	$A_g t$	E_g	E_g	
		$\begin{pmatrix} a & 0 & 0 \\ 0 & a & 0 \\ 0 & 0 & b \end{pmatrix}$	$\begin{pmatrix} c & 0 & 0 \\ 0 & -c & d \\ 0 & d & 0 \end{pmatrix}$	$\begin{pmatrix} 0 & -c & -d \\ -c & 0 & 0 \\ -d & 0 & 0 \end{pmatrix}$	
	32	A	$E(X)$	$E(Y)$	
	$3m$	$A_1(Z)$	$E(Y)$	$E(-X)$	
	$\bar{3}m$	A_{1g}	E_g	E_g	
Tetragonal		$\begin{pmatrix} a & 0 & 0 \\ 0 & a & 0 \\ 0 & 0 & b \end{pmatrix}$	$\begin{pmatrix} c & d & 0 \\ d & c & 0 \\ 0 & 0 & 0 \end{pmatrix}$	$\begin{pmatrix} 0 & 0 & e \\ 0 & 0 & f \\ e & f & 0 \end{pmatrix}$	$\begin{pmatrix} 0 & 0 & -f \\ 0 & 0 & e \\ -f & e & 0 \end{pmatrix}$

(continued)

Continued.

System	Class	Raman tensors				
	4	$A(Z)$	B	$E(X)$	$E(Y)$	
	$\bar{4}$	A	$B(Z)$	$E(X)$	$E(-Y)$	
	$4/m$	A_g	B_g	E_g	E_g	
		$\begin{pmatrix} a & 0 & 0 \\ 0 & a & 0 \\ 0 & 0 & b \end{pmatrix}$	$\begin{pmatrix} c & 0 & 0 \\ 0 & -c & 0 \\ 0 & 0 & 0 \end{pmatrix}$	$\begin{pmatrix} 0 & d & 0 \\ d & 0 & 0 \\ 0 & 0 & 0 \end{pmatrix}$	$\begin{pmatrix} 0 & 0 & e \\ 0 & 0 & 0 \\ e & 0 & 0 \end{pmatrix}$	$\begin{pmatrix} 0 & 0 & 0 \\ 0 & 0 & e \\ 0 & e & 0 \end{pmatrix}$
	$4mm$	$A(Z)$	B_1	B_2	$E(X)$	$E(Y)$
	422	A	B_1	B_2	$E(X)$	$E(X)$
	$\bar{4}2m$	A_g	B_1	$B_2(Z)$	$E(Y)$	$E(X)$
	$4/mmm$	A_{1g}	B_{1g}	B_{2g}	E_g	E_g
		$\begin{pmatrix} a & 0 & 0 \\ 0 & a & 0 \\ 0 & 0 & b \end{pmatrix}$	$\begin{pmatrix} 0 & 0 & c \\ 0 & 0 & d \\ c & d & 0 \end{pmatrix}$	$\begin{pmatrix} 0 & 0 & 0 \\ 0 & 0 & c \\ 0 & c & 0 \end{pmatrix}$	$\begin{pmatrix} e & f & 0 \\ f & -e & 0 \\ 0 & 0 & 0 \end{pmatrix}$	$\begin{pmatrix} f & -e & 0 \\ -e & -f & 0 \\ 0 & 0 & 0 \end{pmatrix}$
Hexagonal						
	6	$A(Z)$	$E_1(X)$	$E_1(Y)$	E_2	E_2
	$\bar{6}$	A'	E''	E''	E'	$E'(Y)$
	$6/m$	A_g	E_{1g}	E_{1g}	E_{2g}	E_{2g}
		$\begin{pmatrix} a & 0 & 0 \\ 0 & a & 0 \\ 0 & 0 & b \end{pmatrix}$	$\begin{pmatrix} 0 & 0 & 0 \\ 0 & 0 & c \\ 0 & c & 0 \end{pmatrix}$	$\begin{pmatrix} 0 & 0 & -c \\ 0 & 0 & 0 \\ -c & 0 & 0 \end{pmatrix}$	$\begin{pmatrix} 0 & d & 0 \\ d & 0 & 0 \\ 0 & 0 & 0 \end{pmatrix}$	$\begin{pmatrix} d & 0 & 0 \\ 0 & -d & 0 \\ 0 & 0 & 0 \end{pmatrix}$

622	A_1	$E_1(X)$	$E_1(Y)$			
6mm	$A_1(Z)$	$E_1(Y)$	$E_1(-X)$			
$\bar{6}m2$	A_1'	E''	E''			
6/mmm	A_{1g}	E_{1g}	E_{1g}			
	$\begin{pmatrix} a & 0 & 0 \\ 0 & a & 0 \\ 0 & 0 & a \end{pmatrix}$	$\begin{pmatrix} b & 0 & 0 \\ 0 & b & 0 \\ 0 & 0 & b \end{pmatrix}$	$\begin{pmatrix} b & 0 & 0 \\ 0 & b & 0 \\ 0 & 0 & b \end{pmatrix}$			

Cubic

23	A	E	E	E_2	E_2
m3	A_g	E_g	E_g	E_2'	E_2'
432	A_1	E	E	$E'(X)$	$E'(Y)$
$\bar{4}3m$	A_1	E	E	E_{2g}	E_{2g}
m3m	A_{1g}	E_g	E_g		
		$\begin{pmatrix} 0 & 0 & 0 \\ 0 & 0 & d \\ 0 & d & 0 \end{pmatrix}$	$\begin{pmatrix} 0 & 0 & d \\ 0 & 0 & 0 \\ d & 0 & 0 \end{pmatrix}$	$\begin{pmatrix} 0 & d & 0 \\ d & 0 & 0 \\ 0 & 0 & 0 \end{pmatrix}$	
		$F(X)$	$F(Y)$	$F(Z)$	
		F_g	F_g	F_g	
		F_2	F_2	F_2	
		$F_2(X)$	$F_2(Y)$	$F_2(Z)$	
		F_{2g}	F_{2g}	F_{2g}	

Polarizability for A_g and H_g Raman active modes under icosahedral symmetry.[b]

A_g	$\begin{pmatrix} a & 0 & 0 \\ 0 & a & 0 \\ 0 & 0 & a \end{pmatrix}$			
H_g	$\begin{pmatrix} b & 0 & 0 \\ 0 & b & 0 \\ 0 & 0 & -2b \end{pmatrix}$	$\begin{pmatrix} \sqrt{3}b & 0 & 0 \\ 0 & -\sqrt{3}b & 0 \\ 0 & 0 & 0 \end{pmatrix}$	$\begin{pmatrix} 0 & \sqrt{3}b & 0 \\ \sqrt{3}b & 0 & 0 \\ 0 & 0 & 0 \end{pmatrix}$	$\begin{pmatrix} 0 & 0 & 0 \\ 0 & 0 & \sqrt{3}b \\ 0 & \sqrt{3}b & 0 \end{pmatrix}$	$\begin{pmatrix} 0 & 0 & -\sqrt{3}b \\ 0 & 0 & 0 \\ -\sqrt{3}b & 0 & 0 \end{pmatrix}$
Partner	1	2	3	4	5

[b] Based on data published by J. Menéndez and J. Page, *Top. Appl. Phys.*, 2000, **76**, 27.

Subject Index

π-electron shell 124–6

a_1/a_2 vectors 144, 145, 147
a_5 C–C bonds 123–4
a_6 C–C bonds 123–4
AB_4-type planar molecules 52–3, 54
ABAB stacking 156
achiral tubes 153–5
acidic treatments 161, 162
adiabatic compressibility ($\Delta\kappa_\sigma$) 226, 227
alcohol solubility 219–20
alkane solubility 217–19
allene 58
alumina nanofluids 36
3-aminopropyltriethoxysilane (APTS) 223
ammonia 57
anti-Stokes lines
   Boltzmann equilibrium distribution 83–4
   definition 46, 81, 82
   single-walled carbon nanotubes 209–11, 214
   Stokes/anti-Stokes ratio 84–5
applied electric field effects 205
arc-plasma-jet method *see* Huffman–Krätschmer arc-discharge evaporation method
argon 130
armchair tubes
   atomic displacement vectors 192, 193
   characteristics 145, 146, 148, 149–51, 153–4
   graphene crystals 163
   multi-walled 156
   Raman active modes 190

aromatic solvents 216–17, 225–7
artificial molecular machinary 24, 25, 26
aspect ratio 27–8
asymmetric stretch (ν_3)
   carbon dioxide molecule 48–9, 50, 73
   hydrogen sulfide molecule 88, 89
   water molecule 76–7
atomic degrees of freedom 78, 183
atomic displacement vectors 183–4, 192, 193

B_2 species 72, 77
beauty, symmetry 50
bench-top reactors 130, 131
benzene
   C_{60} fullerene effects on 225–7
   combustion method 131–2, 133
   molecular rotation axes 52
   point group/symmetry operations 59
   Raman spectrum 80
Binnig, Gerd 10
biological systems 3, 23–5
Boltzmann equilibrium distribution 82, 83–5
bond lengths 89, 123–4, 125, 130
bond spring constant 78
boric acid 57
boron trifluoride 58
bowl-shaped 20-carbon graphene 118
Bravais centring of lattices 66, 67
Brillouin scattering 226–7, 229
building block definition 109–10
bulk systems 11–14, 33–5, 229
bundled single-walled carbon nanotubes 211, 214
butterfly wings 6, 26

C_{2v} point group 70, 71–2
 character table 72, 259
 fullerenes 119, 120
 hydrogen sulfide molecule 88
 normal vibrations formula 268
 water molecule 72, 77
C_2 symmetry operation 69–71
C_4 symmetry operation 73–4
C_{20} fullerenes 117–18
C_{60} fullerenes
 effects on solvents 225–9
 peapods (C_{60}@SWCNT) 110, 111
 point group/symmetry operations 60
 Raman scattering 183–9
 self-assembly in solvents 223–5
 solvent interactions 215, 216–20, 236, 238–9
 structure 123–6, 127, 128
 water molecule interactions 221–2
C_{70} fullerenes 59
 mode symmetries 189, 191
 Raman scattering 189–90, 191–2
 solvent interactions 215, 216–20, 239
 structure 126–9
C_{78} fullerenes 123
C_{80} fullerenes 118, 119
cadmium selenide (CdSe) quantum dots 27
capillarity 14–15, 18–19
carbon dioxide molecule
 centre of symmetry 48, 51, 52
 point group/symmetry operations 59
 stretch mode symmetric/asymmetric, Raman activity 48–9, 50, 73
 vibration modes 48–50
carbon monoxide 160
carbon nano-onions (CNOs) 110, 111, 140–3
carbon nanotubes
 see also double-walled carbon nanotubes; multi-walled carbon nanotubes; single-walled carbon nanotubes
 Damascus saber (14th-16th Century) 8, 9
 production 158–61
 tube–tube interactions 194, 200–1
 zipping/unzipping 166–7

carbon tetrachloride 228
carbon-13 isotopes 187
Cartesian axes 96–8
centre of symmetry (i) 48, 51–2
character tables 69–81
 C_{2v} point group 72, 259
 D_{2j+1} point group single-walled nanotubes 154–5
 definition 72
 point groups 259–66
charge-coupled devices (CCDs) 105
chemical potential (μ) 11–12, 20–1, 92, 94–5
chiral angle θ, 144, 145, 147
chiral tubes 145, 146, 148, 153–5, 190
chiral vector C_h 145, 147, 149, 150
chloroaromatic precursors 133
classical electromagnetic theory 44–7
classification 109
 character tables for point groups 259–66
 fullerenes 109–10
 dimensionality 116
 normal vibrations in symmetry species 74, 76–9, 267–71
 polarisability tensors for point groups 272–5
clouds, water properties 7
CNOs (carbon nano-onions) 110, 111, 140–3
cohesive energy 92
colloidal gold, medical use 8, 9
colour 7, 26–32, 137
coloured (magnetic) space groups 69
combination modes/bands 79, 189
combinatorial scattering of light see Raman spectroscopy
computer-controlled stepping motors 105
condensation method of fullerene production 132–3, 159–60
conduction gap, metal clusters 33, 34
configurational entropy 21–6, 94
confinement entropy 21–6, 94
confocal optical microscopy 102
contact arc method 130–1
 see also Huffman–Krätschmer arc-discharge evaporation method

core–shell structures 29, 31, 32
correction terms, Hill's 20–1
correlation length (ξ) 17–19, 125
critical exponent (ν) 18
critical point (T_c) 18
crystallographic restriction theorem 60, 63
crystals
  crystal Raman modes 80
  face-centred cubic C_{60} 186–7
  orthorhombic 96–7
  point groups 60, 62, 63, 64–5, 96
  strain effects 85–6, 87
cubane 59
cubic crystal point group 65, 275
Curl, Robert Jr 3, 10, 11
cyclic alkanes 218
cyclobutane 58
cyclohexane 58
cylindrical giant fullerenes 139–40
  see also carbon nanotubes
C–C bonds 123–7, 130, 144–7

D-band Raman scattering
  D'-bands (overtones) 79, 100
  definition 79
  dispersion phenomenon 100–1
  graphene sheets 203, 209
  single-walled carbon nanotubes 192, 194, 195, 197, 199, 213
  structural defect effects 93
D_{2j+1} point symmetry group 154–5
D_{5d} point symmetry group 118, 119, 121, 126
Damascus saber (14th-16th Century) 7, 8, 9
data acquisition units 105
Debye temperature 36
degenerate symmetry species 73–4
degrees of freedom 78, 183
density functional theory (DFT) 118, 119, 121, 206–8
detectors (Raman) 105
di-vacancy (DV) defects 165, 166, 206–8, 212
diameter of nanotubes 147, 193, 214
diamond single crystals 87
diatomic molecules 45–6

1,2-dichloroethane 57
dimanganese decacarbonyl 59
dimensionality of fullerenes 116
$D_{[infinity]h}$ point group 74, 75
dipole moment (μ) 45–6
Dirac fermions 205
directed assembly 223–5, 240
directional properties 69–73
discovery of fullerenes 3, 10–11
dispersion effect 99–101, 195, 196
double resonance Raman scattering 195, 196
double-layer graphene (DLG) 203, 208
double-walled carbon nanotubes (DWCNTs)
  configurations 199
  definition 143
  properties 157
  Raman spectra 197–202, 203
doubly degenerate electronic orbitals 73–4
Drexler, Eric 2–3
drop-drying process 223–4
DWCNTs *see* double-walled carbon nanotubes

edge filters 102
efficiency capability 5
Egyptian gold medicines 8
Einstein, Albert 5, 18
El-Sayed, Mostafa 8
elastic scattering of light *see* Rayleigh scattering
electric field effects 205, 211
electromagnetic theory 44–7
electronic nanophenomena 30, 32–4
electronic orbitals 69–70, 125, 126, 144, 153
electronics industry 2, 86
end caps 148–9, 157
endo-fullerenes 125–6, 127, 128
energy
  bandgap in semiconductor clusters 34
  conversions in natural molecular machinary 23–5
  energy density (φ) 12–13, 14
  molecular transfer and Raman activity 81–5

Subject Index

enthalpy 35
entropy, configurational 21–6, 94
equation of state (EOS) 90, 91
equilibrium 21–3, 139–40
ethane molecule 53–5
ethylene 58
ethylene glycol (EG) 36, 37
ethyne 51, 52
Euler's theorem for polyhedrons 117
evaporation
  see also Huffman–Krätschmer arc-discharge evaporation method
  self-assembly 223–5
excitation *see* laser excitation energy
excited energy states 81–3
exfoliated graphite 114
extraction (Soxhlet) method 133–6, 137–8

face-centred cubic (FCC) crystals 186–7
facetted structures 121
Faraday, Michael 8, 9
ferrocene 59
Feynman, Richard P. 1–2, 3, 10, 20
fingerprinting (Raman) 79, 80
fluctuations (Hill's thermodynamics) 21
fluorescence/scattering distinction 82–3
fly wing 229–30
four-dimensional space 69
free energy density (ψ) 12
free energy (F) 11
freedom, degrees of 78, 183
Frieze groups 68
full-width at half maximum (FWHM) 83, 200, 202, 204–5, 210
Fuller, Buckminster 112
fullerene generator *see* Huffman–Krätschmer arc-discharge evaporation method
fullerene onions 140–3
Fullerene Rush controversies 187

G' band Raman scattering
  double-walled carbon nanotubes 200, 202, 203
  graphene monolayer on various substrates 210

single-walled carbon nanotubes 192, 194, 195, 196, 199
  pressure effects 230–1, 232
G-band Raman scattering
  combination bands 79
  dispersion effect 100–1
  double-walled carbon nanotubes 200, 202, 203
  graphene 201–8, 209, 210, 211, 212
  single-walled carbon nanotubes 192, 194, 195, 199, 213, 214
  solvent MW effects 236, 237
  structural imperfections effect 91, 93
  thermal effects 208, 213, 214
geodesic dome design 112
gerade 73
giant fullerenes 118–23, 139–40
Gibbs free energy (G) 20–1, 23–5
glass substrates 32
glide plane symmetry element 65–6
GNRs (graphene nano-ribbons) *see* graphene sheets
gold
  coloured solutions 26–7
  Lycurgus Cup (AD 400) 7, 8
  nanorods 8, 27–9, 30
  optical nanophenomena 29, 31, 32
grains 37, 91, 93, 98–9
graphene sheets 161–8
  crystals 163
  definition 109
  first isolation 114
  Raman scattering 201–8, 209, 210, 211, 212
    number of layers (n) 202, 205, 206, 207
    perturbation effects 202, 203–4, 205
    structural defects 205–8, 212
  single-layered sheet structure 164–8
  single-walled carbon nanotubes formation 143–4, 155
  structure 111
  typical hexagonal structure 144
graphite
  family branches 115–16
  graphene comparison 109, 201, 204

Raman scattering
   combination bands 79, 81
   dispersion effects 100–1
   strain profile 87
   structural imperfections 91, 93
gravity 6, 7, 14, 18
ground state 81

Hall–Petch effect 37, 38
haloalkanes 218–19
hardness, mechanical 230
helical axis 63, 65
helium 130–1, 158–9
Hermann–Mauguin notation 60, 62, 66
Herzberg's formulae 77, 78
hexagons
   C_{60} fullerenes 123–4
   C_{70} fullerenes 126, 127, 128
   hexagonal crystal point group 64, 274
   point defects 164–7, 206–8, 212
   zero-dimensional fullerenes 117
Hill, Terrell L. 19–21
HiPco process 160
history 110–16
   early use of nanotechnology 1–4, 7–11
   Fullerene Rush 187
   Hill's studies 20
   Raman spectroscopy 43–4
hoop stresses 156
Huffman, D.R. 112, 113
Huffman–Krätschmer arc-discharge evaporation method
   carbon nanotubes 132, 158–9, 161
   history 113
   instrumentation 130, 132
   procedure 129–31
   products 123
   Raman studies potential 185
hydrogen chloride 58
hydrogen peroxide 57
hydrogen sulfide 88–9, 90
hydrostatic pressure 88–90, 91
hypochlorous acid 57

icosahedral giant fullerenes 139–40
icosahedral symmetry 275
ideal gas entropy 21–6, 94

identity (E) 51, 52, 53
I_h point C_{80} symmetry group 118, 119, 121
Iijima, Sumio 113, 114, 115
imaging spectrographs 103
improper rotation 53–5
incar-fullerenes 125–6, 127, 128
incarcerated atoms 125–6, 127
inelastic scattering of light see Raman spectroscopy
inhomogeneity scales 14–19, 25
instrumentation
   benzene combustion method 133
   Huffman–Krätschmer arc-discharge evaporation method 130–1
   microwave chemical vapour deposition 165
   plasma arc fullerene generator 132
   Raman spectroscopy 101–5
   Soxhlet extraction/purification 134, 138
integrated intensity 96
interactions
   see also correlation length
   nanoparticle monolayers, optical effects 29, 32
   range
      correlation length 17–19, 125
      fullerene effects on solvents 227–8
      nanotube–nanotube 194, 200–1
      types 94
      within/outside small systems 13
interfacial dislocations 156
internal energy (U) 11
international (Hermann–Mauguin) notation 60, 62, 66
intramolecular C_{60} vibration modes 183–4
inverse Hall–Petch effect 37, 38
inversion centre 48, 51–2
isolated pentagon rule 117–18, 119, 120
isomers 118, 119, 121, 123

Jay, Coneyl 4

Koch, Robert 8, 9
Krätschmer, W. 112, 113
   see also Huffman–Krätschmer arc-discharge evaporation method

Subject Index

Krishnan, K.S. 43
Kroto, Harry 3, 10, 11

Landsberg, G.S. 43
Langmuir–Blodgett films 16–17
Laplace's equation 15
large fullerenes 118–23
large system definition 11–14
laser excitation energy (E_L)
   Raman spectra
      multi-walled carbon nanotubes 198, 201
      resonant Raman effect 95
      single-walled carbon nanotubes 194, 195, 197, 209, 211
laser TOF mass spectra 135, 136, 139
lattices
   see also crystals
   Bravais centring 66, 67
   face-centred cubic C_{60} crystals 186–7
   lattice Raman mode 79–80
   single-layered graphene sheets 164–7
length scales 25, 125, 182–3
length units 5
length-to-diameter (aspect) ratio 27–8
Lennard–Jones fluids 228
Leonardo Da Vinci 18
librational modes 187
light
   see also colour; optical nanophenomena; Raman scattering
   visible/UV 27
line tension/surface tension ratio (τ/σ) 16–17
linear molecule symmetry species 74
linear objects *see* one-dimensional space groups
liquid chromatography 137, 138, 139
liquid nanoscale droplets 15–16
liquid–gas systems 14–15
local thermodynamic functions 12–13
lowest energy structures 119, 122
Lycurgus Cup (AD 400) 7, 8
lysergic acid 57

magnetic space groups 69
Mandelstam, L.I. 43, 44

MarkIII(k) nanoscale planetary gear 24
mathematical formulations
   character tables for point groups 259–66
   normal vibrations in symmetry species 74, 76–9, 267–71
   point groups, four rules 56
   polarisability tensors for point groups 272–5
   Raman band positions 95–6
   symmetry 51–6
   species, normal vibrations 267–71
mechanical hardness 230
mechanical principles 3, 37–8
medical applications 8–9
mesoscopic systems, length-scales 182–3
mesostructures 110, 111
metal clusters 33, 37, 38
metallic nanotubes
   double-walled 199, 200, 203
   semiconducting nanotubes discrimination 195
   single-walled carbon 144, 148
methane 59
methanol 94, 232–3, 235–6
microwave chemical vapour deposition 165
miniaturisation, Feynman's 2
mirror planes 53, 54
mixing state probabilities 22–3
molecular dynamics simulations 151–2, 228, 236, 238
molecular interaction range *see* correlation length
molecular machines 3, 23–5
molecular monolayers 16–17
   see also graphene sheets
molecular Raman mode 79–80
molecular structural mechanics 152–3
molecular weight of solvent 236, 237
mono-vacancy (MV) defects 165, 166, 206–8, 212
monochromators 103
monoclinic crystal point group 64, 273
morphic effects on Raman bands 85–6, 87
Mueller, Hans 85
multi-channel detectors 105

multi-walled carbon nanotubes (MWCNTs) 143, 155–7
  configurations 199
  end caps 157
  first isolation 114
  Raman spectra 197–201, 204
  structure 110, 111, 152
multilayered particles 29, 31

nano prefix meaning 2, 5–6
nano-buds 110, 111
nano-hybrid structures 110, 111
nanobots 4
nanoparticles 29, 31, 32
nanophenomena
  electronic 30, 32–4
  mechanical 37–8
  optical 26–32
  thermal 35–7
nanospheres 29, 31, 32
nanotechnology
  configurational entropy 21–6
  definitions 4–11
  history 2, 3
  nanophenomena 26–32
  origins 1–4
  property changes with size reduction 7
  realisation of importance 9–11
  scales of inhomogeneity 14–19
  science of 11–14
  size of nanosystems 6–7
  usefulness 5
nanotorus 111
nanotubes see carbon nanotubes
naphthalenes 220
Neumann's principle 69
Newton, Isaac 6
nonequivalent carbon sites 129
nonsymmorphic space groups 153, 154
normal vibration modes 47–8
  C_{60} molecule 183–4
  fullerenes 183–5, 187–92
  symmetry species
    correlation 72
    general formula 267–71
  symmetry-based classification 74, 76–9
  water molecule 76–7

notch filters 102
number density (ρ) 12, 13
number of layers (n) 202, 205, 206, 207

one-dimensional fullerenes 143–61
  chiral angle 144, 145, 147
  chiral/achiral 153–5
  diameter estimation 147
  end caps 148–9
  metallic behaviour 144, 148
  molecular structural simulation 151–2
  multi-walled carbon nanotubes 155–7
  pseudo molecules 223–4
  single-walled carbon nanotubes 143–55
  smallest 152–3
  symmetry 153–5
  transitional unit cell length 149–51
  types 145
one-dimensional space groups 66, 67–8
onions (carbon nano-onions) 110, 111, 140–3
opals 6, 26
optical guiding systems 105
optical microscopes 102
optical nanophenomena 7, 26–32, 137
orientation
  gold nanorods, optical effects 28–9
  orientation ordering temperature 186
  polarised Raman intensity 98–9
  single-walled carbon nanotubes 195–7, 199, 200
orthorhombic crystals 64, 96–7, 273
Osawa, Eiiji 111
overtone modes 79, 189
oxidation treatment 166, 167
O–H transformations 76–7

p_χ electronic orbital 69–71
peapods (C_{60}@SWCNT) 110, 111
pentagons
  C_{60} fullerene 123–4, 188
  isolated pentagons rule 117–18, 119, 120
  pentagon shear mode 234–5
  pentagonal pinch mode 188
  zero-dimensional fullerenes 117

Subject Index

perturbation effects
 C_{60} molecule 186
 graphene 202, 203–4, 205
 Raman bands 85–95
 single-walled carbon nanotubes
  194, 233–4
phase transitions 186–7
phonons *see* lattice vibration modes;
 normal vibration modes
physical laws, size of nanosystems 6–7
planar AB_4-type molecules 52–3, 54
planar defects 91
plane space groups 66–7, 68–9
planes of symmetry (σ) 53, 54
plasma arc fullerene generator *see*
 Huffman–Krätschmer arc-
 discharge evaporation method
plateau starting pressures (PSP) 235–6
point defects 164–7, 206–8, 212
point symmetry groups
 assignment flow chart 61
 character tables 69–81, 259–66
 crystals 60, 62, 63, 64–5, 96
 fullerenes 118, 119, 121
 polarisability tensors 272–5
 Raman spectroscopy 56–62
 reasons for 62
 stereographic representation 60, 63–5
 symmetry operations 57–60
 symmetry species, normal vibrations
  calculation 267–71
point thermodynamic approximation
 11–12
polar solvents 219–20
polarisability ellipsoid 49, 50
polarisability tensor (α)
 carbon dioxide molecule 49
 point groups 272–5
 Raman band intensity 96
 Raman scattering 45
 vibration modes 47–8
polarisation direction 28–9
polarised Raman spectra 96–9, 185,
 195–7, 200
polychromators 103
polycyclic aromatic hydrocarbons 132–3

polyhedrons, Euler's theorem 117
polymer molecule Frieze groups 68
poly(*para*-phenylene) (PPP) 100, 101
potential distribution theorem
 (Widom) 12
pressure (*p*)
 excess in liquid droplets 15–16
 fullerene behaviour 229–36
 fullerene onions treatment 140–2
 large one-component system 20–1
 large/small systems 11–12
 plasma arc fullerene generator 158
 pressure–temperature phase diagram
  89, 90
 pressure–volume equation of state
  90, 91
pressure transmission fluid (PTF)
 230, 232–6, 237
probabilities of mixing 22–3
production methods 129–33
 see also Huffman–Krätschmer arc-
  discharge evaporation method
 carbon nanotubes 158–61
 graphene 165, 167
proper rotation axis 52
pseudo three/two/one-dimensional
 molecules 223–4
public perceptions 5
purification methods 137–40, 161, 162
pyrocarbon 79, 81
pyrolytic production methods 114, 160–1

quantum bubbles 31
quantum dots 27
quantum theory 2, 44, 95–6
quasi-thermodynamic assumption 11–12
quenching 83

radial breathing mode (RBM)
 C_{60} molecule 183
 double-walled carbon nanotubes
  200, 202, 203
 multi-walled carbon nanotubes 198
 single-walled carbon nanotubes 192–4,
  199, 209, 210–11, 213, 214
 pressure effects 230–1, 232–3

radius of curvature 15–16
rain formation 7
Raman, Chandrasekhara Venkata 43, 44
Raman scattering 43–105, 183–215
  anti-Stokes lines
    Boltzmann equilibrium distribution 83–4
    definition 46, 81, 82
    radial breathing mode, single-walled carbon nanotubes 209–11, 214
    Stokes/anti-Stokes ratio 84–5
  Boltzmann equilibrium distribution 83–5
  C_{60} molecule 183–9
  C_{70} molecule 189–90, 191–2
  character table 69–81
  dispersion effect 99–101
  double-walled carbon nanotubes 197–200, 203
  electric field effects 205, 211
  energy transfer 81–3
  general theory 44–7
  graphene 201–8, 209, 210, 211, 212
  instrumentation 101–5
  perturbation effects 85–95
  point groups 56–62, 64–5
  polarised/Raman band intensity 96–9
  Raman band position calculations 95–6
  selection rules 47–50
  single-walled carbon nanotubes 190, 192–7, 198, 199, 200, 204
  space groups 62–3, 65–9
  Stokes lines 46, 81, 82, 83–5, 210–11, 214
  symmetry 50–6
  thermal effects 208–15
Rayleigh scattering 44, 46–7, 82, 102
regular truncated icosahedrons 124
relative Raman intensity 96
representations 69–79
  see also symmetry species
  definition 71
  degenerate 73–4
resonant Raman effect 95
ring-shaped 20-carbon structure 118
Rohrer, Heinrich 10
rotation axes (C_n) 52–3
rotation matrix (φ) 98
rotation reflection axes (S_n) 53–5
ruby glass 7, 8

sapphire single crystals 86
scales of inhomogeneity 14–19
  artificial molecular machinary 25
  capillary length 14–15
  correlation length 17–19
  line tension 16–17
  thermal gravitational scale 14
  Tolman length 15–16
scanning monochromators 103
scanning tunnelling microscopy (STM) 10, 145, 146, 164–8
scattering of light see Raman scattering
screw axis ($n[p]$) 63, 65
selection rules 47–50
self-assembly 16–17, 29, 32, 223–5
self-healing systems 240
semiconductors
  clusters 34
  nanoparticles 27
  nanotubes 195, 199, 200, 203
shape of particles 27–8
shear stress 140–2
Shönflies notation rule 57–60, 62, 66
silica-coated gold nanoparticles 32
silicon
  nanowires 35, 36
  silicon carbide 103, 105
  single crystals 85
  substrates
    C_{60} molecule 185
    C_{70} molecule 189, 190
    residual stress 104, 105
single-channel detectors 105
single-walled carbon nanotubes (SWCNTs) 115, 143–55
  chiral vector C_h 145, 147, 149
  C–C bonds 144–7
  diameter estimation 147
  discovery 113–14
  effects on solvents 228
  end caps 148–9

Subject Index

formation 143–4, 155
Raman scattering 190, 192–7, 198, 199, 200, 204, 213, 214
  diameter/RBM correlation 193, 214
  pressure effects 229, 230–6
  thermal effects 209–11, 213–14
structure 110, 111, 143–55
unit cell construction 149–51
six-fold rotation axis 52
size of clusters 33, 34
size of nanosystems 6–7, 11–14
Smalley, Richard 3, 10, 11, 112, 113, 160
solid state C_{60} 185–8
solubility 187, 215, 216–20
solution state 210–11, 214
solvents
  effects on C_{60} fullerenes 187, 215, 216–20
  effects on C_{70} fullerenes 215, 216–20
  fullerene effects on solvents 225–9
  fullerene Raman spectra 215–29
  molecular weight effects 236, 237
  Raman spectra 94–5
  Soxhlet extraction method 134–5, 136
Soxhlet extraction method 133–6, 137–8
sp^2 hybridisation
  C_{60} fullerenes 125, 126
  graphene sheet 144
  single-walled carbon nanotubes 153
sp^3 carbon replacement in regular nanotubes 153
space groups 62–3, 65–9
specific heat capacity 36
spectrographs 103
spherical giant fullerenes 139–40
spring constant 78
stability 120–1, 161
stationary LC phase 137
stereographic representations 60, 63–5
Stokes lines 46, 81, 82, 83–5
  Stokes/anti-Stokes intensities 210–11, 214
Stone–Wales (SW) rearrangement
  density functional theory 167, 206–8, 212
  isomerisation in higher fullerenes 119
  optimised atomic structure of graphene 165, 166, 167

strain effects 85–6, 87, 91, 104
stretch modes 48–9, 50, 73, 94
structural defects/imperfections 90–2, 205
  di-vacancy 165, 166, 206–8, 212
  mono-vacancy 165, 166, 206–8, 212
  Stone–Wales rearrangement 119, 165, 166, 167, 206–8, 212
sublimation method 135, 137
subscripts in symmetry designations 73
substrate interactions 204, 210
surface plasmon resonance (SPR) 29
surface tension (σ) 15–17, 19
surface-decorated particles 29, 31
SWCNTs *see* single-walled carbon nanotubes
symbols *see* terminology
symmetric stretch (v_1)
  carbon dioxide molecule 48–9, 50, 73
  hydrogen sulfide molecule 88, 89
  water molecule 76–7
symmetry
  C_{60} molecule Raman active modes 185, 186
  elements/operations distinction 55–6
  groups *see* point symmetry groups
  one-dimensional fullerenes 153–5
  Raman scattering 50–6, 85–6, 87
symmetry operations
  C_4 symmetry operation 73–4
  point groups 57–60
  symmetry elements distinction 55–6
  transformation of directional properties 69–73
symmetry species 69–79
  C_{60} molecule 183
  linear molecules 74
  normal vibration formulae 267–71
  normal vibration modes correlation 72
  single-walled carbon nanotubes 190
  water molecule, nine degrees of freedom 78
symmorphic space groups 153

tangential Raman scattering modes 192, 193, 194
Taniguchi, Norio 2, 3

temperature (T)
  C_{60} phase transitions 186, 208–15
  conductivity 35, 36
  critical point 18
  fullerene stability 120–1
  graphene sheets Raman scattering 211
  large one-component system 20–1
  large/small systems 11–12
  Raman scattering 84, 85, 86–8
    hydrogen sulfide 89, 90
  single-walled carbon nanotubes
    Raman scattering 213, 214
  specific heat capacity size
    dependence 36
  Stokes/anti-Stokes lines 46
  thermal effects on Raman spectra
    208–15
tensor *see* polarisability tensor
terminology 2, 5–6, 11
  Bravais lattice centring symbols 67
  character table 69–81, 259–66
  fullerenes atomic numbering 119, 122
  Hermann–Mauguin notation 60, 62, 66
  nonmacroscopic/nanosystems 20
  one-dimensional space groups 68
  point group symbols 56, 57–60, 62
  representations 69–73
  S symbol derivation 53
  Shönflies notation rule 57–60, 62
  symmetry designations 73, 74
  symmetry element symbols 63
  symmetry species 267–71
  two-dimensional objects 68
tetragonal crystal point group 64, 273–4
tetrahydrofuran (THF) 135
thermal effects
  see also temperature
  C_{60} phase transitions 186, 208–15
  conductivity 35, 36
  nanophenomena 35–7
  Raman spectroscopy 86–8
thermal gravitational scale 14
thermodynamic functions
  cofigurational entropy 21–6
  large/small systems 11–14, 35–7
  small systems 19–21

three-dimensional *pseudo* molecules
  223–4
Tolman length 15–16
toluene 80, 225–7
topological defects, graphene 205–8, 212
transformation of directional properties
  69–73
transitional SWCNT unit cell 149–51
translation operation 63
triangular symmetries 164
triclinic crystal point group 64
trigonal crystal point group 64, 273
triphenylphosphine 57
tris(ethylenediamine) cobalt(III) cation 58
tube–tube interactions 194, 200–1
tubular giant fullerenes 139–40
twist axis 63, 65
two small systems in contact, mixing
  state probabilities 22–3
two-dimensional space groups 66–7, 68–9
two-dimensional systems 223–4
  see also graphene sheets
two-fold rotation axis 52
two-fold screw axis 65, 66
two-phase liquid–gas systems 14–15

ungerade 73
uniaxial tensile strain 86
unit SWCNT cell construction 149–51
units for Raman/Rayleigh lines 46–7
Universe-scale correlation length 18, 19
usefulness of nanotechnology 5
UV light 27

van Hove singularity 194
vibration modes
  see also normal vibration modes
  C_{60} molecule 183–9
    symmetry assignment 185
  C_{70} molecule, symmetry
    assignment 189, 191–2
  carbon dioxide molecule 48–50
  Raman lines 47–8
virtually excited species 81, 83
visible light 27
volume (V) 12, 13

Subject Index

wallpaper groups 66–7, 68–9
water
  bulk/nano-domain properties 7
  molecules
    C_{60} fullerene 124, 125, 221–2
    nine degrees of freedom 78
    normal vibration classification by symmetry 74, 76–8
    O–H transformations 76–7
    single-walled carbon nanotube interactions 228, 235, 236
  point group 57
Weyl, Hermann 50–1
Widom's potential distribution theorem 12

xenon oxytetrafluoride 58
xenon tetrafluoride 59
xylene 225–7

YBCO single crystal high-temperature superconductor 98–9
yield stress 37
Yoshida, Z. 111

Zeldin's method 56
zero-dimensional fullerenes 117–29, 130
zigzag tubes
  atomic displacement vectors 192, 193
  graphene crystals 163
  multiwalled carbon nanotubes 155–6
  Raman active modes 190
  single-walled carbon nanotubes 145, 146, 148, 149–51, 153–4
zipping/unzipping of single-layered graphene sheets 166–7